BIOTECHNOLOGY FOR THE ENVIRONMENT:
STRATEGY AND FUNDAMENTALS
VOLUME 3A

FOCUS ON BIOTECHNOLOGY

Volume 3A

Series Editors
MARCEL HOFMAN
Centre for Veterinary and Agrochemical Research, Tervuren, Belgium

JOZEF ANNÉ
Rega Institute, University of Leuven, Belgium

Volume Editors
SPIROS N. AGATHOS
Université Catholique de Louvain,
Louvain-la-Neuve, Belgium

WALTER REINEKE
Bergische Universität,
Wuppertal, Germany

COLOPHON

Focus on Biotechnology is an open-ended series of reference volumes produced by Kluwer Academic Publishers BV in co-operation with the Branche Belge de la Société de Chimie Industrielle a.s.b.l.

The initiative has been taken in conjunction with the Ninth European Congress on Biotechnology. ECB9 has been supported by the Commission of the European Communities, the General Directorate for Technology, Research and Energy of the Wallonia Region, Belgium and J. Chabert, Minister for Economy of the Brussels Capital Region.

Biotechnology for the Environment:
Strategy and Fundamentals
Volume 3A

Edited by

SPIROS N. AGATHOS
Université Catholique de Louvain,
Louvain-la-Neuve, Belgium

and

WALTER REINEKE
Bergische Universität,
Wuppertal, Germany

KLUWER ACADEMIC PUBLISHERS
DORDRECHT / BOSTON / LONDON

A C.I.P. Catalogue record for this book is available from the Library of Congress.

ISBN 1-4020-0529-6

Published by Kluwer Academic Publishers,
P.O. Box 17, 3300 AA Dordrecht, The Netherlands.

Sold and distributed in North, Central and South America
by Kluwer Academic Publishers,
101 Philip Drive, Norwell, MA 02061, U.S.A.

In all other countries, sold and distributed
by Kluwer Academic Publishers,
P.O. Box 322, 3300 AH Dordrecht, The Netherlands.

Printed on acid-free paper

Printed in the Netherlands.

EDITORS PREFACE

At the dawn of the 21st century we are witnessing an expanding human population in quest of survival and continued well-being in harmony with the environment. Many segments of society are increasingly preoccupied with the battle against both diffuse and concentrated pollution, the remediation of contaminated sites, the restoration of damaged areas due to anthropogenic activities and the re-establishment of functioning biogeochemical cycles in vulnerable ecosystems. There is an enhanced awareness of the value of pollution prevention and waste minimization in industrial, urban and agricultural activities, as well as an increased emphasis on recycling. Faced with these major contemporary challenges, biotechnology is emerging as a key enabling technology, and, frequently, as the best available technology for sustainable environmental protection and stewardship.

Although the activities of microorganisms and their subcellular agents have been recognized, studied and harnessed already for many years in the environmental arena, there is a new dynamism in the in-depth understanding of the molecular mechanisms underlying the functioning of microorganisms and their communal interactions in natural and polluted ecosystems, as well as an undeniable expansion of practical applications in the form of the new industry of bioremediation. A number of distinct but increasingly overlapping disciplines, including molecular genetics, microbial physiology, microbial ecology, biochemistry, enzymology, physical and analytical chemistry, toxicology, civil, chemical and bioprocess engineering, are contributing to major insights into fundamental problems and are being translated into practical environmental solutions and novel economic opportunities.

The book set «Biotechnology for the Environment», based on a compilation of some of the outstanding presentations made at the 9th European Congress on Biotechnology (Brussels, Belgium, July 11-15, 1999), captures the vitality and promise of current advances in the field of environmental biotechnology and is charting emerging developments in the beginning of the new millennium. This first volume, subtitled 'Strategy and Fundamentals' offers an exciting bird's-eye view on carefully selected topics of basic science and methodology in the area of pollutant biodegradation and bioremediation, illustrating both the richness of the mechanistic insights and the importance of the multidisciplinary approaches for years to come. After an opening account on the robustness of several biotechnological solutions for waste minimization and pollution control in the chemical industry, a number of contributions offer state-of-the-art descriptions of the genetics and biochemistry of degradation of important recalcitrant molecules, such as halogenated aliphatic and aromatic compounds, polycyclic aromatic hydrocarbons, polychlorinated biphenyls, pesticides, detergents and chelating agents. Pesticides and chlorinated or polycyclic hydrocarbons are well-known priority pollutants and continue to be fertile areas of study given the hazards resulting from their abundance and ubiquitous presence due to historical pollution from past industrial and agricultural practices. On the other hand, detergents, chelating agents and other organic molecules at low concentrations (micropollutants, endocrine disruptors) are emerging as equally if not more insidious for the health of humans and other organisms and, thus, studies on their biodegradation are starting to gather momentum. Molecular approaches to track microbial diversity with impact on global change, novel vector-based bioaugmentation strategies for complex environmental matrices (soils, sludges, etc.) and, last but not least, transgenes as new-generation sensors for ecotoxicological monitoring, complete the picture.

The Editors hope that the integration of the depth of scientific fundamentals with the breadth of current and future environmental applications of biotechnology so evident in these selected contributions will be of value to microbiologists, chemists, toxicologists, environmental scientists and engineers who are involved in the development, evaluation or implementation of biological treatment systems. Ultimately, a new generation of environmental scientists should take these lessons to heart so that new catalysts inspired from the biosphere can be designed for safe, eco- and energy-efficient manufacturing and environmental protection.

Spiros N. Agathos   Walter Reineke

TABLE OF CONTENTS

4

PART 1
STRATEGIC VIEWS

APPLICATION OF THE BIOTECHNOLOGICAL POTENTIAL FOR ENVIRONMENTAL CHALLENGES IN THE CHEMICAL INDUSTRY

The biotechnological potential is ecoefficient and thus will become increasingly applied in future

GERHARD STUCKI AND PATRICK STEINLE
Ciba Specialty Chemicals Inc., WS-2090, CH-4133 Pratteln, Switzerland
Present address for Gerhard Stucki, BMG Engineering AG,
Grammetstrasse 14 CH-4410 Liestal Switzerland.
Email:gerhard.stucki@bmgeng.ch Fax:+41-619232580

Abstract

Nature has an enormous potential to cycle materials and energy. This potential, and specifically that of biological processes can be applied in the chemical industry to recycle or treat waste, wastewater, and off-gas, and thus reintegrate the man-made, synthetic chemicals into the natural cycles.

This contribution describes some powerful and innovative biotechnical processes investigated in and partly applied at Ciba Specialty Chemicals Inc., Switzerland for the breakdown of C-, Cl-, N-, S-, and P-containing molecules for water, air and soil treatment. Among these are laboratory investigations on the degradation of atrazine or chlorophenols in soils, the large-scale application of microorganisms capable of mineralising aromatic or chlorinated aliphatic compounds in fixed-film bioreactors for groundwater treatment, the anaerobic treatment of chemical effluents, or the removal of solvents from off-gas using biological trickling filters.

Although the biotechnological potential available is vast, its applicability in the environmental field is slow. Possible reasons and potential barriers to overcome are discussed from a perspective of the chemical industry. Chances for optimal solutions including biological processes increase if the approach to solve environmental challenges is holistic, if the wastewater or off-gas is well characterised, if cheap and reliable biodegradation tests are used coupled with professional interpretation of the test results, and if interdisciplinary know-how transfer occurs.

1. Introduction

Excellent solutions in the environmental field are those which avoid problems from the start. Therefore, if the principle to avoid, reduce and dispose of the wastes end of the

9

S.N. Agathos and W. Reineke (eds.),
Biotechnology for the Environment: Strategy and Fundamentals, 9–20.
© 2002 *Kluwer Academic Publishers. Printed in the Netherlands.*

process is strictly followed, very few environmental problems would occur, and thus not many environmental engineers and biotechnologists would be required. In reality, we are far from such an ideal situation. In fact, applying too strict a principle would make neither economical nor ecological sense in many situations. Example: it will mostly be better to collect small effluents and treat them together in a cheap and environmentally friendly way than to treat them individually, or to incinerate the wastewater with the result of a "zero-discharge facility", which is "zero discharge" in relation to wastewater but not to energy or air. Therefore, in practice, rather the optimal than the maximal solutions are looked for. Biotechnological solutions are especially advantageous in the long run, they generally require more volume (higher investment cost) than chemical and physical options, but they are often far more advantageous regarding resources including energy requirements and operation costs.

In the following chapter, case studies are presented of biological processes investigated and implemented in the chemical industry to solve environmental challenges. Some of them have been transferred to full scale, while others still wait to be applied some day. Very often, the biological process used was just one of a chain of process steps to reach required goals.

2. Biotechnological processes to remove/degrade chemicals

During the last 12 years, many biological processes were studied and partially applied at Ciba Specialty Chemicals, Switzerland, to eliminate waste chemicals (Table 1).

2.1. REMOVAL OF ORGANIC CARBON

In contrast to the food industry, where high loads and concentrations of a more or less constant composition of organic materials are converted by applying anaerobic processes, larger quantities of carbonaceous compounds are most often distilled, recycled or used as fuel replacement in the chemical industry. Cases with high volumes and loads of organic materials to be degraded are rare, and are restricted to wastewaters containing waste carbohydrates or organic acids. Such type of wastewater was generated in a dyestuff factory. The large amounts of dilute aqueous acetic acid and ethylene glycol among trace amounts of aromatic chemicals could neither be avoided nor replaced nor recovered economically and thus was pre-treated anaerobically in a 170 m^3 reactor. For the successful application, it was important to identify toxic substances inhibiting the anaerobic microorganisms. The major "toxic chemical" was sulphuric acid which was replaced by an alternative inorganic acid to avoid inhibition due to H_2S formation.

Table 1. Environmental biotechnological processes tested and applied at Ciba Specialty **Chemicals** *Inc. during the last 10 years*

Element	Examples studied	Phase of application (1999)
C	Removal of high organic loads and colour from wastewater (acetic acid, diethylene glycol, dyestuffs) by an anaerobic end of process treatment	Cheap, anaerobic pre-treatment processes (up-flow-anaerobic sludge blanket process) successfully in operation since several years
Cl	Removal of chlorinated chemicals determined as AOX (adsorbable organic halogens)	Several successful full scale end-of-pipe process applications in operation using a combination of fixed-film bioreactors and activated carbon adsorbers
Cl	Degradation of low concentrations of chlorinated hydrocarbons (chlorinated aromatic compounds, 1,2-dichloroethane [DCA], tetrachloroethene [PER]) in groundwater	Aerobic full scale process to remove DCA using laboratory strains; an anaerobic-aerobic full scale process followed by activated carbon adsorption to remove PER, pilot studies eliminate aerobically chlorinated aromatic compounds at low temperatures, low concentrations, and high hydraulic loading rates
Cl	Degradation of chlorinated phenols or dichloromethane in industrial waters	Laboratory and pilot studies, potential not yet applied
Cl	Removal of chlorinated pesticides from soil (e.g. atrazine) and wastewater (substituted phenols)	Laboratory, bench, and field scale trials using aerobic microorganisms with special degradation capabilities, concept developed to remediate a contaminated site including *in-situ* remediation and soil washing
Cl	Removal of chlorinated solvents from off-gas	Pilot studies using aerobic microorganisms to remove chloromethane, 2-chloropropane, DCA and other synthetic compounds
N	Nitrification and denitrification of ammonia and nitrate, respectively	Classical processes implemented at several production sites
N	*s*-Triazine herbicides (atrazine, simazine, etc.) removed from process streams as N- sources	Laboratory approach worked; field trials failed due to the presence of nitrogen fixing bacteria competing for C
S	Biological recycling of dilute waste sulphuric acid using an anaerobic process	Excellent laboratory performance, process not followed up further due to unfavourable economics
S	Hydrogen sulphide (H_2S) and mercaptan removal from process air	Excellent performance in laboratory and pilot scale, first biological trickling filters under construction ($20\ m^3$ working volume)
P	Biological P fixation	Either no P available or too high P loads are discharged, the bio-P process is applicable as "polishing treatment" only (in the chemical industry)
P	Phosphonates as P- sources	Degradation potential applicable at P limiting conditions
P	Organic phosphates as C- and energy sources	Conventional biological elimination possible

Wastewater containing high quantities of sulphuric acid were excluded from the anaerobic treatment. Methanol as emergency feed stock helped to run the anaerobic process reliably, because it guaranteed a continuous methane production rate, which allowed some permanent removal of toxic H_2S by stripping. The biological process was much cheaper than the previously applied chemical wastewater treatment. In addition to the carbon load, the colour was removed very efficiently by this process.

Another application of the anaerobic potential was the pre-treatment of a highly concentrated leachate of an old dumping site. The H_2S produced from the trace sulphates precipitated the heavy metals efficiently [1].

2.2. CHLORINATED COMPOUNDS

In industrial wastewaters, chlorinated compounds are either analysed as individual substances (e.g. pesticides like atrazine) or as sum of halogenated chemicals (AOX), which adsorb onto activated carbon. We applied microorganisms mineralising chlorinated chemicals at full scales (10, 30 and 500 m^3 aerobic reactors) using the following concept: The chemical wastewater is treated in a series of fixed-film bioreactors followed by activated carbon adsorption. Slowly growing bacteria, known to have a broad degradation potential developed preferably in the last reactor. With the broader degradation potential available, more chlorinated chemicals were mineralised and thus more compounds contributing to AOX were biologically eliminated. As a consequence, less AOX had to be removed by adsorption, which resulted in a longer service life of the activated carbon columns and thus in lower operation costs.

The aerobic degradation of chlorinated compounds serving as C- and energy source such as atrazine [2] or dichloromethane [3] has been investigated at laboratory and pilot scale. Although the chemicals were efficiently removed, the biological processes were not applied at full scale due to different reasons such as economical considerations, production stops, alternative chemical or physical environmental processes.

Low concentrations of chlorinated hydrocarbons in groundwater and leachate of solid waste disposal sites were successfully treated at laboratory and pilot scales, using the site specific microbial degradation potential, enrichment cultures, or pure cultures isolated and grown under laboratory conditions. Crucial for the successful transfer of the biotechnological know-how was not only the microbiological knowledge but also the engineering competence in reactor technology. Intensive pilot studies revealed that low contaminated groundwater can be treated aerobically at similar organic loading rates as domestic or industrial wastewater treatment plants, even though the feed and effluent concentrations are up to 100 times lower than those of industrial wastewater treatment plants ([3, 4], Table 2). In contrast to the typical industrial or domestic wastewater treatment, the hydraulic residence time for the groundwater treatment was as low as 15 to 20 min, since all the pollutants were dissolved and thus were rapidly taken up by the bacteria (Table 2). At higher hydraulic rates or shorter residence times, the conversion possibly became diffusion limited. The removal of DCA from groundwater at full scale was described in detail [5].

Tab. 2. Comparison of the hydraulic retention time, the organic loading rate, and the elimination efficiencies in different water treatment systems

Parameter	Units	Domestic wastewater	Chemical wastewater	Groundwater
Feed concentration (as organic compound)	mg/L	10 - 100	100 - 1000	1 - 10 [3]
Effluent concentration	mg/L	< 10	50	< 10 µg/L
Hydraulic retention time	H	2	12	0.3
Specific hydraulic loading rate [1]	$M^3/(m^3 d)$	12	2	72
Temperature	°C	10 - 35	10 - 40	8 - 15
Organic loading rate [2]	$Kg/(m^3 d)$	0.06 - 0.6	0.2 - 2.0	0.07 -0.7
Sludge concentration	g SS/L	2	6	20 [4]
Sludge loading rate	g organics/ (g SS d)	0.03 - 0.3	0.03 - 0.33	0.004 - 0.035 [5]
Elimination efficiency	%	80 - 90 [6]	ca. 90 [6]	99 - 99.9 [7]

(1) related to the empty bed volume; (2) related to the amount of pollutant per unit reactor volume; (3) chlorinated aromatic and aliphatic compounds (4) estimate for fixed film reactor; (5) a considerable part of the sludge is mineralised under these conditions; (6) based on dissolved organic carbon, (7) based on single compound analysis

The presence of bacteria supplied with fertiliser and H_2O_2 allow the reduction of the synthetic chemical from the mg/l-level to levels below 10 µg/L at temperatures of 8 to 12 °C and at hydraulic residence times below 1 h. As a result of the innovative biotechnical approach, the operation costs fell seven fold compared to conventional treatment options.

Factors limiting the application window of biological processes were also investigated in soils, because many excellent *in-situ* applications to decontaminate gasoline contaminated land demonstrate the powerful biotechnological potential. An example of the full scale application of chlorophenol contaminated soil in biopiles has been described by Häggblom and Valo [6]. Our initial soil studies ware carried out using atrazine as model compound. This chemical showed only weak adsorption properties in soils with a moderate amount of organic matter, thus atrazine was fully available for biodegradation. In addition, the soil required to be inoculated by special microorganisms to achieve atrazine mineralisation. In bench scale trials, a low number of bacteria (0.1 mg/kg soil) able to mineralise atrazine removed this chemical so rapidly from soil that its herbicidal activity against atrazine-sensitive plants was lost [7, 8]. Thus the kinetics of the atrazine activity versus degradation will finally determine certain fields of application of the atrazine degrading organisms.

Another model compound used for our studies was 2,6-dichlorophenol (DCP), which was one of the major pollutants of a former disposal site. A pure microbial culture was found and isolated from soil and water samples from that site [9]. One of the remediation concepts developed included i) the incineration of the highly contaminated soils, ii) an alkaline soil washing treatment for the moderately contaminated soils up to 1 g substance/kg soil with a subsequent biological treatment of the washing medium in a fixed-film bioreactor, and iii) an in-situ biological polishing treatment for concentrations below 100 mg/kg soil [15]. The aqueous extraction solution of the moderately contaminated soil contained 1 g/l DCP. This solution was efficiently purified in a fixed-film aerobic bioreactor to levels below the detection limit of DCP (0.3 mg/l) and what made it possible to recycle the extraction liquid [10]. Although the environmentally more friendly concept that included soil washing and biological treatment of the extraction liquid for the moderately contaminated soil looked promising, it did not compete successfully with the conventional alternatives: a thermal desorption process followed by a thermal treatment of the off-gas was finally chosen to clean up the soils.

Off-gas treatment has a large potential for industrial applications. So far, compost filters have been widely applied to treat odorous gases. In the chemical industry, these filters could rarely be applied, because many compounds yield acid degradation products. As a result of the biological breakdown, the pH in the compost filters fell and the air cleaning efficiency decreased. A very promising tool is the biotrickling filter, in which bacteria fixed on an inert surface clean the air by mineralising the organic off-gas components. Chemical and physical processes limiting the biological turnover are a technical challenge to overcome to make the biological process competitive and reliably applicable. In the trickling filter, metabolites such as inorganic acids can be easily washed off by the aqueous recycle stream. Pilot studies showed that chlorobenzene together with dioxane and toluene were eliminated at rates between 20 and 100 g $C_{org}/(m^3_{reactor\ volume}h)$. Similar removal rates were obtained for 1,2-dichloroethane (DCA) using a 1 m^3 pilot reactor. The temporal variation in load and concentration of the chemicals and the absence of oxygen require process modifications (humidifiers, buffers such as activated carbon adsorbers, peroxide feeding options). At the present stage of process development, the technology is often not cheap enough to compete with chemical and physical alternatives. Therefore, the field of application is still small. Nevertheless, we investigated the use of this technology for the removal of methyl chloride, DCA, different acrylic esters, 2-chloropropane, and aliphatic amines. So far, a 20 m^3 trickling filter prototype has been taken into operation in 1999. It is treating off-gas which was incinerated before and which contains amines, isopropanol and traces of

2-chloropropane. A second biological trickling filter is under evaluation for the removal of 3 kg/h DCA. In an adjacent chemical company in Pratteln, Switzerland, a 350 m^3 trickling filter has recently been erected to remove chlorinated benzenes and toluene.

2.3. NITROGENEOUS COMPOUNDS

Bulk amounts of most nitrogen-containing compounds are generally recycled or disposed off by other than biological unit operations. In the normal case, only low amounts and low concentrations of N-containing compounds are discharged into effluent treatment plants. These chemicals serve as N-source for the bacteria, mineralising the carbonaceous chemicals. Examples for compounds discharged in larger amounts as chemical effluent are waste ammonia, nitrates, solvents difficult to recycle such as dimethylformamide, or secondary and tertiary amines. The many new regulations issued during the last 10 years required the modification of numerous conventional wastewater treatment plants into nitrifying and denitrifying treatment plants.

Nitrifying conditions in industrial wastewater are obtained in treatment plants where a high sludge age prevails. These conditions, however, also allow the development of slowly growing microorganisms known to have a broader degradation potential than the fast growing strains. As a result, difficult to degrade compounds are often co-oxidised in nitrifying plants, whereas they seem to be refractory in conventionally operated plants. An example is the disappearance of *cis*-1,2-dichloroethene (unpublished) in a two stage fixed-film nitrifying bioreactor converting 300 mg/l ammonia to nitrate [11].

Nitrate as a waste product in chemical effluents can hardly be recycled economically. A very efficient disposal technology is the denitrification process. If used in an industrial wastewater treatment plant, nitrate replaces oxygen, and thus may lead to considerable savings in (aeration) energy. This potential will be applied soon in one of our pigment plants, where tons of waste nitrate have to be disposed off, daily. The site itself does not discharge carbon compounds, which could have served as electron donors. Therefore, waste nitrate will be transported via pipeline to the effluent treatment plant of an adjacent paper works to replace pure oxygen used to increase the aeration capacity and part of the blower capacity.

Certain pesticides were tried to be mineralised as nitrogen source using the biological potential of strong oxidising enzymes produced by the microbial cell under N- (or P- or S-) limiting conditions. Unfortunately, N-limiting conditions in the chemical industry are exceptional, and the biological processes are difficult to run reliably due to the competition of N-fixing bacteria competing for the carbon compounds available. As a result, the window of application of this degradation potential is considered very small.

2.4. SULPHUROUS COMPOUNDS

One of the bulk sulphur compounds used in the chemical industry is sulphuric acid. Whereas economic high temperature processes exist to recycle concentrated waste acid, no solution other than neutralisation or precipitation and disposal as waste gypsum exists for the dilute acids below 15 to 20 % H_2SO_4. For this purpose, a few years ago, we combined a biotechnological with a chemical process to recycle sulphuric acid via

H_2S and elemental sulphur. The biological process was run anaerobically using acetic acid, another abundant waste in the chemical industry, as reductant. Both, sulphuric (H_2SO_4) and acetic (CH_3COOH) acid were converted to the two weaker corresponding acids H_2S and H_2CO_3 with the biological process run at neutral conditions [12]. So far, the process was never applied at full scale due to the cheap price for sulphur and the still abundant sites for gypsum disposal.

Mercaptans and H_2S are often the source of odor problems. They were usually oxidised in scrubbers using strong oxidising chemicals. In future, off-gases containing pollutants such as sulphur containing compounds as well as smelly chlorinated or N-containing volatile chemicals will be decomposed more and more in low-energy consuming systems such as compost filters and biotrickling filters. Especially the latter allow to control the biological activity much better than the simple compost filter, which is prone to clogging, acidification and drying out.

2.5. PHOSPHOROUS COMPOUNDS

Whereas numerous domestic wastewater treatment plants have been converted/erected/modified to remove phosphates using the biological potential of microorganisms to fix and release increased amounts of inorganic phosphate at different redox conditions, the chemical wastewater treatment plants are unlikely to undergo such modifications. The main reasons are that substrate concentrations determined as total organic carbon (TOC) or biological oxygen demand (BOD) are so high, that phosphate often must be added to support the unlimited growth of activated sludge. To remove high phosphate quantities, the chemical precipitation of concentrated P-containing effluents at the source of the pollution is usually much more efficient than a bio-P-treatment process at the end of the pipe, where only traces (a few mg/l) of P are eliminated.

The removal of phosphonates under laboratory conditions has been studied intensively. This class of substances can certainly be degraded under P-limiting conditions. The removal of phosphonates in presence of inorganic phosphates has been reported [13] but could not be obtained in pilot scale studies (Stucki, unpublished). The biological elimination of phosphonates is assumed to proceed at one of Ciba's effluent treatment plants in Switzerland where low amounts of phosphates are available and considerable quantities of phosphonates "disappear". In contrast to phosphonates, organic phosphates are easy to degrade aerobically. The resulting phosphate has to be precipitated subsequently.

3. Technical barriers delaying the application of environmental biotechnological processes in the chemical industry

Although the biotechnological potential to solve environmental challenges is huge, its applicability in the chemical industry is slow. Barriers preventing any new technology to be applied exist on different levels, e.g. at the level of regulation and legislation, at the economic level, and last but not least on the technical level. In this chapter, some technical barriers are identified and strategies to overcome these hurdles are suggested (Table 3).

A technical barrier to apply innovative biological solutions often exists already in the approach to solve an environmental challenge. Many engineers and chemists responsible for the management and realisation of environmental projects in the chemical industry are obviously more familiar with chemical and physical processes, which seem to be more predictable. There is a lack of know-how to estimate the biological degradation potential and how to improve the reliability of the biological processes.

Table 3. Technical barriers delaying the application of biotechnological solutions to solve environmental challenges in the chemical industry

Challenge	In the past	In future
Ecoefficiency	An holistic perspective frequently missing, problems shifted from water to off-gas, or from off-gas to water, etc.	Optimal environmental processes in relation to economics and ecology are looked for - a big chance for biotechnological processes
Indicator for biodegradability	COD/BOD$_5$ ratio	DOC (dissolved organic carbon) biodegradation
Testing time to determine the biodegradation potential in a given wastewater treatment plant	New tests requiring shorter testing times are developed and used	Cheap test set-up, low maintenance work, long incubation times (14 days and more)
Microbial growth rate	Fast growing strains are considered as those with the better application potential	Slowly growing bacteria are equally applicable but require sophisticated engineering, good reactor design, etc.
New degradation abilities	Isolation of new strains	Application of sludge with special degradation competence to evaluate the extent of degradation and/or speed up initial degradation rates
Off-gas treatment	Quite expensive energy- intensive technologies are implemented to remove biologically degradable chemicals	Field of application will be widened, biotrickling filters will clean highly and weakly contaminated off-gases, a requirement for success is an interdisciplinary engineering approach

Until today, environmental processes are often implemented without a holistic perspective of the problem, and usually under a big time pressure. As a result, many environmental activities are, if strictly analysed, counter-productive, thus polluting the environment more than if nothing was done. Examples are the moving of contaminated soils from one site to a distant site for disposal, the use of a high-energy technology for the removal of trace amounts of biodegradable chemicals, or the stripping of water contaminants to the atmosphere. The most important chance for biotechnological processes is if ecoefficient processes were preferred to solve environmental problems. Ecoefficient processes might be defined as those showing the highest environmental gain at the lowest costs. We have used several modified methods originally developed for life cycle analysis to compare and assess environmental processes. We realised that biological processes correctly applied often outperform chemical and physical alternatives, which require more energy than the biological options. Nature continuously optimises energy consumption of its processes. Natural processes are very energy efficient, and thus, the applicability of biological processes is very promising if ecoefficient options are favoured.

The biodegradative potential to treat water, air and/or soil is still underestimated in the chemical industry. The most widely used rule of thumb to estimate the extent of biodegradation of a certain wastewater is by the ratio of the chemical oxygen demand and the biological oxygen demand within 5 days: COD/BOD_5 ratio. Whereas this term is successfully applicable in the field of domestic wastewater treatment, the chemical environment may influence both parameters COD and BOD too much, and thus the ratio is very unreliable. Compounds at a high oxidation level like glyoxylate yield very low COD/BOD_5 ratios and thus pessimistic degradation forecasts even though they are excellently biodegradable. Chemicals such as H_2S, hydrazine, chloride ions in the g/l-range, SO_2 or NO_2 have a big impact on the COD/BOD_5 ratios, and thus lead to false negative conclusions. The evaluation of biodegradability of certain wastewaters or chemicals based on the extent of TOC elimination is considered as more reliable, and is used at most sites of the Ciba Corporation for the internal wastewater management.

Many new tests to determine the extent of biodegradation in shorter time periods are developed. Often, the newly designed tests are used to determine the degradation potential of certain compounds in wastewater treatment plants. Again, the application in the domestic field is acknowledged due to the more or less constant wastewater composition. However, the biodegradation potential in the chemical industry based on batch processes is likely to be underestimated with these short-time tests, since chemicals whose degradation requires time for adaptation, induction of certain catabolic enzymes, or growth of slowly growing microorganisms until degradation is detectable will show negative results. Therefore, too short biodegradation tests are likely to yield too low degradability predictions. At many sites of our corporation, tests to determine the inherent biodegradability within 2 to 4 weeks are carried out. A simple test set-up consisting of a laboratory shaker and some Erlenmeyer flasks allows to run many tests cheaply. The extent of biological elimination is determined on the basis of DOC. Analysis is done at the test start and after each week.

Microbiologists mostly favour fast growing bacteria and ascribe to these organisms a higher application potential. Whereas this might be true for production processes, the growth rate in environmental biotechnology is often of secondary importance. Even slow growing strains such as e.g. the acetotrophic methanogens contribute to powerful and extremely efficient biological processes, provided that the reactors are designed such that the bacteria are kept in the system where they can bring about their degradation potential.

New bacteria or sludge capable of degrading chemicals known as non-biodegradable are detected every year [9, 14]. In many cases, not enough attention is paid to the fact, that the degradation potential of sludge is usually restricted to a certain site, an off-gas filter, or a specific wastewater treatment plant receiving the substances under consideration for a long period of time. In a few cases, we have detected that the purification efficiency of wastewater of a certain production process decreased considerable when the production was moved from one site to another. Thus, the competence to degrade certain chemicals was lost and had to be recovered at the new site. We consider the use of sludge able to eliminate the substances under consideration as crucial to estimate the extent of biodegradation at the start of a project, simply to keep the biological option viable. Later, the transfer of sludge with special degradation capabilities form one site to another might speed up the degradation competence of a given environmental system such as wastewater treatment plants, off-gas filters or remediation sites.

In the chemical industry, low loads of odorous compounds are often cleaned using huge amounts of energy and expensive equipment. Although different systems to treat off-gas biologically are on the market, their application in the chemical industry is small. Process limitations are not primarily seen in the biological field, but also regarding chemical (mass transfer), physical (humidity) and engineering (nozzles) issues. The field of application will be widened with the trickling filter, where low volumes/high concentrations and high volumes/low concentrations will be economically treatable. Success in the field of biological off-gas treatment requires an interdisciplinary approach.

4. Conclusions

Biotechnological processes offer ecoefficient solutions for environmental challenges in the chemical industry. Case studies performed during the last twelve years at Ciba Specialty Chemicals, where this potential was tried to be applied wherever possible show that the field of application becomes wider. To apply biological processes more frequently, a holistic approach to solve the environmental issues is required. In addition, simple, cheap, and reliable evaluation methods to determine the extent of biological elimination and degradation of chemicals in given treatment facilities should be applied.

References

1. Matter, B. and Gschwind, N.: Abwasserreinigung bei der Sanierung der Sondermülldeponie Bonfol. gwa/Gas Wasser Abwasser 69 (1989), 381-388.
2. Stucki, G., Yu, C.W., Baumgartner, T. and Gonzalez-Valero, J.F.: Microbial atrazine mineralisation under carbon limiting and denitrifying conditions. Water Res. 29 (1995), 291-296.
3. Stucki, G.: Biological decomposition of dichloromethane from a chemical process effluent. Biodegradation 1 (1991), 221-228.
4. Thüer, M., Reisinger, M. and Stucki, G.: Biologischer CKW-Abbau zur Grundwasserbehandlung. Vergleich unterschiedlicher Bioreaktoren im Pilotversuch und erste Erfahrungen aus der Praxis. gwa/Gas Wasser Abwasser 74 (1994), 1-9
5. Stucki, G. and Thüer, M.: Experiences of a large-scale application of 1,2-dichloroethane degrading microorganisms for groundwater treatment. Environ. Sci. Technol. 29 (1995), 2339-2345.
6. Häggblom, M.M. and Valo, R.J. (1995) Bioremediation of chlorophenol wastes, in L.Y. Young and C.E. Cerniglia (eds.), Microbial transformation and degradation of toxic organic chemicals. Wiley-Liss Inc., New York, pp. 389-434.
7. Wenk, M., Bourgeois, M., Allen, J. and Stucki G.: Effects of atrazine-mineralising microorganisms on weed growth in atrazine-treated soils. J. Agric. Food Chem. 45 (1997), 4474-4480.

8. Wenk M., Baumgartner, T., Dobovšek, J., Fuchs, T., Kuscera, J., Zopfi, J. and Stucki, G.: Rapid atrazine mineralisation in soil slurry and moist soil by inoculation of an atrazine degrading Pseudomonas sp. strain. Appl. Microbiol. Biotechnol. 49 (1998), 624-630.
9. Steinle, P., Stucki, G., Stettler, R. and Hanselmann, K.W.: Aerobic mineralisation of 2,6-dichlorophenol by Ralstonia sp. Appl. Environ. Microbiol. 64 (1998), 2566-2571.
10. Steinle, P., Stucki, G., Bachofen, R. and Hanselmann, K.W.: Alkaline soil extraction and subsequent mineralisation of 2,6-dichlorophenol in a fixed-bed bioreactor. Bioremediation J. 3 (1999), 223-232.
11. Siegel, O. and Lais, P.: Schmutzwasserbehandlung der Sondermülldeponie Koelliken. gwa/Gas Wasser Abwasser 78 (1998), 193-200.
12. Stucki, G., Hanselmann, K.W. and Hürzeler, R.A.: Biological sulfuric acid transformation: Reactor design and process optimization. Biotechnol. Bioengin. 41 (1993), 303-315.
13. Schowanek, D. and Verstraete, W.: Phosphonate utilization by bacteria in the presence of alternative phosphorous sources. Biodegradation 1 (1990), 43-53.
14. Gisi, D., Stucki, G. and Hanselmann, K.W.: Biodegradation of the pesticide 4,6-dinitro-ortho-cresol by microorganisms in batch cultures and in fixed bed column reactors. Appl. Microbiol. Biotechnol. 48 (1997), 441-448.
15. Steinle, P., Thalmann, P., Höhener, P., Hanselmann, K.W. and Stucki, G.: Effect of environmental factors on the degradation of 2,6-dichlorophenol in soil. Environm. Sci. Technol. 34 (2000), 771-775.

PART 2
MOLECULAR ASPECTS OF XENOBIOTIC DEGRADATION

MOLECULAR CHARACTERISATION OF KEY ENZYMES IN HALORESPIRATION

HAUKE SMIDT, ANTOON D. L. AKKERMANS, JOHN VAN DER OOST AND WILLEM M. DE VOS

Laboratory of Microbiology, Wageningen University, Hesselink van Suchtelenweg 4, NL-6703 CT Wageningen, The Netherlands
Email: hauke.smidt@algemeen.micr.wag-ur.nl
tel: +31-317-483118; fax: +31-317-483829
Present addressHauke Smidt:University of Washington, Dept. of Civil and Environmental Engineering, 268 Wilcox Hall,Box 352700, Seattle, WA 98195-2700, USA,Phone: +1-206-543-2094 (office) +1-206-685-6657 (lab)FAX : +1-206-685-3836 email: smidt@u.washington.edu

Abstract

Halorespiring bacteria are able to couple the reductive dechlorination of halogenated aliphatic and aromatic compounds to energy conservation and hence to microbial growth. Isolation of these strains and their expected potential for application in *in situ* biodegradation of haloorganic compounds also have led to an increased interest in the molecular basis of the halorespiratory pathway. Integrated physiological, biochemical and molecular genetic approaches have provided deeper insights in the structure, function and regulation of the halorespiratory electron transfer chain. The identification of reductive dehalogenases as the key enzymes in this process was followed by their detailed molecular characterisation. This revealed considerable similarities at both the mechanistic and structural level, suggesting that these enzymes constitute a novel class of corrinoid containing reductases. Our current knowledge on the phylogeny of halorespiring bacteria and on the molecular characterisation of their dehalogenating systems provides a sound basis for the further exploitation of these microorganisms as dedicated degraders in polluted environments.

1. Introduction

Halogenated hydrocarbons are present in the environment in high quantities due to their past and present application in industry and agriculture e.g. as solvents, pesticides and preservatives, compromising environmental integrity and health [1, 25, 33]. However, as more than 2000 haloorganic compounds are naturally produced at considerable

23

S.N. Agathos and W. Reineke (eds.),
Biotechnology for the Environment: Strategy and Fundamentals, 23–46.
© 2002 *Kluwer Academic Publishers. Printed in the Netherlands.*

levels, they should not be regarded as of solely anthropogenic origin. Rather, the abundance of natural halogenated compounds has been the selective pressure that has resulted in the evolution of microbial dehalogenating populations. This might explain the unexpectedly high microbial capacity to dehalogenate different classes of xenobiotic haloorganics [23, 24].

The biodegradability of halogenated hydrocarbons largely depends on their chemical structure and the environmental conditions. The degradation of lower halogenated compounds, which proceeds relatively efficiently under aerobic conditions has been studied in considerable detail at the physiological, biochemical and genetic level [18, 32, 73]. However, with the exception of oxidative pentachlorophenol dehalogenation [18], dehalogenating systems that depend on molecular oxygen are only suitable for the attack of haloorganic compounds carrying a limited number of highly electronegative halogen-substituents, resulting in the persistence of polyhalogenated compounds in aerobic environments. Whereas e.g. the co-oxidation of partially chlorinated mono-, di-, and trichloroethene is fortuitously catalysed by mono- and dioxygenases in various bacteria, the fully halogenated tetrachloroethene is not degradable under these conditions [2, 39]. During the last two decades, it has been shown for a large variety of halogenated aliphatic and aromatic compounds, that reductive dehalogenation is the crucial step, by which the degradation of these pollutants can be initiated in anoxic environments. The abiotic or co-metabolic conversion under different redox conditions by numerous anaerobic mixed and pure cultures is proposed to be catalysed in most cases by metal ion-containing heat stable tetrapyrroles or enzymes, in which these compounds are incorporated as cofactors [17, 19, 28]. In addition, an increasing number of anaerobic bacteria has recently been isolated that are able to couple the reductive dehalogenation of halogenated aliphatic and aromatic compounds to energy conservation and hence to microbial growth ([17, 30], Fig. 1, 2). In contrast to the above-mentioned co-metabolic conversions, in these microorganisms dehalogenation is catalysed by specific enzymes with high specific activity and affinity. This novel respiratory process has previously been described as *halorespiration* (*halo*genated hydrocarbons as terminal electron acceptor in anaerobic *respiration*, [68]). Although the alternative term *dehalorespiration* has been proposed to be more appropriate (*dehalo*genation as terminal electron accepting process in anaerobic *respiration*, [30]), we refer to this process as halorespiration for its analogy with fumarate respiration and because of the obvious direction of the conversion.

The different mechanisms, by which bacteria are able to dehalogenate haloaliphatic and haloaromatic compounds, has been recently reviewed in several excellent communications and their detailed exhaustive description goes far beyond the scope of this review (see [6, 18, 30, 38, 52, 89] and references therein).

This review will discuss the current knowledge on the key characteristics of halorespiring bacteria and the structure, function and regulation of halorespiratory systems active in these microorganisms.

2. Halorespiring bacteria –thermodynamic rationale, phylogeny and key characteristics

Estimation of Gibbs free energies and redox potentials have indicated that halogenated aliphatic and aromatic compounds should potentially be good electron acceptors in anaerobic environments [15, 16]. As an example, the Gibbs free energy available from the reductive dehalogenation of 2,3-dichlorophenol to 2-chlorophenol has been calculated to −147.9 kJ, whereas the standard redox potential of this couple amounts to +353 mV. This value is significantly higher than those calculated for the redox couples SO_4^{2-} / H_2S (E_0' = -217 mV) and fumarate/succinate (E_0' = +30 mV), and comparable to the redox potential of NO_3^- / NO_2^- (E_0' = +433 mV). Even when taking into account that H_2-partial pressures in anaerobic environments are orders of magnitude below standard conditions, reductive dehalogenation seems energetically still highly competitive with other occurring anoxic terminal electron accepting processes [15, 78].

Over the past decade, a rapidly growing number of bacteria has been isolated based on their ability to use chloroalkenes, such as tetrachloroethene (PCE) and trichloroethene (TCE), or chloroaromatic compounds like chlorophenols and chlorobenzoates as the terminal electron acceptor (Fig. 1). Strains have been isolated from various polluted and pristine environments, ranging from activated- and anaerobic granular sludge to freshwater- and estuarine sediments. These microorganisms have gained increasing attention because of their potential in bioremediation of contaminated anoxic environments, the novel respiratory pathways they possess and the capacity of various isolates to dechlorinate both chloroaromatics and chloroalkenes.

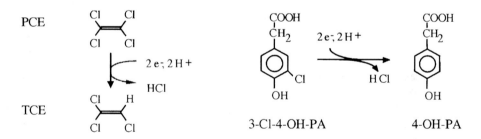

Figure 1. Examples of reductive dehalogenation reactions as they are performed by halorespiring microorganisms. (PCE = tetrachloroethene; TCE = trichloroethene; PA = phenylacetic acid)

With one exception, halorespirers have been affiliated with distinct phylogenetic branches of the bacterial domain, namely the groups of low G+C Gram-positives, δ- and ε-proteobacteria (Fig. 2), and their main characteristics have recently been summarised in two exhaustive reviews [17, 30]. Most of the isolates are rather versatile with respect to their ability to use, besides fermentative growth at the expense of e.g. pyruvate, a whole range of different electron donors and acceptors for growth. However, also a small number of apparently obligate halorespiring isolates has been identified to date.

The phylogenetically deeply branching *Dehalococcoides ethenogenes* couples the oxidation of H_2 to the reduction of PCE, TCE, cis-dichloroethene (cis-DCE), 1,1-DCE and dichloroethane (DCA) to vinylchloride (VC) and ethene. It is currently the only pure culture known to completely dechlorinate and, hence, detoxify the abundant bulk contaminant PCE.

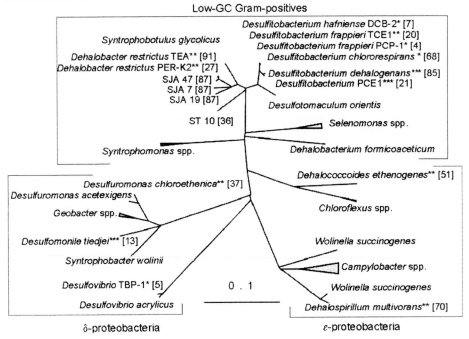

Figure 2. Phylogenetic tree based on bacterial SSU rRNA sequences. The reported capacity to dehalogenate chloroaromatics (*), chloroalkenes (**) or both (***) is indicated. Alignment and phylogenetic analysis were performed with the ARB software [77], and the tree was constructed following maximum parsimony criteria with nearest neighbour optimisation. No SSU rRNA sequences were available for Desulfitobacterium strains PCE-S and Viet1, Desulfuromonas strain BB1, and strains 2-CP1 and 2-CPC. The reference bar indicates 10 nucleotide changes per 100 nucleotides.

While the reductive dehalogenation of the above mentioned compounds is coupled to growth, the conversion of trans-DCE and VC occurs at comparably slow rates with first-order kinetics, indicating co-metabolic conversion [50, 51]. Two closely related isolates of the low G+C Gram-positive *Dehalobacter restrictus* have been described that are both strictly dependent on anaerobic respiration, coupling the oxidation of H_2 to the reductive dehalogenation of PCE and TCE to predominantly cis-DCE [27, 91]. Moreover, the *Dehalobacter*-related strain ST-10 could be identified as one of two dominant species in a thermophilic (65°C) PCE-dehalogenating enrichment culture using a 16S rRNA gene-based molecular approach [36]. Using general bacterial 16S rRNA gene-specific fingerprinting techniques, amplicons have been isolated in several cases that showed highest sequence similarity with obligate halorespiring isolates. In one recent study, a clone family could be identified from an anaerobic trichlorobenzene degrading community that showed 98.8 to 99.4 % sequence identity with the 16S rRNA gene of *Dehalobacter restrictus* (SJ7, SJ19, SJ47) [87]. Interestingly, a *Dehalococcoides*-like strain was shown to be specifically stimulated in an actively 2,3,5,6-tetrachlorobiphenyl-*ortho*-dechlorinating microbial consortium [31].

The first bacterial isolate, for which halorespiration was unambiguously proven, is the sulphate-reducing 3-chlorobenzoate (3-CB) degrading *Desulfomonile tiedjei* [13, 58]. The organism is able to couple the reductive dehalogenation of 3-CB to pyruvate-, formate- and H_2-oxidation [12, 57]. With hydrogen as the electron donor, chemiosmotic coupling of reductive dehalogenation and ATP-synthesis could be demonstrated, as uncouplers and ionophores reduced the ATP-pool relative to dehalogenation activity [58].

Among the halorespiring pure cultures that have been reported to date, the Gram-positive genus *Desulfitobacterium* comprises a major group of isolates. All of *Desulfitobacterium* spp. are rather versatile with respect to their metabolic properties. In most cases, the coupling of reductive dehalogenation to the oxidation of H_2 and/or formate could be demonstrated, indicating energy conservation via electron-transport-coupled phosphorylation [20, 40, 47, 68]. Like other halorespiring isolates, most strains of *Desulfitobacterium* also couple the oxidation of other substrates (e.g. pyruvate, lactate) to reductive dehalogenation. However, as these substrates also support energy conservation via substrate-level phosphorylation, it cannot be excluded that under these conditions reductive dehalogenation merely serves as an electron sink rather than supporting electron-transport-coupled phosphorylation ([30], van de Pas *et al.*, to be published). Haloorganic compounds that are used as terminal electron acceptor by *Desulfitobacterium* strains, include chlorinated ethenes (either PCE [21, 40, 90] or PCE / TCE [20, 54]) and halogenated phenolic compounds (in most cases *ortho*-chlorinated compounds [4, 7, 21, 68, 85], but also *meta*- and *para*-substituted isomers [4]), including several hydroxylated polychlorinated biphenyls [90]. Interestingly, two of the isolates, namely strain PCE1 and *Desulfitobacterium frappieri* PCP-1, were shown to have two independent activities, indicating the presence of multiple enzyme systems, as was also reported for *Dehalococcoides ethenogenes* (see below) [4, 20, 48].

Additional microorganisms that have been isolated for their halorespiring capacity include the ε-proteobacterium *Dehalospirillum multivorans*, which uses chlorinated ethenes as the terminal electron acceptor [70]. The study of physiology, bioenergetics and, to some extend, molecular biology of halorespiration in this organism has significantly contributed to our understanding of this process (recently reviewed in [30]). The first example of a halorespiring microorganism from an estuarine environment is the recently isolated *Desulfovibrio* strain TBP-1, which couples the reduction of *ortho*-and *para*-brominated phenols to the oxidation of lactate [5]. Recently, two strains of *Desulfuromonas* have been isolated based on their unique ability to couple the oxidation of acetate, but not hydrogen, to the reductive dehalogenation of PCE and TCE to cis-DCE, namely *Desulfuromonas chloroethenica* and *Desulfuromonas* strain BB1 [37, 42]. The only facultative anaerobic halorespiring microorganisms reported to date are the closely related 2-CP1 and 2-CPC, which use *ortho*-chlorinated phenols and fumarate as e-acceptors. 16S rRNA gene-based phylogenetic analysis revealed that they are most closely related to the myxobacteria within the δ-proteobacteria [9, 43].

3. Key components of halorespiratory chains

Isolation of halorespiring bacteria and their expected potential for application in *in situ* biodegradation of haloorganic pollutants has also led to an increased interest in the molecular basis of this novel anaerobic respiratory pathway. To date, investigations have mainly focused on reductive dehalogenase as the key enzyme in halorespiration. However, efforts have also been made to identify additional structural and regulatory components of the halorespiratory electron transport chain (see section 3.2.).

It is known from physiological experiments that in several halorespiring bacteria described to date, reductive dehalogenase activity is induced in the presence of a halogenated substrate. Moreover, the influence of alternative electron acceptors on the activity of the dehalogenating system has been investigated, indicating that particularly sulphur oxy-anions are potential inhibitors of halorespiration. However, insight in the regulatory circuits involved in the induction and repression of the halorespiration process is still very limited. Evidence is now emerging that at least partly, regulation takes place at the level of transcription (see section 4.).

3.1. REDUCTIVE DEHALOGENASES - ENZYMES AND GENES

3.1.1. Major characteristics of reductive dehalogenases

To date, several haloaryl- and haloalkyl reductive dehalogenases have been at least partially purified and characterised on the biochemical, and in some cases, on the genetic level. As one would expect for a respiratory complex, all enzymes were shown to be membrane-associated. Only the PCE reductive dehalogenase from *Dehalospirillum multivorans* was isolated from the cytoplasm [63]. However, evidence is available from the molecular characterisation of the encoding gene that the catalytic subunit might be anchored to the cytoplasmic membrane through an integral membrane protein ([64], see below). In some cases, the electron-accepting site of the enzymes was inaccessible for reduced methyl viologen (MV) in whole cells. As MV is not able to permeate the cytoplasmic membrane [34], it was concluded that the enzyme is facing the cytoplasm. Nonetheless, molecular analysis has revealed that the proteins are produced as pre-proteins, in which the mature polypeptides are preceded by a *twin arginine*-type signal sequence characteristic for periplasmic respiratory complexes [3]. Obviously, additional experiments are required to solve the topology of the enzymes (see below). Major characteristics of the enzymes are summarised in Table 1.

culture and characterised on the biochemical level, was the inducible 3-CB reductive dehalogenase from the sulphate reducing *Desulfomonile tiedjei* [65]. As for all reductive dehalogenases isolated since then, the enzymatic activity could be measured *in vitro* using reduced MV (E_0' = -446 mV) rather than benzyl viologen (BV, E_0' = -360 mV) as artificial electron donor. The enzyme was purified from the membrane fraction as a heterodimer, which was insensitive to oxygen. A yellowish chromophore, proposed to be a heme, was present in the small 37-kDa subunit of the enzyme. *In vivo*, the reductive dehalogenation of *m*-chlorinated phenols and chlorinated ethenes (PCE, TCE) is co-induced by 3-CB, suggesting co-metabolic conversion of these compounds by the 3-CB reductive dehalogenase. However, no data are available for the purified enzyme [56, 82].

In contrast to the reductive dehalogenase from *Desulfomonile tiedjei*, all other proteins characterised to date have been isolated as monomers, probably containing Fe-S clusters as well as a corrinoid as cofactors, indicating a common mode of catalytic action (Table 1).

3.1.2. Involvement of transition metal cofactors and Fe-S clusters in catalysis

Photoreversible inactivation by iodoalkanes indicated that cob(I)alamin is probably involved in the reductive dehalogenation of chlorinated aliphatic and aromatic compounds in most halorespiring bacteria ([48, 54, 62, 71]; van de Pas, personal communication). The 3-CB reductive dehalogenase from *Desulfomonile tiedjei* was not inhibited by 250 mM 1-iodopropane, and it has been concluded that no corrinoid is involved in its catalytic activity [44]. Rather, heme (yet another transition metal cofactor) seems to be present in the small subunit of this enzyme [65]. However, in two cases, namely the 3-Cl-4-OH-phenylacetic acid (Cl-OHPA)- and PCE reductive dehalogenases from *Desulfitobacterium hafniense* and *Dehalococcoides ethenogenes*, respectively, the involvement of a corrinoid was demonstrated despite the fact that no inhibition by 1-iodopropane was observed [8, 48]. This indicates that it cannot be unambiguously excluded that a corrinoid plays a role in the *Desulfomonile tiedjei* enzyme. It is noteworthy that for different classes of transition metal cofactors, including corrinoids and hemes, abiotic reductive dehalogenation activity has been demonstrated [19, 22], indicating their potential role as cofactors also in enzyme catalysed dechlorination.

For two enzymes, the presence of cofactors and their involvement in catalytic activity has been demonstrated by optical and electron paramagnetic resonance (EPR) spectroscopic analysis. Studies on the purified *o*-CP reductive dehalogenase of *Desulfitobacterium dehalogenans* revealed the presence of a cobalamin ($Em(Co^{1+/2+})$ = -370 mV and $Em(Co^{2+/3+}) > 150$ mV), one [4Fe4S] cluster ($Em \approx -440$ mV) and one [3Fe4S] cluster ($Em \approx +70$ mV). The reoxidation of fully (light/deazaflavin/EDTA) reduced enzyme with Cl-OHPA yielded base-off cob(II)alamin [86]. Similar results were obtained with purified PCE reductive dehalogenase from *Dehalobacter restrictus*. However, this enzyme contains two [4Fe4S] clusters with $Em \approx -480$ mV, rather than one high- and one low-potential cluster [72]. The observed redox potentials of the corrinoid are significantly higher than those found for other corrinoid containing

The first dehalogenating enzyme that has been purified from a halorespiring pure enzymes, indicating that activation is probably not required prior to reduction to the Co(I)-state. Rather, this could easily be accomplished by the low-potential Fe-S centres. It was therefore suggested that a single-electron transfer occurs from a yet unknown electron donor via the two cubanes and the corrinoid to the chlorinated substrate [72]. Nevertheless, experimental proof for a single-electron transfer yielding an alkyl-radical intermediate is still lacking.

Table 1. Main characteristics of reductive dehalogenases from halorespiring microorganisms (Part1)

Organism	*Desulfomonile tiedjei*	*Desulfitobacterium dehalogenans*	*Desulfitobacterium hafniense*	*Desulfitobacterium chlororespirans*
Substrates	*m*-CB, *m*-CP, PCE/TCE	*o*-CP	3-Cl-4-OH-PA	*o*-CB, *o*-CP
Inducers	*m*-CB, *m*-substituted benzamides and -benzylalcohols	3-Cl-4-OH-PA, 2,4,6-TCP, 2,4-DCP, 2,3-DCP, 2-Br-4-CP, 2-Br-4-CH$_3$ -P, 2-Cl-4-R-P (R=F, Br, CO$_3^-$, CH$_3$, COOCH$_3$)	3-Cl-4-OH-PA	3-Cl-4-OH-BA
Inhibitors	sulphite, thiosulphate	1-iodopropane[d]		sulphite, thiosulphate
Electron donor	MV	MV	MV	MV
Specific activity in cell extracts (U / mg)[e]	0.15×10^{-3}	0.38	0.55	0.009
Specific activity of purified protein (U / mg)	0.018	28	6	0.019 (in membranes)
Size (SDS-PAGE)	64 & 37 kDa	48 kDa	47 kDa	ND
Cofactors (mol/mol enzyme)	heme[c] Fe-S	1 cobalamin[a, b] 1 [3Fe4S][b] 1 [4Fe4S][b]	1 cobalamin[c] 12Fe/13S	ND
N-terminus	ND	AETMNYVPG	AETMNYVPG	ND
O$_2$-sensitivity (t$_{1/2}$)	insensitive	≈ 90 min[d]	≈ 100 min (in membranes)	insensitive (in membranes)
References	[14, 59, 65, 83]	[84, 86]	[8]	[41]

[a] Indicated by photoreversible inactivation of the reduced enzyme by iodoalkanes; [b] Indicated by EPR spectroscopic analysis; [c] Indicated by optical spectroscopic analysis; [d] van de Pas, personal communication; [e] One unit (U) of enzymatic activity is defined as 1 μmol Cl⁻ released or 2 μmol MV oxidised per minute; abbreviations: ND, not determined; CB, chlorobenzoate; CP, chlorophenol; PCE, tetrachloroethene; TCE, trichloroethene; PA, phenylacetic acid; BA, benzoic acid; MV, methylviologen (reduced).

Table 1. Main characteristics of reductive dehalogenases from halorespiring microorganisms (Part 2)

Organism	Dehalobacter restrictus	Dehalospirillum multivorans	Desulfitobacterium PCE-S	Dehalococcoides	ethenogenes
Substrates	PCE, TCE	PCE, TCE, Cl-propenes	PCE, TCE, Cl-propenes	TCE, cis-DCE	PCE
Inducers	ND	PCE (also constitutive and non-inducible strain isolated)	constitutive	ND	constitutive
Inhibitors	1-iodopropane, cyanide	1-iodopropane, Cl-methanes, EDTA sulphite, cyanide	1-iodopropane, EDTA, cyanide, sulphite, azide,	1-iodopropane cyanide, sulphite, dithionite	iodoethane, cyanide, sulphite, dithionite
Electron donor	MV	viologens with $E_0' < -360$ mV	viologens with $E_0' < -360$ mV	MV	MV
Specific activity in cell extracts (U / mg)[e]	0.18	1.5	0.24	0.51	0.27
Spec. activity of purified protein (U / mg)	14	158	39	12	21
Size (SDS-PAGE)	60 kDa	58 kDa	65 kDa	61 kDa	51 kDa
Cofactors (mol/mol enzyme)	1 cobalamin[a-c] 2 [4Fe4S][b]	1 cobalamin[a, c] 8Fe/8S	1 cobalamin[a, c] 8Fe/8S	1 cobalamin[a] Fe-S	1 cobalamin[a] Fe-S
N-terminus	19/20 residues identical to PCE-S	GVPGANAAE K	ADIVAPITESF	ND	ND
O_2-sensitivity ($t_{1/2}$)	280 min	120 min	50 min	ND	ND
References	[30,71,72]	[54,62,44]	[54,55]	[48]	[48]

[a] Indicated by photoreversible inactivation of the reduced enzyme by iodoalkanes; [b] Indicated by EPR spectroscopic analysis; [c] Indicated by optical spectroscopic analysis; [d] van de Pas, personal communication; [e]One unit (U) of enzymatic activity is defined as 1μmol Cl⁻ released or 2 μmol MV oxidised per minute; abbreviations: ND, not determined; CB, chlorobenzoate; CP, chlorophenol; PCE, tetrachloroethene; TCE, trichloroethene; PA, phenylacetic acid; BA, benzoic acid; MV, Methyl viologen (reduced)

31

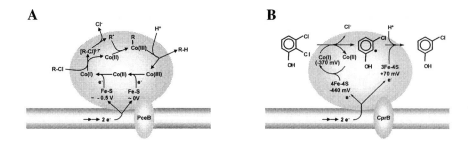

Figure 3. Proposed reaction mechanisms of (A) corrinoid-containing PCE reductive dehalogenases according to Holliger et al. [30] and (B) o-CP reductive dehalogenase from Desulfitobacterium dehalogenans [86].

Alternatively, Wohlfarth & Diekert [92] proposed that the dehalogenation reaction proceeds through addition of cob(I)alamin to a carbon of the halogenated substrate with subsequent β-elimination of a chlorine ion. This model assumes the splitting of electrons into one low-potential and one high-potential electron via the two Fe-S clusters. The low-potential electron would be required for the reduction of cob(II)alamin to cob(I)alamin prior to catalysis, the high potential electron for the reduction of cob(III)alamin to cob(II)alamin after dechlorination [92]. However, it has recently been observed that the PCE reductive dehalogenases from both *Dehalospirillum multivorans* and *Desulfitobacterium frappieri* PCE-S also convert *trans*-1,3-dichloropropene to 1-chloropropene. This favours the involvement of a one-electron transfer mechanism as it was suggested for the *Dehalobacter restrictus* PCE reductase rather than the proposed β-elimination, since only the former would yield the actual dehalogenation product (Fig.3A) [30].

Still, in both enzymes for which the different cofactors have been analysed by EPR spectroscopic analysis, the corrinoid could not be oxidised to the cob(III)alamin form, suggesting that it might not play a role in the catalytic cycle of reductive dehalogenation. Moreover, this would be in agreement with the mechanism proposed for the reductive dehalogenation of PCE by free cobalamin. Aiming at the incorporation of all experimental evidence obtained to date, yet another reaction mechanism can be proposed for the *o*-chlorophenol reductive dehalogenase from *Desulfitobacterium dehalogenans* (Fig. 3B, [86]). This model involves activation of the oxidized cob(II)alamin to cob(I)alamin by a low-potential electron from the low-potential [4Fe4S] cluster. The reduced cob(I)alamin might then donate one electron to the chlorinated substrate, yielding the release of a chlorine ion and an alkyl-radical intermediate, which, however, has not been unambiguously demonstrated. The dehalogenated product would then be released following the transfer of a second electron from the high-potential [3Fe4S] cluster. Future experiments have to reveal, whether both, haloalkene- and haloaryl reductive dechlorination, proceed through identical mechanisms, or whether the various enzymes catalyse the reactions along similar, albeit different pathways.

3.1.3. Molecular biology of reductive dehalogenases

Using a reverse genetics approach based on the N-terminal amino acid sequences of the purified enzymes, the *o*-CP- and PCE reductive dehalogenase encoding genes were isolated from genomic libraries of *Desulfitobacterium dehalogenans*, *Dehalospirillum multivorans* and *Desulfitobacterium* strain PCE-S, respectively, cloned and sequenced [60, 64, 86]. In all cases, sequence analysis revealed the presence of two closely linked genes: i) *cprA* and *pceA*, coding for the catalytic subunit of the respective reductive dehalogenases; and ii) *cprB*, *pceB* and orf1, encoding small integral membrane proteins of 11, 8.4 and 11 kDa, that are composed of two (PceB) or three (CprB, Orf1) transmembrane helices (Fig. 4). As cotranscription of both genes could be demonstrated for *Desulfitobacterium dehalogenans* and *Dehalospirillum multivorans*, it has been proposed that the integral membrane protein acts as a membrane anchor for the reductive dehalogenase [64, 75, 86]. The smaller size and lack of a third transmembrane helix in PceB, as compared to the *Desulfitobacterium* counterparts, might explain the significantly looser attachment of the *Dehalospirillum* catalytic subunit to the cytoplasmic membrane, as it was observed during purification [63].

Furthermore, all reductive dehalogenases share a rather long twin arginine (RR) signal sequence of 30 to 42 amino acids, which is cleaved off in the purified proteins (Fig. 5). These signal peptides (consensus (S/T)-R-R-x-F-L-K) are thought to play a major role in the maturation and translocation of mainly periplasmic proteins binding different redox cofactors by the recently described Twin Arginine Translocation (TAT) system [3]. This is in obvious contradiction to a possibly cytoplasmic orientation of the reductive dehalogenases, and the only other RR-enzyme with a proposed cytoplasmic location is the *E. coli* DMSO reductase [88]. It has been hypothesised that the presence of a membrane anchor might prevent translocation by the TAT-system [64]. Still, the actual role of the TAT-system in maturation and, possibly, translocation of reductive dehalogenases from halorespiring microorganisms deserves further study. Very interesting results have recently been obtained in a study, in which the halorespiration-induced expression of several molecular chaperone-encoding genes flanking the *cprBA* gene cluster in *Desulfitobacterium dehalogenans* has been shown by Northern blot analysis [75]. This might be a further step towards the unravelling of mechanisms of post-translational maturation of dehalogenating complexes in halorespiring bacteria, as it has been suggested that accessory proteins like the GroEL-type chaperonins might play a role in correct assembly and cofactor insertion during the TAT-dependent maturation of RR-signal peptide containing proteins [3, 69].

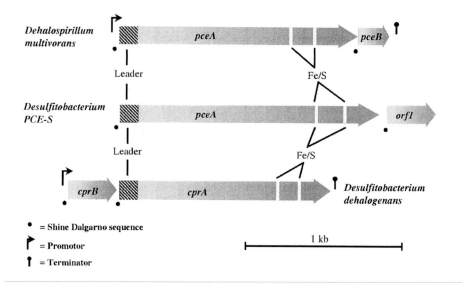

Figure 4: Comparison of o-chlorophenol- and PCE reductive dehalogenase encoding operons from different halorespiring microorganisms [60, 64, 75, 86].

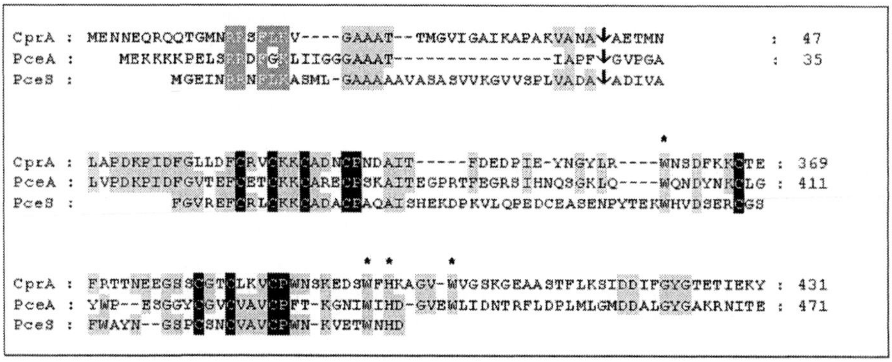

```
CprA :   MENNEQRQQTGMN  S   V----GAAAT--TMGVIGAIKAPAKVANA↓AETMN        :  47
PceA :      MEKKKKPELS  D G LIIGGGAAAT---------------IAPF↓GVPGA        :  35
PceS :          MGEIN N   ASML-GAAAAAVASASVVKGVVSPLVADA↓ADIVA

                                                          *
CprA :  LAPDKPIDFGLLDF RV KK ADN  NDAIT-----FDEDPIE-YNGYLR----WNSDFKK TE : 369
PceA :  LVPDKPIDFGVTEF ET KK ARE  SKAITEGPRTFBGRSIHNQSGKLQ----WQNDYNR LG : 411
PceS :          FGVREF RL KK ADA  AQAISHEKDPKVLQPEDCEASENPYTEKWHVDSER GS

                     *  *      *
CprA :  FRTTNEEGS S GT LKV  WNSKEDSWFHKAGV-WVGSKGEAASTFLKSIDDIFGYGTETIEKY : 431
PceA :  YWP--ESGGY GV VAV FT-KGNIWIHD-GVEWLIDNTRFLDPLMLGMDDALGYGAKRNITE  : 471
PceS :  FWAYN--GSP SN VAV WN-KVETWNHD
```

Figure 5. Primary sequence alignment of reductive dehalogenases from halorespiring bacteria. The alignment was performed using the programs Clustal X and GeneDoc [66, 80]. The apparent sites of leader cleavage, deduced from the N-terminal sequence of the purified proteins, are indicated by (↓). Conserved residue, light grey; RR-motif, dark grey; iron-sulfur cluster binding motif, black; conserved tryptophan and histidine residues, stars. CprA, Desulfitobacterium dehalogenans o-CP reductive dehalogenase (RD) (acc. no. AAD44542); PceA, Dehalospirillum multivorans PCE-RD (acc. no. AAC60788); PceA-S, Desulfitobacterium PCE-S PCE-RD [60].

The presence of two Fe-S clusters, as determined by EPR-analysis, was confirmed by the identification of one ferredoxin-like and one truncated Fe-S cluster binding motif in the sequence of CprA (Fig. 5). The same two motifs are also present in the sequence of the two PCE reductive dehalogenases, indicating a conserved mode of intramolecular transfer of electrons to the enzyme's active site that contains the cobalamin. These enzymes thus might differ in Fe-S cluster contents from the PCE reductive dehalogenase of *Dehalobacter restrictus*, for which EPR analysis had indicated the presence of 2 [4Fe4S] clusters of identical midpoint potential. Similarly, 3 Fe-S clusters have been suggested for the Cl-OHPA reductive dehalogenase from *Desulfitobacterium hafniense*. Nevertheless, these differences still have to be supported by the as yet unidentified primary structures of both proteins.

Although the reductive dehalogenases share highly conserved sequences in the C-terminal part of the proteins, they all lack the consensus sequence for the binding of the corrinoid cofactor, which is common to known methylcobalamin-dependent methyltransferases and mutases [46]. This is, however, not surprising as EPR analysis revealed that the active CprA enzyme contains the corrinoid in the base-off form, which is in contrast to the above mentioned proteins [86]. Of special interest is the presence of highly conserved tryptophan and histidine residues that have been shown or proposed to be involved in the stabilisation of the leaving halide in hydrolytic dehalogenases and dichloromethane dehalogenases, respectively (Fig. 5, [10, 49]).

The elucidation of structure-function relations and the identification of residues involved in catalysis in this novel class of corrinoid-containing reductases will have to await the availability of a structural model as well as heterologous expression systems, which are an obligate prerequisite for site-directed- and other mutagenesis approaches. Recently, heterologous expression of the PCE reductive dehalogenase encoding gene *pceA* from *Dehalospirillum multivorans* could be achieved in *Escherichia coli* in the presence of a helper plasmid, which supplied the expression host with the rare *E. coli* tRNA$_4^{Arg}$, correcting for differences in codon usage between *E. coli* and *Dehalospirillum multivorans*. However, expression of the full-length *pceA* led to the production of unprocessed inactive pre-protein, still containing the RR-signal peptide. Neither truncation of the signal sequence from the construct nor expression in the presence of *Dehalospirillum multivorans* corrinoids led to functional expression [64].

3.1.4. Multiple activities in halorespiring bacteria –enzymes involved and evolutionary aspects

It is known from several halorespiring bacteria that they possess multiple dechlorinating activities. Thus, the question arises, whether this is due to single enzymes with rather broad substrate specificities or a set of highly specific isoenzymes. Support for the latter scenario has been obtained by substrate-dependent differential induction patterns in *Desulfitobacterium frappieri* PCP-1 and *Desulfitobacterium* PCE1. The former has been shown to possess one specific dehalogenating system inducible for *ortho*-dehalogenation and one for *meta*- and *para*-dehalogenation [4]. Moreover, Cl-OHPA is transformed to 2-chlorophenol rather than being dehalogenated, indicating, that the *ortho*-dehalogenating enzyme system differs from the *o*-chlorophenol- and Cl-OHPA reductive dehalogenases isolated from *Desulfitobacterium dehalogenans* and *Desulfitobacterium hafniense* [11].

Desulfitobacterium PCE1 is able to reductively dehalogenate *o*-chlorinated phenolic compounds as well as PCE. However, cells grown in the presence of Cl-OHPA as the electron acceptor, showed only 0.25% of the PCE reductive dehalogenation activity of cells that used PCE as electron acceptor, indicating that two distinct enzyme systems might be involved in the two activities [20]. As the *o*-chlorophenol reductive dehalogenase of the closely related *Desulfitobacterium dehalogenans* does not convert PCE, it is tempting to speculate that the low PCE dechlorinating activity present in this organism could be due to a similar protein as it is active in *Desulfitobacterium* PCE1, of which either catalytic functionality or induction have been drastically impaired. Similarly, the 3,5-DCP *meta*-dechlorinating activity of *Desulfitobacterium hafniense* could be possibly due to a distinct enzyme, which might resemble the highly active catalyst in the closely related *Desulfitobacterium frappieri* PCP1.

From the PCE-detoxifying *Dehalococcoides ethenogenes*, two distinct enzyme systems have been partially purified, a 51 kDa PCE-reductive dehalogenase and a 61-kDa TCE/cis-DCE reductive dehalogenase [48]. Recently, the partial genome sequence of *Dehalococcoides ethenogenes* has been released, and sequence comparison led to the identification of at least sixteen non-identical copies of potentially reductive dehalogenase encoding genes. Moreover, these were all closely linked to genes, coding for CprB/PceB like small hydrophobic proteins of ~10 kDa (Preliminary sequence data were obtained from The Institute for Genomic Research website at http://www.tigr.org). The presence of these multiple alleles in the genome of *Dehalococcoides ethenogenes* might reflect a common strategy that should enable the organism to relatively quickly acquire novel degradation capacities upon exposure to an environmental trigger such as the anthropogenic release of non-natural halogenated hydrocarbons. It has been suggested that the cryptification of dehalogenase genes and their decryptification by environmentally directed mutations might play an important role in adaptive evolution [26, 79]. Hill and co-workers used degenerate primers to PCR-amplify halocarboxylic acid dehalogenase encoding genes (*deh*) from a wide variety of bacterial isolates [26]. Expression studies on the identified genes allowed for the discrimination between cryptic or silent, as well as active *deh* genes, and yielded the identification of several cryptic genes that either encoded non-functional gene products or had been silenced by the absence of an active promoter [26]. It is tempting to speculate that e.g. the high PCE reductive dehalogenase activity in *Desulfitobacterium* PCE1 and its hardly detectable presence in the closely related *Desulfitobacterium dehalogenans* might be another example of such a silencing / unsilencing process.

3.2. THE HALORESPIRATORY CHAIN

Physiological, biochemical and molecular genetic approaches have been used to unravel structure, function and regulation of the halorespiratory electron transfer chain. Studies have mainly focused on halorespiratory chains in which either hydrogen or formate serves as the electron donor as in these cases, energy can only be conserved by electron-transport-coupled phosphorylation.

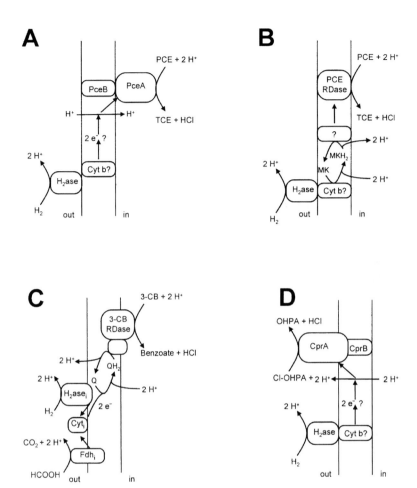

Figure 6. Proposed models of the halorespiratory chains in Dehalospirillum multivorans (A, adapted from [30]), Dehalobacter restrictus (B, adapted from [30]), Desulfomonile tiedjei (C, adapted from [44] and Desulfitobacterium dehalogenans (D). Abbreviations: H₂ase(i), (inducible) hydrogenase; Fdhi, inducible formate dehydrogenase; Cyt (b), cytochrome (b); Rdase, reductase; MK, menaquinone; Q, quinone.

For several organisms, additional evidence for energy conservation via a chemiosmotic coupling mechanism has now arisen from the application of uncouplers, protonophores and oxidant pulses [53, 54, 58]. Based on these and other results that have been collected to date and which are summarised below, basic models for structure and function of halorespiratory chains have been proposed (Fig. 6A-C, [30, 44]). An essential assumption for these models is the cytoplasmic location of the dehalogenating enzyme. In case this is located at the outside of the cell membrane, as is suggested based on the above-mentioned evidence, these models do not apply. An alternative model is proposed, assuming an extracytoplasmic orientation of the reductive dehalogenase (Fig. 6D).

3.2.1. Topology of the halorespiratory chain

The enzymes catalysing the electron donating process have been characterised from different halospiring strains, indicating an extra-cytoplasmic location in all cases. In *Desulfitobacterium* strain PCE-S and *Dehalospirillum multivorans*, membrane-associated extra-cytoplasmic hydrogenases and formate dehydrogenases have been identified [53]. Similarly, hydrogenase activity was localised on the outside of the cytoplasmic membrane of *Desulfomonile tiedjei* and *Dehalobacter restrictus*, as MV-dependent activity was similar in intact and lysed cells. Moreover, proton liberation from H_2 as the electron donor was inhibited by membrane impermeable Cu^{2+} [29, 44, 71]. A periplasmic formate dehydrogenase was induced in *Desulfomonile tiedjei* when grown in the presence of formate as the electron donor [44].

Based on the apparent extra-cytoplasmic location of the electron donating process and the proposed cytoplasmic location of the terminal reductive dehalogenase, a scalar mechanism of proton translocation was suggested, which would theoretically yield a H^+/e^- ratio of 1. However, from yield studies on *Dehalospirillum multivorans* cells grown with H_2 / PCE and acetate as the carbon source, it was calculated that only 0.4 mole ATP were formed per mole Cl^- released, suggesting a H^+/e^- ratio of 0.5 [53, 70]. In view of the proposed reaction mechanism for PCE reductive dehalogenase, which requires the endergonic transfer of a high-potential electron for the reduction of cob(III)alamin to cob(II)alamin, the results point towards the involvement of a reversed electron flow, driven by the electrochemical proton potential (Fig. 3). As reductive dehalogenation was still observed in the presence of protonophores in *Dehalobacter restrictus* and *Desulfomonile tiedjei*, whereas proton liberation was abolished, reverse electron flow does not seem to be required for *in vivo* reductive dehalogenase activity in these systems [58, 71]. On the contrary, fast proton liberation during electron transport from H_2 to PCE and TCE in *Dehalobacter restrictus* resulted in a H^+/e^- ratio of 1.25 ± 0.2. Similarly, an $H^+/3$-CB ratio of 2.1 was observed during halorespiration in *Desulfomonile tiedjei*, indicating that additional vectorial proton translocation cannot be excluded in these systems. However, experimental proof is still missing.

3.2.2. Electron mediating components of the halorespiratory chain

Membrane-bound menaquinone (MK), which could possibly act as proton pump, was shown to be directly involved in electron mediation from hydrogenase to PCE reductase in *Dehalobacter restrictus*. Nevertheless, oxidant pulse experiments with the MK/MKH$_2$ analogues DMN/DMNH$_2$ indicated that MK is not involved in vectorial proton translocation [71]. Furthermore, MK is presumably not the direct physiological electron donor for the reductive dehalogenase, suggesting the presence of an additional electron-mediating component, which remains to be identified [30]. In contrast, it was assumed that MK is not involved in electron transfer in *Dehalospirillum multivorans*, as PCE reduction was not affected by 2-heptyl-4-hydroxyquinoline-N-oxide (HOQNO), which is an inhibitor of many quinone-dependent redox reactions. However, the insensitivity of normally MK-dependent fumarate reduction might suggest that the inhibition experiment should be interpreted with caution [53]. Neither ubi- nor menaquinone could be detected in *Desulfomonile tiedjei*. However, a yet unidentified quinoid could be extracted from the organism. An essential function of this component in halorespiration was proposed based on the observation that reductive dehalogenation in whole cells of *Desulfomonile tiedjei* is severely inhibited by HOQNO. Moreover, reductive dehalogenation of this organism is dependent on 1,4-naphthoquinone, which might be an essential quinoid-precursor. However, the isolated quinoid failed to replace MV in *in vitro* activity measurements, indicating that it is not the direct electron donor for reductive dehalogenation [44].

Yet another class of potent electron-mediating components are the cytochromes. Both b- and c-type cytochromes have been isolated from virtually all halorespiring bacteria. Nevertheless, involvement in halorespiratory electron transport has not been investigated. Interestingly, a high spin c-type cytochrome was co-induced with dehalogenation activity in *Desulfomonile tiedjei*, and the corresponding gene has been cloned and sequenced [45]. The 50-kDa gene product exhibited two conserved c-type heme binding motifs (CxxCH), but substantially differed from known cytochromes and also no homologues could be identified from the *Dehalococcoides ethenogenes* partial genome sequence. However, as with the above-mentioned quinoid, also the induced cytochrome presumably is not the physiological electron donor for the dehalogenating system [44].

3.2.3. Halorespiration-deficient mutants in Desulfitobacterium dehalogenans

A molecular genetic approach was taken in order to get deeper insights in structure, function and regulation of halorespiration in the *o*-chlorophenol dehalogenating *Desulfitobacterium dehalogenans*. In this study, halorespiration-deficient mutants were isolated after insertional mutagenesis by the conjugational transposon Tn*916*. Characterisation at the physiological level indicated a pleiotropic phenotype for most of the mutants, as nitrate respiration, and occasionally fumarate respiration, was impaired in addition to the inability to use Cl-OHPA as the terminal electron acceptor with lactate as the electron donor. From these results, at least partial integration of both respiratory chains has been suggested [74]. Similarly, sulphite reduction and dehalogenation with H_2 as electron acceptor in *Desulfomonile tiedjei* were impaired by the same respiratory inhibitors, indicating a partially shared electron transport chain [12]. The biochemical and molecular genetic analysis of the halorespiration deficient mutants suggested a major role of various respiratory key enzyme complexes (formate dehydrogenase, hydrogenase, possibly proton translocating formate hydrogen lyase) and regulatory proteins in halorespiration (see below).

4. Regulation of halorespiration

It is known from physiological experiments that in several halorespiring bacteria described to date, reductive dehalogenase activity is induced by the presence of a halogenated substrate (Table 1). In most cases, the inducer is also substrate for the dehalogenating system, as might be expected. Occasionally, as it was observed for the *meta*-dehalogenating enzyme in *Desulfomonile tiedjei*, reductive dehalogenase activity is not only inducible by the substrate, but also by a whole variety of non-halogenated compounds [14, 59].

In strains of *Desulfitobacterium* that contain multiple dehalogenating activities, independent induction of the various enzyme systems has been observed [20]. In at least *Desulfitobacterium dehalogenans* and *Desulfitobacterium frappieri* PCP-1, induction is dependent on *de novo* protein biosynthesis, as dehalogenating activity did not develop in case chloramphenicol had been added prior to the inducer [11, 84]. Only for a few strains, constitutively present dehalogenating activity has been described to date [20, 50, 54, 61] One of these is the versatile *Dehalospirillum multivorans*, for which the isolation of three different variants has been reported: one of the isolates was found to be inducible for its dehalogenating activity, one expressed this system constitutively and one isolate could not be induced by the addition of the chlorinated substrate [61].

A better understanding of the regulatory circuits that are involved not only in the induction, but also the occasional repression of halorespiration activity by various environmental factors, is believed to be highly relevant for the application of halorespiring microorganisms for the *in situ* bioremediation of contaminated environments and the optimisation of such processes [20]. Being one of the relevant parameters, the influence of alternative electron acceptors on the activity of the dehalogenating system has been investigated for a few isolates, indicating that particularly sulphur oxy-anions are potential inhibitors of halorespiration [14, 20, 83]. Extensive chemostat studies on the PCE dehalogenating *Desulfitobacterium frappieri* TCE-1 showed that dehalogenation was especially sensitive to the presence of various alternative electron acceptors under electron donor-limiting conditions, whereas only sulphite significantly inhibited PCE reduction when the culture was PCE-limited [20]. On the contrary, the presence of equimolar concentrations of sulphite did not inhibit Cl-OHPA dechlorination in non-acclimated cultures of *Desulfitobacterium dehalogenans* [47].

Only little is known on the underlying mechanisms of control and the level at which regulation takes place. Recently, however, evidence has been gained from the detailed molecular analysis of the *cpr* gene cluster in *Desulfitobacterium dehalogenans* that induction of the *o*-chlorophenol reductive dehalogenase is strongly regulated at the transcriptional level [75]. Transcription of the bi-cistronic *cprBA* operon was 15-fold induced already 30 minutes after addition of the chlorinated substrate, with almost simultaneous appearance of dehalogenating activity. Comparison of the mapped *cprB*- and two additional putative halorespiration-specific promoters within the *cpr* gene cluster revealed the presence of conserved putative regulatory binding motifs that resemble the FNR box, suggesting the involvement of an FNR-like transcriptional regulator in the control of halorespiration-specific gene expression. *E. coli* FNR (fumarate- / nitrate-reductase regulator) is the archetype of a family of transcriptional regulators, mainly involved in modulation of expression of genes that encode (anaerobic) respiratory complexes [76]. Interestingly, one such regulatory protein might be encoded by the *cprK* gene in the *cpr* gene cluster [75].

The possible involvement of two-component regulatory systems in the control of dehalogenation in halorespirers has been suggested from the molecular analysis of halorespiration deficient mutants of the same *Desulfitobacterium dehalogenans*. In two of the mutants, the transposon was found to be inserted in the vicinity of genes encoding putative sensor histidine kinases. It was hypothesised that such two-component regulators could play a role in sensing the haloorganic compound [74]. Interestingly, genes encoding such two-component regulators were also found to be closely linked with putative reductive dehalogenase encoding genes in the *Dehalococcoides ethenogenes* genome [Preliminary sequence data were obtained from The Institute for Genomic Research website at http://www.tigr.org].

5. Outlook

We now know that both biological and non-biological factors contribute to the persistence of biohazardous compounds in the environment. There is obviously an urgent need not only for optimised natural attenuation approaches, but also a need for new experimentally designed degradation pathways [35, 81]. The present knowledge on the phylogeny of halorespiring bacteria and on the molecular characteristics of their reductive dehalogenases provides a powerful basis for the further development of culture-independent molecular tools for the detection of potential and actual halorespiring activity, and will enable the fine-tuning of current *in situ* bioremediation strategies. Moreover, it will allow the development of innovative strategies including the design of dedicated degraders with improved and / or novel metabolic capacities.

The remarkable phylogenetic diversity of halorespiring microorganisms has brought up the question of evolutionary origin of the intriguing ability to respire haloorganic compounds. The molecular characterisation of the key enzymes of this process, the reductive dehalogenases, indicated the presence of highly conserved potentially function-related sequence motifs. From this finding, it is tempting to speculate that halorespiration itself is an evolutionary ancient process involved in the degradation of naturally produced chlorinated compounds, rather than a recent development triggered by anthropogenic release of halogenated hydrocarbons into the environment. Nevertheless, spreading by lateral gene transfer might also be a possible explanation for the ubiquitous occurrence of halorespiration. Several phenotypes that have been under recent evolutionary pressure, like dehalogenation in aerobic microorganisms, antibiotic resistance and heavy metal resistance, are encoded on mobile genetic elements including extrachromosomal plasmids and transposable elements [67]. Such a mechanism of horizontal spreading cannot be excluded for halorespiration. It might be interesting to note that one of the putative reductive dehalogenase encoding genes present on the genome of *Dehalococcoides ethenogenes* is flanked by sequences with some similarity to bacterial integrase- and transposase-encoding genes. The isolation and characterisation of reductive dehalogenase encoding genes from pure and mixed cultures as well as from contaminated and pristine environments might provide deeper insight in their evolutionary history.

Acknowledgements

This work was partly supported by a grant of the Studienstiftung des Deutschen Volkes and contract BIO4-98-0303 of the European Union.

The authors would like to thank Bram van de Pas, Fred Hagen, Gosse Schraa and Fons Stams for frequent discussions on structure and function of halorespiratory chains and mechanisms of reductive dehalogenation.

Sequencing of *Dehalococcoides ethenogenes* was accomplished with support from the DOE Microbial Genome Program.

References

1. Ahlborg, U.G. and Thunberg, T.M. (1980) Chlorinated phenols: occurrence, toxicity, metabolism, and environmental impact, Crit. Rev. Toxicol. 7, 1-35
2. Arp, D.J. (1995) Understanding the diversity of trichloroethene co-oxidation, Curr. Opin. Biotechnol. 6, 352-358.
3. Berks, B.C., Sargent, F. and Palmer, T. (2000) The Tat protein export pathway, Mol. Microbiol. 35, 260-274.
4. Bouchard, B., Beaudet, R., Villemur, R., McSween, G., Lepine, F. and Bisaillon, J.G. (1996) Isolation and characterization of *Desulfitobacterium frappieri* sp. nov., an anaerobic bacterium which reductively dechlorinates pentachlorophenol to 3-chlorophenol, Int. J. Syst. Bacteriol. 46, 1010-1015.
5. Boyle, A.W., Phelps, C.D. and Young, L.Y. (1999) Isolation from estuarine sediments of a *Desulfovibrio* strain which can grow on lactate coupled to the reductive dehalogenation of 2,4,6-tribromophenol, Appl. Environ. Microbiol. 65, 1133-1140.
6. Bradley, P.M. (2000) Microbial degradation of chloroethenes in groundwater systems, Hydrogeology journal 8, 104-111.
7. Christiansen, N. and Ahring, B.K. (1996) *Desulfitobacterium hafniense* sp. nov., an anaerobic, reductively dechlorinating bacterium, Int. J. Syst. Bacteriol. 46, 442-448.
8. Christiansen, N., Ahring, B.K., Wohlfarth, G. and Diekert, G. (1998) Purification and characterization of the 3-chloro-4-hydroxy-phenylacetate reductive dehalogenase of Desulfitobacterium hafniense, FEBS Lett. 436, 159-162.
9. Cole, J.R., Cascarelli, A.L., Mohn, W.W. and Tiedje, J.M. (1994) Isolation and characterization of a novel bacterium growing via reductive dehalogenation of 2-chlorophenol, Appl. Environ. Microbiol. 60, 3536-3542.
10. Damborský, J. and Koca, J. (1999) Analysis of the reaction mechanism and substrate specificity of haloalkane dehalogenases by sequential and structural comparisons, Protein Eng. 12, 989-998.
11. Dennie, D., Gladu, I., Lepine, F., Villemur, R., Bisaillon, J.G. and Beaudet, R. (1998) Spectrum of the reductive dehalogenation activity of *Desulfitobacterium frappieri* PCP-1, Appl. Environ. Microbiol. 64, 4603-4606.
12. DeWeerd, K.A., Concannon, F. and Suflita, J.M. (1991) Relationship between hydrogen consumption, dehalogenation, and the reduction of sulfur oxyanions by *Desulfomonile tiedjei*, Appl. Environ. Microbiol. 57, 1929-1934.
13. DeWeerd, K.A., Mandelco, L., Tanner, R.S., Woese, C.T. and Suflita, J.M. (1990) *Desulfomonile tiedjei*, new genus new species, a novel anaerobic, dehalogenating, sulfate-reducing bacterium, Arch. Microbiol. 154, 23-30.
14. DeWeerd, K.A. and Suflita, J.M. (1990) Anaerobic aryl reductive dehalogenation of halobenzoates by cell extracts of *Desulfomonile tiedjei*, Appl. Environ. Microbiol. 56, 2999-3005.
15. Dolfing, J. and Harrison, B.K. (1992) Gibbs free energy of formation of halogenated aromatic compounds and their potential as electron acceptors in anaerobic environments, Environ. Sci. Technol. 26, 2213-2218.
16. Dolfing, J. and Janssen, D.B. (1994) Estimation of Gibbs free energies of formation of chlorinated aliphatic compounds, Biodegradation 5, 21-28.
17. El Fantroussi, S., Naveau, H. and Agathos, S.N. (1998) Anaerobic dechlorinating bacteria, Biotechnol. Prog. 14, 167-188.
18. Fetzner, S. (1998) Bacterial dehalogenation, Appl. Microbiol. Biotechnol. 50, 633-657.
19. Gantzer, C.J. and Wackett, L.P. (1991) Reductive dechlorination catalyzed by bacterial transition-metal coenzymes, Environ. Sci. Technol. 25, 715-722.

20. Gerritse, J., Drzyzga, O., Kloetstra, G., Keijmel, M., Wiersum, L.P., Hutson, R., Collins, M.D. and Gottschal, J.C. (1999) Influence of different electron donors and acceptors on dehalorespiration of tetrachloroethene by *Desulfitobacterium frappieri* TCE1, Appl. Environ. Microbiol. 65, 5212-5221.
21. Gerritse, J., Renard, V., Pedro Gomes, T.M., Lawson, P.A., Collins, M.D. and Gottschal, J.C. (1996) *Desulfitobacterium* sp. strain PCE1, an anaerobic bacterium that can grow by reductive dechlorination of tetrachloroethene or ortho-chlorinated phenols, Arch. Microbiol. 165, 132-140.
22. Glod, G., Angst, W., Holliger, C. and Schwarzenbach, R.P. (1997) Corrinoid-mediated reduction of tetrachloroethene, trichloroethene, and trichlorofluoroethene in homogeneous aqueous solution: Reaction kinetics and reaction mechanisms, Environ. Sci. Technol. 31, 253-260.
23. Gribble, G.W. (1996) Naturally occurring organohalogen compounds--a comprehensive survey, Fortschr. Chem. Org. Naturst. 68, 1-423.
24. Häggblom, M.M., Knight, V.K. and Kerkhof, L.J. (2000) Anaerobic decomposition of halogenated aromatic compounds, Environmental pollution 107, 199-207.
25. Hileman, B. (1993) Concerns broaden over chlorine and chlorinated hydrocarbons. Chem. Eng. News 71, 11-20.
26. Hill, K.E., Marchesi, J.R. and Weightman, A.J. (1999) Investigation of two evolutionarily unrelated halocarboxylic acid dehalogenase gene families, J. Bacteriol. 181, 2535-2547.
27. Holliger, C., Hahn, D., Harmsen, H., Ludwig, W., Schumacher, W., Tindall, B., Vazquez, F., Weiss, N. and Zehnder, A.J.B. (1998) *Dehalobacter restrictus* gen. nov. and sp. nov., a strictly anaerobic bacterium that reductively dechlorinates tetra- and trichloroethene in an anaerobic respiration, Arch. Microbiol. 169, 313-321.
28. Holliger, C. and Schraa, G. (1994) Physiological meaning and potential for application of reductive dechlorination by anaerobic bacteria, FEMS Microbiol. Rev. 15, 297-305.
29. Holliger, C. and Schumacher, W. (1994) Reductive dehalogenation as a respiratory process, Antonie Van Leeuwenhoek 66, 239-246.
30. Holliger, C., Wohlfarth, G. and Diekert, G. (1999) Reductive dechlorination in the energy metabolism of anaerobic bacteria, FEMS Microbiol. Rev. 22, 383-398.
31. Holoman, T.R.P., Elberson, M.A., Cutter, L.A., May, H.D. and Sowers, K.R. (1998) Characterization of a defined 2,3,5,6-tetrachlorobiphenyl-*ortho*-dechlorinating microbial community by comparative sequence analysis of genes coding for 16S rRNA, Appl. Environ. Microbiol. 64, 3359-3367.
32. Janssen, D.B., Pies, F. and van der Ploeg, J.R. (1994) Genetics and biochemistry of dehalogenating enzymes, Annu. Rev. Microbiol. 48, 163-191.
33. Jensen, J. (1996) Chlorophenols in the terrestrial environment, Rev. Environ. Contam. Toxicol. 146, 25-51.
34. Jones, R.W. and Garland, P.B. (1977) Sites and specificity of the reaction of bipyridylium compounds with anaerobic respiratory enzymes of *Escherichia coli*. Effects of permeability barriers imposed by the cytoplasmic membrane, Biochem. J. 164, 199-211.
35. Keasling, J.D. and Bang, S.-D. (1998) Recombinant DNA techniques for bioremediation and environmentally-friendly synthesis., Curr. Opin. Biotechnol. 9, 135-140.
36. Kengen, S.W., Breidenbach, C.G., Felske, A., Stams, A.J., Schraa, G. and de Vos, W.M. (1999) Reductive dechlorination of tetrachloroethene to cis-1,2-dichloroethene by a thermophilic anaerobic enrichment culture, Appl. Environ. Microbiol. 65, 2312-2316.
37. Krumholz, L.R., Sharp, R. and Fishbain, S.S. (1996) A freshwater anaerobe coupling acetate oxidation to tetrachloroethylene dehalogenation, Appl. Environ. Microbiol. 62, 4108-4113.
38. Lee, M.D., Odom, J.M. and Buchanan, R.J. (1998) New perspectives on microbial dehalogenation of chlorinated solvents: Insights from the field, Annu. Rev. Microbiol. 52, 423-452.
39. Leisinger, T. (1996) Biodegradation of chlorinated aliphatic compounds, Curr. Opin. Biotechnol. 7, 295-300.
40. Löffler, F.E., Ritalahti, K.M. and Tiedje, J.M. (1997) Dechlorination of chloroethenes is inhibited by 2-bromoethanesulfonate in the absence of methanogens, Appl. Environ. Microbiol. 63, 4982-4985.
41. Löffler, F.E., Sanford, R.A. and Tiedje, J.M. (1996) Initial characterization of a reductive dehalogenase from *Desulfitobacterium chlororespirans* Co23, Appl. Environ. Microbiol. 62, 3809-3813.
42. Löffler, F.E., Sun, Q., Li, J.R. and Tiedje, J.M. (2000) 16S rRNA gene-based detection of tetrachloroethene-dechlorinating *Desulfuromonas* and *Dehalococcoides* species, Appl. Environ. Microbiol. 66, 1369-1374.
43. Löffler, F.E., Tiedje, J.M. and Sanford, R.A. (1999) Fraction of electrons consumed in electron acceptor reduction and hydrogen thresholds as indicators of halorespiratory physiology, Appl. Environ. Microbiol. 65, 4049-4056.

44. Louie, T.M. and Mohn, W.W. (1999) Evidence for a chemiosmotic model of dehalorespiration in *Desulfomonile tiedjei* DCB-1, J. Bacteriol. 181, 40-46.
45. Louie, T.M., Ni, S.S., Xun, L.Y. and Mohn, W.W. (1997) Purification, characterization and gene sequence analysis of a novel cytochrome c co-induced with reductive dechlorination activity in *Desulfomonile tiedjei* DCB-1, Arch. Microbiol. 168, 520-527.
46. Ludwig, M.L. and Matthews, R.G. (1997) Structure-based perspectives on B12-dependent enzymes, Annu. Rev. Biochem. 66, 269-313.
47. Mackiewicz, N. and Wiegel, J. (1998) Comparison of energy and growth yields for *Desulfitobacterium dehalogenans* during utilization of chlorophenol and various traditional electron acceptors, Appl. Environ. Microbiol. 64, 352-355.
48. Magnuson, J.K., Stern, R.V., Gossett, J.M., Zinder, S.H. and Burris, D.R. (1998) Reductive dechlorination of tetrachloroethene to ethene by two-component enzyme pathway, Appl. Environ. Microbiol. 64, 1270-1275.
49. Marsh, A. and Ferguson, D.M. (1997) Knowledge-based modeling of a bacterial dichloromethane dehalogenase, Proteins 28, 217-226.
50. Maymó Gatell, X., Anguish, T. and Zinder, S.H. (1999) Reductive dechlorination of chlorinated ethenes and 1,2-dichloroethane by "*Dehalococcoides ethenogenes*" 195, Appl. Environ. Microbiol. 65, 3108-3113.
51. Maymó Gatell, X., Chien, Y.T., Gossett, J.M. and Zinder, S.H. (1997) Isolation of a bacterium that reductively dechlorinates tetrachloroethene to ethene, Science 276, 1568-1571.
52. Middeldorp, P.J.M., Luijten, M.L.G.C., van de Pas, B.A., van Eekert, M.H.A., Kengen, S.W.M., Schraa, G. and Stams, A.J.M. (1999) Anaerobic microbial reductive dehalogenation of chlorinated ethenes, Bioremediation J. 3, 151-169.
53. Miller, E., Wohlfarth, G. and Diekert, G. (1996) Studies on tetrachloroethene respiration in *Dehalospirillum multivorans*, Arch. Microbiol. 166, 379-387.
54. Miller, E., Wohlfarth, G. and Diekert, G. (1997) Comparative studies on tetrachloroethene reductive dechlorination mediated by *Desulfitobacterium* sp. strain PCE-S, Arch. Microbiol. 168, 513-519.
55. Miller, E., Wohlfarth, G. and Diekert, G. (1998) Purification and characterization of the tetrachloroethene reductive dehalogenase of strain PCE-S, Arch. Microbiol. 169, 497-502.
56. Mohn, W.W. and Kennedy, K.J. (1992) Reductive dehalogenation of chlorophenols by *Desulfomonile tiedjei* DCB-1, Appl. Environ. Microbiol. 58, 1367-1370.
57. Mohn, W.W. and Tiedje, J.M. (1990) Strain DCB-1 conserves energy for growth from reductive dechlorination coupled to formate oxidation, Arch. Microbiol. 153, 267-271.
58. Mohn, W.W. and Tiedje, J.M. (1991) Evidence for chemiosmotic coupling of reductive dechlorination and ATP synthesis in *Desulfomonile tiedjei*, Arch. Microbiol. 157, 1-6.
59. Mohn, W.W. and Tiedje, J.M. (1992) Microbial reductive dehalogenation, Microbiol. Rev. 56, 482-507.
60. Neumann, A., Reinhardt, S., Wohlfarth, G. and Diekert, G. (1999) Biochemical and genetic comparison of PCE-dehalogenases from a Gram-positive and a Gram-negative bacterium, Biospektrum, special issue VAAM P-D 20.
61. Neumann, A., Scholz-Muramatsu, H. and Diekert, G. (1994) Tetrachloroethene metabolism of *Dehalospirillum multivorans*, Arch. Microbiol. 162, 295-301.
62. Neumann, A., Wohlfarth, G. and Diekert, G. (1995) Properties of tetrachloroethene and trichloroethene dehalogenase of *Dehalospirillum multivorans*, Arch. Microbiol. 163, 276-281.
63. Neumann, A., Wohlfarth, G. and Diekert, G. (1996) Purification and characterization of tetrachloroethene reductive dehalogenase from *Dehalospirillum multivorans*, J. Biol. Chem. 271, 16515-16519.
64. Neumann, A., Wohlfarth, G. and Diekert, G. (1998) Tetrachloroethene dehalogenase from *Dehalospirillum multivorans*: Cloning, sequencing of the encoding genes, and expression of the *pceA* gene in *Escherichia coli*, J. Bacteriol. 180, 4140-4145.
65. Ni, S., Fredrickson, J.K. and Xun, L. (1995) Purification and characterization of a novel 3-chlorobenzoate-reductive dehalogenase from the cytoplasmic membrane of *Desulfomonile tiedjei* DCB-1, J. Bacteriol. 177, 5135-5139.
66. Nicholas, K.B. and Nicholas, H.B.J. (1997) GeneDoc: a tool for editing and annotating multiple sequence alignments, Distributed by the author.
67. Ochman, H., Lawrence, J.G. and Groisman, E.A. (2000) Lateral gene transfer and the nature of bacterial innovation, Nature 405, 299-304.

68. Sanford, R.A., Cole, J.R., Löffler, F.E. and Tiedje, J.M. (1996) Characterization of *Desulfitobacterium chlororespirans* sp. nov., which grows by coupling the oxidation of lactate to the reductive dechlorination of 3-chloro-4-hydroxybenzoate, Appl. Environ. Microbiol. 62, 3800-3808.
69. Santini, C.L., Ize, B., Chanal, A., Muller, M., Giordano, G. and Wu, L.F. (1998) A novel sec-independent periplasmic protein translocation pathway in *Escherichia coli*, EMBO J. 17, 101-12.
70. Scholz-Muramatsu, H., Neumann, A., Messmer, M., Moore, E. and Diekert, G. (1995) Isolation and characterization of *Dehalospirillum multivorans* gen. nov., sp. nov., a tetrachloroethene-utilizing, strictly anaerobic bacterium, Arch. Microbiol. 163, 48-56.
71. Schumacher, W. and Holliger, C. (1996) The proton/electron ratio of the menaquinone-dependent electron transport from dihydrogen to tetrachloroethene in *"Dehalobacter restrictus"*, J. Bacteriol. 178, 2328-2333.
72. Schumacher, W., Holliger, C., Zehnder, A.J.B. and Hagen, W.R. (1997) Redox chemistry of cobalamin and iron sulfur cofactors in the tetrachloroethene reductase of *Dehalobacter restrictus*, FEBS Lett. 409, 421-425.
73. Slater, J.H., Bull, A.T. and Hardman, D.J. (1997) Microbial dehalogenation of halogenated alkanoic acids, alcohols and alkanes, Adv. Microb. Ecol. 38, 133-176.
74. Smidt, H., Song, D.L., van der Oost, J. and de Vos, W.M. (1999) Random transposition by Tn*916* in *Desulfitobacterium dehalogenans* allows for isolation and characterization of halorespiration-deficient mutants, J. Bacteriol. 181, 6882-6888.
75. Smidt, H., van Leest, M., van der Oost, J. and de Vos, W.M. (2000) Transcriptional regulation of the *cpr* gene cluster in the *ortho*-chlorophenol respiring *Desulfitobacterium dehalogenans*, *J. Bacteriol. 182, 5683-5691.*
76. Spiro, S. (1994) The FNR family of transcriptional regulators, *Antonie Van Leeuwenhoek 66*, 23-36.
77. Strunk, O. and Ludwig, W. (1995) A software environment for sequence data, Department of Microbiology, Technical University of Munich, Munich, Germany.
78. Thauer, R.K., Jungermann, K. and Decker, K. (1977) Energy conservation in chemotrophic anaerobic bacteria, Bacteriol. Rev. 41, 100-180.
79. Thomas, A.W., Lewington, J., Hope, S., Topping, A.W., Weightman, A.J. and Slater, J.H. (1992) Environmentally directed mutations in the dehalogenase system of Pseudomonas putida strain PP3, Arch. Microbiol. 158, 176-182.
80. Thompson, J.D., Gibson, T.J., Plewniak, F., Jeanmougin, F. and Higgins, D.G. (1997) The CLUSTAL-X windows interface: flexible strategies for multiple sequence alignment aided by quality analysis tools, Nucleic Acids Res. 25, 4876-4882.
81. Timmis, K.N. and Pieper, D.H. (1999) Bacteria designed for bioremediation, Trends Biotechnol 17, 200-204.
82. Townsend, G.T. and Suflita, J.M. (1996) Characterization of chloroethylene dehalogenation by cell extracts of *Desulfomonile tiedjei* and its relationship to chlorobenzoate dehalogenation, Appl. Environ. Microbiol. 62, 2850-2853.
83. Townsend, G.T. and Suflita, J.M. (1997) Influence of sulfur oxyanions on reductive dehalogenation activities in *Desulfomonile tiedjei*, Appl. Environ. Microbiol. 63, 3594-3599.
84. Utkin, I., Dalton, D.D. and Wiegel, J. (1995) Specificity of reductive dehalogenation of substituted *ortho*-chlorophenols by *Desulfitobacterium dehalogenans* JW/IU-DC1, Appl. Environ. Microbiol. 61, 346-351.
85. Utkin, I., Woese, C. and Wiegel, J. (1994) Isolation and characterization of *Desulfitobacterium dehalogenans* gen. nov., sp. nov., an anaerobic bacterium which reductively dechlorinates chlorophenolic compounds, Int. J. Syst. Bacteriol. 44, 612-619.
86. van de Pas, B.A., Smidt, H., Hagen, W.R., van der Oost, J., Schraa, G., Stams, A.J.M. and de Vos, W.M. (1999) Purification and molecular characterization of *ortho*-chlorophenol reductive dehalogenase, a key enzyme of halorespiration in *Desulfitobacterium dehalogenans*, J. Biol. Chem. 274, 20287-20292.
87. von Wintzingerode, F., Selent, B., Hegemann, W. and Gobel, U.B. (1999) Phylogenetic analysis of an anaerobic, trichlorobenzene transforming microbial consortium, Appl. Environ. Microbiol. 65, 283-286.
88. Weiner, J.H., Bilous, P.T., Shaw, G.M., Lubitz, S.P., Frost, L., Thomas, G.H., Cole, J.A. and Turner, R.J. (1998) A novel and ubiquitous system for membrane targeting and secretion of cofactor-containing proteins, Cell 93, 93-101.
89. Wiegel, J. and Wu, Q. (2000) Microbial reductive dehalogenation of polychlorinated byphenyls, FEMS Microbiol. Ecol. 32, 1-15.

90. Wiegel, J., Zhang, X.M. and Wu, Q.Z. (1999) Anaerobic dehalogenation of hydroxylated polychlorinated biphenyls by *Desulfitobacterium dehalogenans*, Appl. Environ. Microbiol. 65, 2217-2221.
91. Wild, A., Hermann, R. and Leisinger, T. (1997) Isolation of an anaerobic bacterium which reductively dechlorinates tetrachloroethene and trichloroethene, Biodegradation 7, 507-511.
92. Wohlfarth, G. and Diekert, G. (1997) Anaerobic dehalogenases, Curr. Opin. Biotechnol. 8, 290-295.

DIVERSITY AND ACTIVITY OF MICROBES OXIDIZING METHANE AND AMMONIUM IN NORTHERN ORGANIC SOILS UNDER CHANGING ENVIRONMENTAL CONDITIONS

**PERTTI J. MARTIKAINEN [1], RITVA E.VASARA [2],
MARI T. LIPPONEN [3], JAANA TUOMAINEN [2],
MERJA H.SUUTARI [3] AND K. SERVOMAA [2]**
[1] *Department of Environmental Sciences, Bioteknia 2, University of
Kuopio, P.O. Box 1627, FIN-70211 Kuopio, Finland, [2] North Savo
Environment Centre, P.O. Box 1049, e-mail: Ritva.Vasara@vyh.fi, fax.
+358-17-281 1461 FIN-70101 Kuopio, Finland, [3] Laboratory for
Environmental Microbiology, National Public Health Institute, P.O. Box
95, FIN-70701 Kuopio, Finland*

Abstract

Microbial methane and ammonium oxidation in soil has a great importance in the atmospheric gas composition. Peat soils, typical in northern latitudes, are major sources of atmospheric methane as a result of methane producing microbes they harbour. Only a part of the methane produced in anaerobic peat is emitted to atmosphere because of microbial methane oxidation in the uppermost aerobic peat profile. Lowering of water table e.g., for agricultural or forestry purposes, changes the relative activity of methane producing and oxidizing microbes. Drainage generally decreases methane emissions. Nitrogen-rich peat soils are potential sources of nitrous and nitric oxides. Microbes oxidizing ammonium are among the key organisms responsible for the production of nitrogenous oxides. In natural water-logged peat, the production of nitrous and nitric oxides is negligible because activities of nitrifying and denitrifying bacteria are low. Drainage of peat soils for agriculture largely increases decomposition of organic matter and activity of nitrifying and denitrifying bacteria. Among soil ecosystems, organic agricultural soils have the highest nitrous oxide production. Drainage for forestry also induces some nitrous oxide production, especially in the most nitrogen-rich sites. There is a risk that global warming will change the hydrology of northern peatlands and their methane and nitrous oxide dynamics because the microbial processes are closely associated to the hydrological characteristics of peatlands. The global importance of methane and ammonium oxidation in northern organic soils is known but the microbes responsible for these processes are poorly characterized. The isolation of these microbes has proven to be difficult, there are only few isolates. However, PCR amplification and sequencing of DNA extracted from natural and manipulated organic

47

S.N. Agathos and W. Reineke (eds.),
Biotechnology for the Environment: Strategy and Fundamentals, 47–57.
© 2002 *Kluwer Academic Publishers. Printed in the Netherlands.*

soils suggest a great diversity among these microbial populations. These studies are based on the gene sequences of 16S rDNA, or the genes encoding for the ammonia monooxygenase (AMO) of ammonium-oxidizing bacteria, and the methane monooxygenase (MMO) of methane-oxidizing microbes. The results obtained by molecular biological techniques strongly suggest that in acidic organic soils there are novel organisms responsible for ammonium and methane oxidation. The effect of environmental changes on the diversity of these microbes is unknown. Some recent results indicate that there are high similarities in microbial populations oxidizing ammonium and methane in various northern organic soils although their hydrological and nutritional conditions differ greatly.

1. Introduction

Peat soils have received increasing attention during the last ten years because of their great importance for the atmospheric gas composition. In the atmosphere the contents of carbon dioxide (CO_2), methane (CH_4), and nitrous oxide (N_2O) have increased, and are increasing all the time causing warming of our planet. Warming is the most serious global environmental problem, which can destroy e.g. the possibilities for agriculture in some regions as a result of increasing dryness. The increase in the concentration of CO_2 is associated mainly to burning of fossil fuels. However, some of the greenhouse gases mostly originate from biological processes. Of methane and nitrous oxide 70-80 % originate from bacterial processes (Svensson, 1996).

Microbial methane production requires anaerobic environment and organic substrates. Water-saturated peat soil is such an environment where anaerobic microbial processes can take place. As a result of slow decomposition of plant derived organic matter in anaerobic peat, peatlands accumulate carbon, i.e. they are sinks for atmospheric carbon dioxide. However, as anaerobic environments they are net sources for methane. The amount of carbon accumulated since the last glacial period in the northern peatlands is 300-455 Pg corresponding 20-30 % of the global soil-carbon pool (Gorham, 1991). Northern peatlands are responsible for about 10 % of the total atmospheric methane load of 400-600 Tg y^{-1} (Bartlett and Harriss, 1993). Only part of the methane produced in anaerobic environments escapes to the atmosphere because a large part of it, about 60 % globally (Nedwell, 1996), is oxidized by methane-oxidizing microbes. Microbial nitrification and denitrification are the most important sources for nitrous oxide (Davidson, 1991). In nitrification, ammonium is oxidized via nitrite to nitrate by aerobic chemolithotrophic bacteria. The first step, oxidation of ammonium to nitrite, is carried out by ammonium oxidizers, which also liberate some nitrous oxide. Nitrite and nitrate produced by nitrifying bacteria are used as electron acceptors in microbial denitrification where these compounds are reduced via NO and N_2O to N_2. Denitrifying bacteria are facultative anaerobes replacing oxygen with nitrate or nitrite in anaerobic environment (Davidson, 1991). Because of the high nitrogen content peat soils are potential sources for N_2O. The release of methane and nitrous oxide from soil highly depends on the soil physical and chemical conditions. Anthropogenic activities including water table manipulation for forestry or agricultural purposes are known to affect microbial processes responsible for the CH_4, N_2O and NO dynamics. Similarly, global warming would change hydrology and microbial processes in natural peat soils.

48

We present here a brief overview on the microbes and their capacities to oxidize ammonium and methane in northern organic soils, and discuss the effects of environmental stresses on these microbial activities highly affecting atmospheric gas composition.

2. Methane-oxidizing microbes

Methane content in organic soils varies from below the atmospheric concentration (about 2 ppmv) in well-drained soils to several percentages in water-saturated soils. Experiments with soil samples have shown that there are low-affinity methane oxidizing microbes with high K_m and V_{max}, and high-affinity microbes with low K_m and V_{max} (Bender and Conrad, 1992). Well-drained soils can have both low-affinity and high-affinity methanotrophs (Bull *et al.*, 2000). The low-affinity methane-oxidizers have been isolated in pure cultures, but there are no pure cultures of the high-affinity microbes isolated at methane concentration close to the atmospheric methane content. These slowly growing microbes have been proven to be very difficult to cultivate. Dunfield *et al.* (1999) enriched from an organic soil using methane concentration of 275 ppmv a mixed culture, which had similar K_m for methane as the soil itself. The enrichment took four years, and the enrichment culture probably contained only one methanotroph species (type II methanotroph, *Methylocystis/Methylosinus*). Acidity and low nutrient conditions are characteristic for natural peatlands. Dedysh *et al.* (1998a) developed an innovative enrichment technique with acidic medium having very low nutrient content, and were able to enrich low-affinity acidophilic methane-oxidizing microbes from northern peatlands in Russia. Later they isolated several strains from these enrichments. The pH optimum of the strains was about 5. Similar pH optimum was found for methane oxidation in soil of a forested peatland in Finland (Saari *et al.*, unpublished results). The acidophilic strains isolated by Dedysh *et al.* (1998b) had soluble methane monooxygenase and were close to the heterotrophic bacterium *Beijerinckia indica* according to the 16S rDNA sequence analysis. However, the isolated strains were unable to use organic substrates (Dedysh *et al.*, 1998b).

The difficulties with the isolation would indicate a mixed metabolism of the microbes living at low methane concentration. There is evidence that at least some of the high affinity methane oxidizers in soil are promoted by methanol (Jensen *et al.*, 1998, Benstead *et al.*, 1998), acetate, formate (West and Schmidt, 1999), or some other organic compounds (Roslev *et al.*, 1997). It has been suggested that ammonia-oxidizing chemolithotrophic bacteria could participate in methane oxidation in soil. However, recent results have shown that ammonium oxidizers do not have importance in methane oxidation (Jiang and Bakken, 1999, Klemedtsson *et al.*, 1999, Bodelier and Frenzel, 1999).

Molecular biological techniques have been applied to show the occurrence of methane Molecular biological techniques have been applied to show the occurrence of methane oxidizers in soil samples. These techniques where isolation of the targeted organisms is not needed, have a great potential in the ecology of methane oxidizers. 16S rDNA probes designed for the known methanotrophs have shown the occurrence of genera Methylococcus and Methylosinus in peat profile oxidizing methane (McDonald *et al.*, 1996). The differences in the methane oxidizers in natural and manipulated organic soils have been studied recently in Finland (Vasara *et al.*, submitted). PCR amplification and partial sequencing of *pmo*A genes from DNA extracted from the soils showed that they had microbes with *pmo*A sequences clustering with *pmo*A of *Methylococcus capsulatus, Methylobacter, Methylomonas* and *Methylomicrobium*. These bacteria belong to the type I methanotrophs. In the Finnish organic soils there were also sequences related to *pmo*A of type II methanotrophs, and sequences not related to any known methanotrophs. There are recent reports on novel methanotrophs in soils of various geographical regions (Holmes *et al.*, 1999, Henckel *et al.*, 2000); some of these bacteria have similarity to the type II methanotrophs (Holmes *et al.*, 1999). There is evidence that acidophilic methanotrophs, belonging to the type II, are widely distributed in organic soils. Finnish peat soils had sequences (Vasara *et al.*, submitted) highly identical with the *pmo*A sequences found in a peatland at Pennine Hills, England (McDonald and Murrell, 1997). 16S rDNA sequences retrieved from Pennine Hills (McDonald *et al.*, 1996) had 95 % identity value with the Russian acidophilic isolates discussed above (Dedysh *et al.*, 1998b). An interesting finding is that natural peatlands and organic soils drained for forestry or agriculture have very similar *pmo*A gene sequences (Vasara *et al.*, submitted). This indicates that some methanotrophs have a high ability to survive in variable environmental conditions.

3. Methane oxidation

3.1. VERTICAL DISTRIBUTION OF METHANE OXIDATION IN PEAT PROFILE

In natural peatlands the peat profile can be divided in two parts (Clymo, 1992*).* *Acrotelm* is the zone above the water table and is responsible for the uptake of CO_2 by plant photosynthesis. After some decomposition organic matter is transported below the water table to the anaerobic *catotelm*. Catotelm is the site for long-term peat accumulation. The boundary layer between the acrotelm and catotelm is the most intensive site for CH_4 production in peat profile. The maximum methane production occurs 10-20 cm below the water table (Roulet *et al.*, 1993, Sundh *et al*, 1994, Kettunen *et al.*, 1999) because this layer provides the maximum supply of substrates for methanogenic bacteria. We would assume that methane-oxidizing bacteria have their maximum activity above the water table in unsaturated peat where availability of atmospheric oxygen is good. However, experiments with peat samples have shown that maximum methane oxidation potentials, probably also the maximum population density of methane oxidizers, are located 0 -15 cm below the water table (Sundh *et al.*, 1994, 1995, Kettunen *et al.*, 1999). The availability of both CH_4 and oxygen are crucial for methane oxidation. The low diffusion rate of oxygen in water highly limits the aerobic processes below the water table. However, there are vascular plants able to transport gases in their aerenchymal tissues. By this mechanism oxygen can penetrate into the water-logged peat deeper than just by diffusion. The aerobic rhizosphere of vascular plants is the probable reason for the vertical overlapping of methane production and oxidation in peat (Nedwell, 1996). There are results indicating that in fens where aerenchymal vascular plants are abundant, methane oxidation occurs deeper in peat profile than in bog profile lacking vascular plants (Kettunen *et al.*, 1999). The aerenchymal plants also transport CH_4 from the water-logged peat, and highly enhance the release of CH_4 to the atmosphere because that methane is not oxidized by microbes (Torn and Chapin, 1993, Schimel, 1995).

In profiles of boreal peatlands the layers showing maximum methane production or oxidation fluctuate with water table (Kettunen *et al.* 1999). As discussed above both the production and oxidation of methane in peat profile occur in the saturated layer. During a dry summer there can be a long-term lowering in water table. There is evidence that methane producers and oxidizers retain their activity 2-6 weeks in unsaturated peat, evidently methane oxidation is less affected than methane production (Kettunen *et al.*, 1999). It is known that at least part of the microbial community oxidizing methane can survive for a long period in anaerobic conditions (King *et al.*, 1990, Roslev and King 1994, 1995), an important capacity if there is a long-term increase from the mean water table level.

3.2. EFFECTS OF LONG-TERM HYDROLOGICAL CHANGES ON METHANE OXIDATION IN PEAT SOIL

As discussed above methane oxidation in peat profiles of natural peatlands is located in saturated peat, at or below the water table. However, in peatlands where water table has been lowered for forestry maximum methane oxidation occurs in the uppermost few centimetres of the soil profile, far from the water table (Crill *et al.* 1994). The reason is that as a result of peat compaction oxygen diffusion is limited to the uppermost soil. The activity of methane oxidizers keeps methane concentration below the atmospheric level causing methane flow from atmosphere to the peat (Crill *et al.* 1994). There are changes in peatland vegetation after water table lowering for forestry. Most of the species typical for wet peat surfaces are replaced by forest species (Vasander *et al.*, 1997) which apparently limits oxygen diffusion into peat and affects the distribution of methane-oxidizing populations in peat profile.

Methane oxidizers in soil profile of upland forests use atmospheric methane similarly to methane oxidizers in drained peat. However, they are not located in the soil organic horizon but deeper in the uppermost part of the mineral soil (*e.g.* Crill, 1991, Bender and Conrad, 1994, Saari *et al.*, 1997). The reason for the distribution of methane oxidation in forest soil is not known. It has been suggested that in mineral soil the methane-oxidizing microbes are protected from atmospheric acidic load and nitrogen deposition (King and Schnell, 1994a), or that phenolic compounds abundant in the organic horizon inhibit methane oxidation there (Amaral and Knowles, 1997). However, as discussed above the uppermost soil layer of forested peatlands is the most active in methane oxidation, and the hypotheses presented for mineral forest soils do not fit there.

3.3. EFFECT OF NITROGEN ON METHANE OXIDATION

Methane oxidation in soils is restricted by ammonium addition but nitrate does not have such an inhibitory effect (Nesbit and Breitenbeck, 1992, Hutsch *et al.*, 1994). Also non-ammonium salts can disturb methane oxidation in soil (King and Schnell, 1998). There is strong evidence that ammonium is competing with methane in the active site of the methane monooxygenase (MMO) (e.g. King and Schnell, 1994b). Moderate concentrations of non-nitrogenous inorganic salts can have indirect inhibitory effect on soil methane oxidation by desorbing bound ammonium from the soil matrix. Therefore, ammonium would be the inhibitory agent, not the added ions by themselves (see King, 1996). When extra ammonium has been consumed from soil, the inhibition of methane oxidation still continues, showing that there are also some other inhibitory mechanisms than the competition with methane. It has been suggested that nitrite or hydroxylamine produced from ammonium by methane oxidizers disturb the metabolism of methanotrophs (King and Schnell, 1994b, Schnell and King, 1994, Dunfield and Knowles, 1995). When there is a slight increase in the methane concentration from the ambient atmospheric level, the inhibition of methane oxidation by ammonium is more abundant. This would have importance for the global methane budget because atmospheric methane concentration is increasing which can cause higher inhibition of methane oxidation by ammonium in future atmospheric conditions (King and Schnell, 1994a). In addition to the agricultural land use where nitrogen is added as fertilizers, atmospheric acid load and nitrogen deposition decrease soil capacity to oxidize atmospheric methane (Hutsch et al. 1994, Goldman et al., 1995, Sitaula et al., 1995, Willison et al., 1995, Dobbie et al., 1996, Saari et al., 1997).

Nitrogen fertilization has decreased methane oxidation and uptake also in forested peat soil. Ammonium chloride and potassium nitrate are more inhibitory than urea suggesting that not only ammonium but ions in general (see above) can inhibit methane oxidation in drained peat (Crill et al., 1994). There is evidence that methane oxidation in drained peat can tolerate higher ammonium concentration than methane oxidation in well-drained soils generally (Crill et al., 1994) supporting the observations that there are methane oxidizers showing insensitivity against ammonium inhibition (King, 1996). Nitrogen treatments have had variable effects on methane oxidation in profiles of natural peatlands. With ammonium nitrate there are results showing either slight inhibition (Saarnio and Silvola, 1999) or stimulation (Nykänen et al., submitted) in methane oxidation. It can be that methane oxidation in natural peatlands tolerates nitrogen load better than methane oxidation in upland soils. The probable reason, in addition to the difference in the methane-oxidizing populations, would be the efficient nitrogen uptake by vegetation in uppermost peat profile. Therefore, a minor part of the added nitrogen reaches the methane-oxidizing peat layer (Nykänen et al., submitted). A reason for the increase in methane oxidation in peat after nitrogen addition would also be the increased availability of methane as a result of stimulation of methane production by the nitrogen treatment (Nykänen et al., submitted).

4. Ammonium-oxidizing bacteria

The molecular biological methods applied to various soil types (Stephen et al. 1996, Hastings et al., 1997, Kowalchuk et al., 1997, Stephen et al., 1998, Bruns et al., 1999, Mendum et al., 1999) have confirmed the results from isolation studies (Hankinson and Schmidt, 1984, Martikainen and Nurmiaho-Lassila, 1985, De Boer et al., 1989, Allison and Prosser, 1991, Utåker et al., 1995) that Nitrosospira is a common ammonium-oxidizing bacterium in soils, also in the acidic ones. There are only few studies on the diversity of ammonium-oxidizing bacteria in organic soils. MPN enumerations often fail to show any ammonium-oxidizing bacteria from soils of natural peatlands but drainage has greatly increased their occurrence (Regina *et al.* 1996, Vasara *et al.*, submitted). Utåker *et al.*, (1995) and Regina (1998) isolated *Nitrosospira* strains from a forested Finnish peat soil. Primers designed for ammonium monooxygenase gene (*amo*A) of ammonia oxidizers belonging to the β-Proteobacteria detected *Nitrosospira*-like sequences in Finnish natural and forested peat soils, and also in organic agricultural soil, whereas no *Nitrosomonas*-like *amo*A sequences were detected (Vasara *et al.*, submitted). These results show that also in the northern organic soils *Nitrosospira* dominates over *Nitrosomonas*. With primers designed for both *amo*A and *pmo*A most of the cloned sequences were *pmo*A-like, or had similarity to both *amo*A and *pmo*A genes. However, it is not known whether the microbes these sequences originate from are ammonium oxidizers or methane oxidizers (Vasara *et al.*, submitted). Although drainage for forestry or agriculture had highly changed the soil conditions, *e.g.* increased the nitrification rate, the overall sequence distribution in the natural and manipulated organic soils had great similarity. There is evidence that *Nitrosospira* genotypes in mineral soils of different pHs (Stephen *et al.*, 1998) or treatments (Bruns *et al.*, 1999) can vary. Some *Nitrosospira* types are adapted to acidic soil conditions (Stephen *et al.*, 1998). Experiments with peat samples have shown that nitrifying microbes could have different pH optimum at various depths in the peat profile (Lång *et al.*, 1995).

5. Effects of hydrological and nutritional changes on ammonium oxidation and production of nitrogen oxides in peat soils

Availability of ammonium and oxygen highly regulates the activity of chemolithotrophic ammonium oxidizing bacteria, and thus also nitrate production by nitrite-oxidizing chemolithotrophs. In saturated soil of natural peatlands nitrification activity is low (Rosswall and Granhall, 1980, Rangeley and Knowles, 1988, Regina *et al.* 1996) but lowering of water table for forestry enhances nitrification (Regina *et al.*, 1996). However, in the nitrogen-rich fen peat (minerotrophic peatlands) the increase in nitrification is stronger than in the nitrogen-poor very acid bog peat (ombrotrophic peatlands). These results clearly demonstrate the importance of ammonium availability and pH in regulation of nitrification in natural peat soils (Regina *et al.*, 1996). Increase in nitrification after water table lowering is reflected by increase in the emissions of nitrous oxide (N_2O) (Freeman *et al.*, 1993, Martikainen *et al.* 1993) and nitric oxide (NO) (Lång *et al.*, 1995). Lowering of water table in bogs does not induce nitrous oxide production (Regina *et al.*, 1996). Nitrification and associated increase in nitrous oxide production in northern fens are enhanced within few months after the water table lowering (Regina *et al.* 1999). Production of gaseous nitrogen oxides in peat soil without atmospheric nitrate load or nitrate fertilization always requires activity of nitrifiers, at least nitrite produced by ammonium oxidizers. Gaseous nitrogen oxides are produced either directly by the ammonium oxidizers or by denitrifying bacteria, which reduce nitrite and/or nitrate. The relative importance of ammonium oxidation and denitrification in the production of nitrous and nitric oxides in forested peat soils varies (Regina *et al.*, 1996, Regina *et al.* 1998a).

Drainage of peatlands for agriculture enhances production of nitrous and nitric oxides much more than the drainage for forestry (Nykänen *et al.* 1995, Lång *et al.*,

1995). Nitrogen addition is known to increase nitrification and production of nitrous and nitric oxides also in forested peat soil (Regina et al., 1998b) indicating that there availability of inorganic nitrogen limits the production of the nitrogenous gases.

Agricultural organic soils are ploughed, limed and fertilized, activities which induce high rate of organic matter decomposition, and associated nitrification and denitrification. Ammonium oxidation, nitrite oxidation and denitrification occur in drained nitrogen-rich boreal peat soils also during winter below the snowpack at low temperatures. The emissions of nitrous oxide during winter can correspond to even half of the total annual nitrous oxide release (Regina et al., 1998b, Alm et al., 1999).

6. Global change

Some climatic models predict warmer and drier summers in the north (Manabe and Wetherald, 1986, Mitchell, 1989). If the water table levels in peatlands decrease with change of climate, the changes in dynamics of methane and nitrogen oxides would have similarities with the changes induced by lowering of water table of natural peatlands for forestry, i.e. the global methane emission would decrease whereas the emissions of nitrous and nitric oxides would increase. Also nitrogen deposition and acidification can effect the gas dynamics of northern peatlands. There can also be feedback mechanisms between the increasing atmospheric methane concentration and nitrogen load in the well-aerated soils. To know the diversity and activity of microbes oxidizing methane and ammonium in various soil conditions has a key importance when we try to predict the gas dynamics of northern organic soils in changing environmental conditions.

Acknowledgements

This work was supported by the Academy of Finland, and the Ministry of the Environment.

References

Allison, S.M., and Prosser, J.I. (1991) Urease activity in neutrophilic autotrophic ammonia-oxidizing bacteria isolated from acid soils, Soil Biol. Biochem. 23, 45-51.

Alm, J., Saarnio, S., Nykänen, H., Silvola, J., and Martikainen, P.J. (1999) Winter CO_2, CH_4 and N_2O fluxes on some natural and drained boreal peatlands, Biogeochemistry 44, 163-186.

Amaral, J.A., and Knowles R. (1997) Inhibition of methane consumption in forest soils and pure cultures of methanotrophs by aqueous forest soil extracts, Soil Biol. Biochem. 29, 1713-1720.

Bartlett, K.R., Harriss, R.C. (1993) Review and assessment of methane emissions from wetlands, Chemosphere 26, 261-320.

Bender, M. and Conrad, R (1992) Kinetics of CH_4 oxidation in oxic soils exposed to ambient air or high CH_4 mixing ratios, FEMS Microbiol. Ecol. 101, 261-270.

Bender, M. and Conrad, R. (1994) Methane oxidation activity in various soils and freshwater sediments: occurrence, characteristics, vertical profiles, and distribution on grain size fractions, J. Geophys. Res. 99, 16531-16540.

Benstead, J., King, M. and Williams, H.G. (1998) Methanol promotes atmospheric methane oxidation by methanotrophic cultures and soils, Appl. Environ. Microbiol 64, 1091-1098.

Bodelier, P.L.F., and Frenzel, P. (1999) Contribution of methanotrophic and nitrifying bacteria to CH_4 and NH_4 oxidation in the rhizosphere of rice plants as determined by new methods of discrimination, Appl. Environ. Microbiol. 65, 1826- 1832.

Bruns, M.A., Stephen, J.R., Kowalchuk, G.A., Prosser, J.I., and Paul, E.A. (1999). Comparative diversity of ammonia oxidizer 16S rRNA gene sequences in native, tilled, and successional soils, Appl. Environ. Microbiol. 65, 2994-3000.

Bull, I.D., Parekh, N.R., Hall, G.H., Ineson, P. and Evershed, R.P. (2000) Detection and classification of atmospheric methane oxidizing bacteria in soil, Nature 405, 175-178.

Clymo, R.S. (1992) Models of peat growth, in H. Vasander and M. Starr (eds.), SUO, Mires and peat 4, 127- 136, Proceedings of the International Workshop, Hyytiälä., Finland, 28 September – 1 October 1992. Finnish Peatland Society.

Crill, P.M. (1991) Seasonal patterns of methane uptake and carbon dioxide release by a temperate woodland soil, Global Biogeochem. Cycles 5, 319-334.

Crill, P., Martikainen, P.J., Nykänen, H., and Silvola, J. (1994) Temperature and N fertilization effects on methane oxidation in a drained peatland soil, Soil Biol. Biochem 26, 1331-1339.

Davidson, E.A. (1991) Fluxes of nitrous oxide and nitric oxide from terrestrial ecosystems, in J.E. Rogers and W.B. Whitman (eds.), Microbial Production and Consumption of Greenhouse Gases: Methane, Nitrogen Oxides and Halomethanes, American Society for Microbiology, Washington, D.C., pp. 219-235.

De Boer, W., Duyts, H. and Laanbroek, H.J. (1989) Urea stimulated autotrophic nitrification in suspension of fertilized, acid heath soil, Soil Biol. Biochem. 21, 349-354-

Dedysh, S.N., Panikov N.S., and Tiedje, J.M. (1998a) Acidophilic methanotrophic communities from sphagnum peat bogs, Appl. Environ. Microbiol. 64, 922-929.

Dedysh, S.N., Panikov, N., Liesack, W., Grosskopf, R., James, J.Z. and Tiedje, J.M. (1998b) Isolation of acidophilic methane-oxidizing bacteria from northern peat wetlands, Science 282, 281-284.

Dobbie, K.E., Smith, K.A., Prieme, A., Christensen, S., Degorska, A., and Orlanski, P. (1996) Effect of land use on the rate of methane uptake by surface soils in northern Europe, Atmos. Environ. 30, 1005-1011.

Dunfield, P. and Knowles, R. (1995) Kinetics of methane oxidation by nitrate, nitrite, and ammonium in a humisol, Appl. Environ. Microbiol. 61, 3129-3135.

Dunfield, P.F., Liesack, W., Henckel, T. Knowles, R. and Conrad, R. (1999) High.affinity methane oxidation by a soil enrichment culture containing a type II methanotroph, Appl. Environ. Microbiol. 65, 1009- 1014.

Freeman, C., Lock, M.A. and Reynolds, B. (1993). Fluxes of CO_2, CH_4 and N_2O from a Welsh peatland following simulation of water table draw-dawn: Potential feedback to climatic change. Biogeochem. 19, 51-60.

Goldman, M.B., Groffman, P.M., Pouyat, R.V., McDonnell, M.J., and Pickett, S.T.A. (1995) CH_4 uptake and N availability in forest soils along an urban to rural gradient, Soil Biol. Biochem. 27, 281-286.

Gorham, E. (1991) Northern peatlands: Role in the carbon cycle and probable responses to climatic warming, Ecol. Appl. 1, 182-195.

Hankinson, T.R., and Scmidt, E.L. 1984 Examination of an acid forest soil for ammonia- and nitrite-oxidizing autotrophic bacteria, Can. J. Microbiol. 30, 1125-1132.

Hastings, R.C., Ceccherine, M.T., Miclaus, N., Saunders, J.R., Bazzicalupo, M., McCarthy, A.J. (1997) Direct molecular biological analysis of ammonia oxidizing bacteria populations in cultivated soil plots treated with swine manure, FEMS Microbiol. Ecol. 23, 45-54.

Henckel, T., Jäckel, U., Schnell, S., and Conrad, R. (2000) Molecular analyses of novel methanotrophic communities in forest soil that oxidize atmospheric methane, Appl. Environ. Microbiol. 66, 1801-1808.

Holmes, A.J., Roslev, P., McDonald, I.R., Iversen, N., Henriksen, K., and Murrell, J.C. (1999). Characterization of methanotrophic bacterial populations in soils showing atmospheric methane uptake. Appl. Environ. Microbiol. 65, 3312-3318.

Hutsch, B.W., Webster, C.P., and Powlson, D.S. (1994) Methane oxidation in soil as affected by land use, soil pH and N fertilization, Soil Biol. Biochem. 26, 1613-1622.

Jensen, S., Prieme, A., and Bakken, L. (1998) Methanol improves methane uptake in starved methanotrophic microorganisms, Appl. Environ. Microbiol. 64, 1143-1146.

Jiang, Q.Q., and Bakken L. (1999) Nitrous oxide production and methane oxidation by different ammonia.-oxidizing bacteria, Appl. Environ. Microbiol 65, 2679-2684.

Kettunen, A., Kaitala, V., Lehtinen, A., Lohila, AL., Alm, J., Silvola, J., and Martikainen, P.J. (1999) Methane production and oxidation potentials in relation to water table fluctuations in two boreal mires, Soil. Biol. Biochem, 31, 1741-1749.

King, G. (1996) Regulation of methane oxidation: controls between anoxic sediments and oxic soils, in M.E. Lidstrom and F.R. Tabita (eds.), Microbial Growth on C$_1$ Compounds, Kluwer Academic Publisher, Dordrecht, pp. 318-324.

King, G.M., Roslev, P., Skovgaard, H. (1990) Distribution and rate of methane oxidation in sediments of the Florida Everglades, Appl. Environ. Microbiol. 56, 2902-2911.

King G.M., and Schnell S. (1994a) Effect of increasing atmospheric methane concentration on ammonium inhibition of soil methane consumption, Nature 370, 282-284.

King, G.M., and Schnell, S. (1994b) Ammonium and nitrite inhibition of methane oxidation by Methylobacter albus BG8 and Methylosinus trichosporium OB3b at low methane concentration, Appl. Environ. Microbiol. 60, 3508-3513.

King, G.M., and Schnell, S. (1998) Effects of ammonium and non-ammonium salt additions on methane oxidation by Methylosinus trichosporium OB3b and Maine forest soils, Appl. Environ. Microbiol. 64, 253-257.

Klemedtsson, L., Jian, Q., Kasimir-Klemedtsson, Å., and Bakken, L. (1999) Autotrophic ammonium-oxidizing bacteria in Swedish mor humus, Soil Biol. Biochem. 31, 839-847.

Kowalchuck, G.A., Stephen, J.R., De Boer, W., Prosser, J.I., Embley, T.M., and Woldendorp, J.W. (1997) Analysis of ammonia-oxidizing bacteria of the β subdivision of the class proteobacteria in coastal sand dunes by denaturing gradient gel electrophoresis and sequencing of PCR-amplified 16S ribosomal DNA fragments, Appl. Environ. Microbiol. 63, 1489-1497.

Lång, K., Lehtonen, M., Martikainen, P.J. (1993). Nitrification potentials at different pH values in peat samples from various layers of a drained mire, Geomicobiol. J. 11, 141-147.

Lång, K., Silvola, J., Ruuskanen, J., and Martikainen, P.J. (1995) Emissions of nitric oxide from boreal peat soils, J.Biogeography, 22, 359-364.

Martikainen, P.J., and Nurmiaho-Lassila, E-L. (1985) Nitrosospira, an important ammonium-oxidizing bacterium in fertilized coniferous forest soil, Can. J. Microbiol. 31, 190-197.

Martikainen, P.J., Nykänen, H., Crill, P., and Silvola, J. (1993) Effect of a lowered water table on nitrous oxide fluxes from northern peatlands, Nature 366, 51-53.

Manabe, S. and Wetherald, T.R. (1986) Reduction in summer soil wetness induced by an increase in atmospheric carbon dioxide, Science 232, 626-628.

McDonald, I.R., Hall, G.M., Pickup, R.W., and Murrell, J.C. (1996) Methane oxidation potential and preliminary analysis of methanotrophs in blanket bog peat using molecular ecology techniques, FEMS Microbiol. Ecol. 21, 197-211.

McDonald, I.R., and Murrell, J.C. (1997) The particulate methane monooxygenase gene pmoA and its use as a functional gene probe for methanotrophs, FEMS Microbiol. Lett. 156, 205-210.

Mendum, T.A., Sockett, R.E., and Hirsch, P.R. (1999) Use of molecular and isotopic techniques to monitor the response of autotrophic ammonia-oxidizing populations of the β subdivision of the class proteobacteria in arable soils to nitrogen fertilizer, Appl. Environ. Microbiol. 65, 4155-4162.

Mitchell, J.F.B. (1989) The "greenhouse effect" and climate change, Rev. Geophys 27, 115-139.

Nedwell, D.B. (1996) Methane production and oxidation in soils and sediments, in J.C. Murrell and D.P. Kelly (eds.), Microbiology of Atmospheric Trace Gases, NATO ASI Series I, Global Environmental Change, Springer-Verlag, Berlin, pp. 31-49.

Nesbit, S.P., and Breitenbeck, G.A. (1992) A laboratory study of factors influencing methane uptake by soils. Agric. Ecosyst. Environ. 41, 39-54.

Nykänen, H., Alm, J., Lång, K. Silvola, J., and Martikainen, P.J. (1995) Emissions of CH$_4$, N$_2$O and CO$_2$ from a virgin fen and fen drained for grassland in Finland, J. Biogeogr. 22, 351-357.

Nykänen, H., Vasander, H., Huttunen, J., and Martikainen, P.J. Dynamics of methane and nitrous oxide on ombrotrophic boreal peatland receiving experimental nitrogen load (submitted).

Rangeley, A., and Knowles, R. (1988) Nitrogen transformation in a Scottish peat soil under laboratory conditions. Soil Biol. Biochem. 20, 385-391.

Regina, K. (1998) Microbial production of nitrous oxide and nitric oxide in boreal peatlands, PhD thesis, University of Joensuu Publications in Sciences 50, Joensuu, Finland.

Regina, K., Nykänen, H., Silvola, J., and Martikainen, P.J. (1996) Fluxes of nitrous oxide from boreal peatlands as affected by peatland type, water table level and nitrification capacity, Biogeochemistry 35, 401-418.

Regina, K., Silvola, K., and Martikainen, P.J. (1998a) Mechanisms of N$_2$O and NO production in the soil profile of a drained and forested peatland, as studied with acetylene, nitrapyrin and dimethylether. Biol. Fertil. Soil 27, 205-210.

Regina, K., Nykänen, H., Maljanen, M., Silvola, J. and Martikainen, P.J. (1998b) Emissions of N₂O and NO and net nitrogen mineralisation in a boreal forested peatland treated with different nitrogen compounds, Can. J. Forestry 28, 132-140.

Regina, K., Silvola, J., and Martikainen, P.J. (1999) Short-term effects of changing water table on N₂O fluxes from peat monoliths from natural and drained boreal peatlands, Global Change Biol. 5, 183-189.

Roslev, P., and King G.M. (1994) Survival and recovery of methanotrophic bacteria starved under oxic and anoxic conditions, Appl. Environ. Microbiol. 60, 2602-2608.

Roslev, P. and King G.M. (1995) Aerobic and anaerobic starvation metabolism in methanotrophic bacteria, Appl. Environ. Microbiol. 61, 1563-1570.

Roslev, P., Iversen, N., and Henriksen, K. (1997) Oxidation and assimilation of atmospheric methane by soil methane oxidizers, Appl. Environ. Microbiol. 63, 874-880.

Rosswal, T., and Granhall, U. (1980) Nitrogen cycling in a subarctic ombrotrophic mire, Ecol. Bull. 30, 209-234.

Roulet, N.T., Ash, R., Quinton W., and Moore T. (1993) Methane flux from drained northern peatlands: Effect of persistent water table lowering on flux, Global Biogeochem. Cycles 7, 749-769.

Saari, A., Martikainen, P.J., Ferm, A., Ruuskanen, J., De Boer, W., Troelstra, S.R., and Laanbroek, H. (1997) Methane oxidation in soil profiles of Dutch and Finnish coniferous forests with different soil texture and atmospheric nitrogen deposition, Soil Biol. Biochem. 29, 1625-1632.

Saarnio, S., and Silvola, J. (1999) Effects of increased CO₂ and N on CH₄ efflux from a boreal mire: a growth chamber experiment, Oecologia 119, 349-356.

Schimel, J.P. (1995) Plant transport and methane production as controls on methane flux from arctic wet meadow tundra, Biogeochemistry 2, 183-200.

Schnell, S., and King, G.M. (1994) Mechanistic analysis of ammonium inhibition of atmospheric methane consumption in forest soils, Appl. Environ. Microbiol. 60, 3514-3521.

Sitaula, B.K., Bakken, L., and Abrahamsen, G. (1995) CH₄ uptake by temperate forest soil: Effect of N input and soil acidifcation, Soil Biol. Biochem. 27, 871-880.

Stephen, J.R., McCaig, A.E., Smith, Z., Prosser, J.I, and Embley, T.M. (1996) Molecular diversity of soil and marine 16S rRNA gene sequences related to β-subgroup ammonia-oxidizing bacteria, Appl. Environ. Microbiol. 62, 4147-4154.

Stephen, J.R., Kowalchuck, G.A., Bruns, M.-A. V., McCaig, A.E., Phillips, C.J., Embley, T.M., and Prosser, J.I. (1998) Analysis of β-subgroup proteobacterial ammonia oxidizer populations in soil by denaturing gradient gel electrophoresis analysis and hierarchical phylogenetic probing, Appl. Environ. Microbiol. 64, 2958-2965.

Sundh, I., Nilsson, M., Granberg, G., and Svensson, B.H. (1994) Depth distribution of microbial production and oxidation of methane in northern boreal peatlands, Microb. Ecol. 27, 253-265.

Sundh, I., Mikkelä, C., Nilsson, M., and Svensson, B.H. (1995) Potential aerobic methane oxidation in a Sphagnum-dominated peatland – Controlling factors and relation to methane emission, Soil Biol. Biochem., 27, 829-837.

Svensson, B.H. (1996) Contribution of microbial processes to global change, in J.C. Murrell and D.P. Kelly (eds.),

Microbiology of Atmospheric Trace Gases, NATO ASI Series I, Global Environmental Change, Springer-Verlag, Berlin, pp. 255-259.

Torn, M.S., and Chapin, F.S. (1993) Environmental and biotic controls over methane flux from arctic tundra, Chemosphere 26, 357-368.

Utåker, J.B., Bakken, L., Jian, Q.Q., and Nes, I.E. (1995) Phylogenetic analysis of seven new isolates of ammonium-oxidizing bacteria based on 16S rRNA gene sequences, System. Appl. Microbiol 18, 549-559.

Vasander, H., Laiho R., and Laine, J. (1997) Changes is species diversity in peatlands drained for forestry, in C.C. Tretti, M.F. Jurgensen, D.E. Grigal, M.R. Gale, and J.K. Jeglum (eds.), Northern Forested Wetlands: Ecology and Management, CRC Lewis Publisher, Boca Raton, pp. 109-119.

Vasara, R., Suutari, M.H., Lipponen, M.T.T., Martikainen, P.J., Regina, K., Tuomainen, J., Kangasjärvi, J. and Servomaa, K. The diversity of ammonia-oxidizing and methane-oxidizing microbial populations in natural and manipulated northern organic soils (submitted).

West, A.E., and Schmidt, S.K. (1999) Acetate stimulates CH₄ oxidation by an alpine tundra soil, Soil Biol. Biochem 31, 1649-1655.

Willison, T.W., Webster, C.P., Goulding, K.W.T., and Powlson, D.S. (1995) Methane oxidation in temperate soils: Effects of land use and the chemical form of nitrogen fertilizer. Chemosphere 30, 539-546.

THE HALOALKANE DEHALOGENASE GENES *DHLA* AND *DHAA* ARE GLOBALLY DISTRIBUTED AND HIGHLY CONSERVED

GERRIT J. POELARENDS,
JOHAN E. T. VAN HYLCKAMA VLIEG, TJIBBE BOSMA AND
DICK B. JANSSEN
*Department of Biochemistry, Groningen Biomolecular Sciences and
Biotechnology Institute, University of Groningen, 9747 AG Groningen,
The Netherlands. Phone: 31-50-3634209. Fax: 31-50-3634165. E-mail:
d.b.janssen@chem.rug.nl.*

Direct hydrolytic dehalogenation by haloalkane dehalogenases is the most important mechanism involved in biodegradation of synthetic haloalkanes that occur as soil pollutants. Here we show that five Gram-negative 1,2-dichloroethane-utilizing bacteria contain haloalkane dehalogenase genes identical to the *dhlA* gene from *Xanthobacter autotrophicus* GJ10, whereas five Gram-positive chloroalkane degraders, independently isolated from geographically distinct locations, contain genes identical to the *dhaA* gene of *Rhodococcus rhodochrous* NCIMB13064. Furthermore, the *dhaA* gene was detected by PCR amplification in fifteen newly isolated Gram-positive chloroalkane-degrading bacteria. Our results suggest that the *dhlA* and *dhaA* genes recently arose from a single origin and have become distributed globally, most likely as the result of the massive and worldwide use of synthetic chlorinated hydrocarbons in industry and agriculture.

1. Introduction

Synthetic haloalkanes form an important class of environmental pollutants because of their massive and widespread use in industry and agriculture, persistence in the environment, and ability to bioaccumulate. An understanding of how microorganisms and microbial communities adapt to degrade synthetic haloalkanes that occur as soil pollutants is important for the evaluation of the ecological impact and environmental fate of these industrial chemicals. The key step in the microbial degradation of halogenated compounds is the cleavage of carbon-halogen bonds by dehalogenases.

Direct hydrolytic dehalogenation of haloalkanes was first found with the haloalkane dehalogenase of the 1,2-dichloroethane-degrading bacterium *Xanthobacter autotrophicus* GJ10 (Janssen *et al.*, 1985). This haloalkane dehalogenase (DhlA) has been extensively studied and the corresponding *dhlA* gene was cloned, sequenced, and shown to be plasmid localized (Janssen *et al.*, 1989; Tardiff *et al.*, 1991). Another haloalkane dehalogenase (DhaA) with low activity towards 1,2-dichloroethane, but better activity towards long chain mono- and dihalogenated substrates, was found in the

59

S.N. Agathos and W. Reineke (eds.),
Biotechnology for the Environment: Strategy and Fundamentals, 59–66.
© 2002 *Kluwer Academic Publishers. Printed in the Netherlands.*

1-chlorobutane-degrading bacterium *Rhodococcus rhodochrous* NCIMB13064 (Curragh *et al.*, 1994). The corresponding haloalkane dehalogenase gene (*dhaA*) was shown to be located on the autotransmissible plasmid pRTL1 (Kulakova *et al.*, 1995), and was recently cloned and sequenced (Kulakova *et al.*, 1997). Here, we use the availability of the *dhlA* and *dhaA* gene sequences to investigate the presence of homologous genes in other chloroalkane-degrading bacteria. We describe that distinct Gram-positive bacterial strains isolated from different geographical locations with different carbon sources contain identical *dhaA* genes. Gram-negative 1,2-dichloroethane-utilizing bacteria were always found to contain a gene identical to the haloalkane dehalogenase gene *dhlA*.

2. Chloroalkane-utilizing bacteria used in this study

The bacterial strains screened for the presence of the *dhaA* gene are listed in Table 1. Strains Y2, m15-3, HA1, and GJ70 were independently isolated by different research groups (Sallis *et al.*, 1990; Yokota *et al.*, 1986; Scholtz *et al.*, 1987; Janssen *et al.*, 1987), and were all identified by 16S rRNA gene sequence analysis as *Rhodococcus* species (data not shown). Strains NCIMB13064 and 170 (formerly known as *P. cichorii* 170) have been investigated previously (Kulakova *et al.*, 1997; Poelarends *et al.*, 1998), and were shown to possess identical *dhaA* genes. The other strains were isolated in our laboratory for their ability to use 1-chlorobutane or 1,3-dichloropropene as the sole carbon and energy source, and were subjected to minimal subculturing before being analyzed. All newly isolated strains were identified as Gram-positive organisms, but were not further characterized. The relevant catabolic pathways are shown in Fig. 1. The 1,2-dichloroethane-degrading bacterial strains screened for the presence of the *dhlA* gene are listed in Table 2. Five of these bacterial strains were studied here, and strains GJ10, GJ11, AD20, AD25, AD27, and RB8 have been investigated previously (Janssen *et al.*, 1989; Van den Wijngaard *et al.*, 1992). Strains GJ10, GJ11, AD20, and AD25 were shown to possess identical *dhlA* genes, whereas strains AD27 and RB8 were shown by hybridization analysis to contain *dhlA* homologs. The KDE strains (a kind gift of Dr. M. Kästner) were all Gram-negative, non-motile bacteria that were able to grow on methanol and citrate. Their yellow pigmentation and pleomorphic appearance suggest that they might belong to the genus *Xanthobacter* (Wiegel & Schlegel, 1984). The 1,2-dichloroethane catabolic pathway is shown in Fig. 1.

TABLE 1. *List of chloroalkane-degrading strains carrying the dhaA gene*

Strain	Origin	Reference
R. rhodochrous NCIMB13064	UK, industrial site exposed to chlorinated alkanes	Curragh et al., 1994
R. erythropolis Y2	UK, industrial site exposed to haloalkanes	Sallis et al., 1990
Rhodococcus strain m15-3	Japan	Yokota et al., 1986
Rhodococcus strain HA1	Switzerland	Scholtz et al., 1987
Rhodococcus strain GJ70	The Netherlands, sludge contaminated with pesticides	Janssen et al., 1987
P. pavonaceae 170	The Netherlands, soil exposed to 1,3-dichloropropene	Verhagen et al., 1995
Strain TB2	USA, soil exposed to 1,2,3-trichloropropane	This study
Strains KC1, KC2	The Netherlands, industrial site exposed to chlorinated pesticides	This study
Strains AKZ1, C12-1	The Netherlands, industrial site exposed to 1,2-dichloroethane	This study
Strains A7-6, A6-1, A6-2, A12-1, A11-3, A0-7	The Netherlands, agricultural soil exposed to 1,3-dichloropropene	This study
Strains B7-1, B7-2, B12-1, B12-2, B0-3	The Netherlands, agricultural soil with no history of point-source contamination	This study

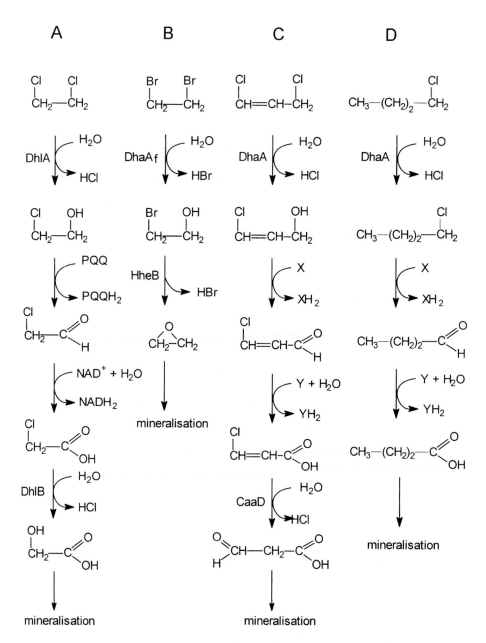

Fig. 1. Haloalkane catabolic pathways. (A) Route for the metabolism of 1,2-dichloroethane in X. autotrophicus and A. aquaticus strains (Janssen et al., 1985). (B) Metabolism of 1,2-dibromoethane in Mycobacterium sp. strain GP1 (Poelarends et al., 1999). (C) Degradation of 1,3-dichloropropene in P. pavonaceae 170 (Poelarends et al., 1998). (D) Proposed route for the degradation of 1-chlorobutane in R. rhodochrous NCIMB13064 (Curragh et al., 1994).

3. Detection of *dhaA* and *dhlA* homologs in chloroalkane-degrading bacteria

Total genomic DNA was isolated from strains Y2, m15-3, HA1, and GJ70 as described elsewhere (Poelarends *et al.*, 1998), and was directly used as template for polymerase chain reaction (PCR) amplification. For amplification of putative dehalogenase genes from the other strains, biomass from a single colony was resuspended in the reaction mixture. Total DNA and synthetic oligonucleotide primers were used at a final concentration of 100 ng per 100 μl in the PCR. The reaction was performed with *Taq* DNA polymerase (Boehringer Mannheim, Mannheim, Germany) using denaturation, annealing and extension temperatures of 94°C, 58°C and 72°C, respectively. Primers were designed specifically for *dhaA* (5'-AAAATC GCCATGGCAGAAATCGGTA-3' and 5'-TGGACATCGGACCATGGCGTGAACC-3', *Nco*I sites underlined), or *dhlA* (5'-GAGGCTCCATGGTAAATGCAAT-3' and 5'-ATAGAATTCCATGGAT CCTCAGTTTTCGTACCGGCACCGG-3', *Nco*I sites underlined), and were synthesized by Eurosequence BV (Groningen, The Netherlands). PCR products were subjected to electrophoresis in 0.8% agarose gels, and were stained with ethidium bromide.

TABLE 2. List of 1,2-dichloroethane-degrading strains carrying the dhlA *gene*

Strain	Origin	Reference
X. autotrophicus GJ10	The Netherlands, soil and sediment from different sites	Janssen *et al.*, 1985
X. autotrophicus GJ11	The Netherlands, sediment from the Rhine River	Van den Wijngaard *et al.*, 1992
A. aquaticus AD20, AD25, AD27	The Netherlands, sediment from the Eems channel	Van den Wijngaard *et al.*, 1992
Pseudomonas sp. RB8	Germany, industrial wastewater-treatment plant	Van den Wijngaard *et al.*, 1992
Xanthobacter-like organisms: strains KDE1, KDE2, KDE3, KDE4, KDE5	Germany, soil from Braunschweig region	Dr. M. Kästner

For all 20 Gram-positive chloroalkane degraders analyzed in this study, PCR amplification with *dhaA*-specific primers consistently produced a 0.9-kb DNA fragment, corresponding to the size of the *dhaA* gene (data not shown). In contrast to the presence of *dhaA* homologs in these Gram-positive chloroalkane degraders, the five Gram-negative 1,2-dichloroethane-utilizers (the KDE strains) were all found to contain *dhlA* homologs (data not shown). Attempts to amplify putative *dhaA* or *dhlA* genes from randomly chosen Gram-positive and Gram-negative laboratory strains that are known not to degrade chloroalkanes, did not result in an amplification product. This indicates that the presence of haloalkane dehalogenase genes is related to the ability to utilize chloroalkanes.

4. Sequence identity among *dhaA* or *dhlA* genes of different chloroalkane-degrading bacteria

The PCR-amplified *dhaA* genes of strains Y2, m15-3, HA1, GJ70, and TB2 were chosen for further analysis, because these strains were independently isolated from geographically distinct locations. Putative *dhlA* genes amplified from the five Gram-negative KDE strains were also further analyzed. The PCR products of strains GJ70 and Y2 were directly cloned in the TA-cloning vector pCR2.1 according to the recommendations of the supplier (Invitrogen). PCR products obtained from the other strains were cloned in the *Nco*I site behind the T7 promoter in the expression vector pGEF+ as described before (Poelarends *et al.*, 1998). The cloned dehalogenase genes were sequenced by the dideoxy chain termination method (Sanger *et al.*, 1977), and nucleotide sequences were aligned by using LALIGN (Institut de Génétique Humaine (IGH), Montpellier, France).

Surprisingly, DNA sequencing revealed that the PCR-amplified haloalkane dehalogenase genes of strains Y2, m15-3, HA1, GJ70, and TB2 were completely identical to the *dhaA* gene that was previously found in *R. rhodochrous* NCIMB13064 and *P. pavonaceae* 170. The PCR-amplified genes of the five KDE strains were completely identical to the *dhlA* gene that was previously found in *X. autotrophicus* GJ10 and GJ11, and in *Ancylobacter aquaticus* AD20 and AD25. This remarkable sequence identity (100%) suggests that each *dhaA* gene, or *dhlA* gene, in the different bacterial strains was recently derived from a common ancestor.

In conclusion, until now only one enzyme (DhlA) that efficiently converts 1,2-dichloroethane to 2-chloroethanol has been found in 1,2-dichloroethane-degrading bacteria, whereas only one enzyme (DhaA) that converts long-chain chloroalkanes to the corresponding alcohols has been found in several Gram-positive chloroalkane degraders and in one Gram-negative strain. Contrary to the haloalkane dehalogenases, many different L-2-monochloropropionic acid specific dehalogenases have been found (Van der Ploeg *et al.*, 1991; Fetzner and Lingens, 1994). This remarkable difference in sequence diversity within these two classes of hydrolytic dehalogenases may reflect the natural occurence of haloacids in nature, whereas the substrates for the haloalkane dehalogenases have been present in the environment for a short period on an evolutionary time scale. The haloacid dehalogenase genes are probably of old evolutionary origin and their distribution is not related to environmental contamination. Our observations argue for a single, recent evolutionary origin of the haloalkane dehalogenase genes *dhaA* and *dhlA* and their subsequent global distribution as a result of the massive and worldwide use of synthetic haloalkanes in industry and agriculture.

In our search for homologous dehalogenase genes in haloalkane-degrading bacteria, we recently found a *dhaA* homolog, called *dhaA$_f$*, in the 1,2-dibromoethane-degrading organism *Mycobacterium* sp. strain GP1 (Poelarends *et al.*, 1999) (Fig. 1). This dehalogenase gene appeared to encode a 307-amino-acid polypeptide, and is the result of a fusion between two known genes which encode dehalogenase enzymes. The first 293 amino acids of DhaA$_f$ are identical to DhaA, except for three amino acid substitutions, whereas the last 14 amino acids are identical to the C-terminal sequence of the haloalcohol dehalogenase HheB from *Corynebacterium* sp. strain N-1074 (Yu *et al.*, 1994). In contrast to the absolutely conserved sequence of the *dhaA* gene in the analyzed *Rhodococcus* strains, the *dhaA* gene is thus not perfectly conserved in strain GP1.

The dehalogenases DhaA and DhlA most likely existed in preindustrial times as enzymes with a closely related function. However, DhaA and DhlA do not share high sequence similarity to each other and there are no known closely related proteins from which they may have recently evolved. Questions concerning the origin of these dehalogenases thus remain to be solved.

The data presented here provide further support for previous studies suggesting that transfer of genes involved in pollutant biodegradation may play an important role in the evolution of bacterial strains and the adaptation of microbial communities to different environmental contaminants (Fulthorpe *et al.*, 1995; Herrick *et al.*, 1997; De Souza *et al.*, 1998). The *dhlA* genes of *X. autotrophicus* GJ10 and GJ11, as well as those of *Ancylobacter aquaticus* AD25 and AD27, were shown to be localized on similar plasmids (Tardiff *et al.*, 1991; Van der Ploeg JR, unpublished results). Haloalkane-catabolic plasmids of different sizes were recently identified in the *Rhodococcus* strains Y2, m15-3, HA1, GJ70, and TB2. Our further research will focus on the identification and characterization of these mobile elements, and their role in gene transfer.

Acknowledgements

The study was supported by the Life Sciences Foundation (SLW), which is subsidized by the Netherlands Organization for Scientific Research (NWO), and by the EC Environmental and Climate Research Program contract ENV4-CT95-0086.

We thank the Czech Collection of Microorganisms (Masaryk University, Brno, Czech Republic) for providing *R. erythropolis* Y2. We also thank Dr. M. Kästner (Tech. University Hamburg-Harburg, Hamburg, Germany), Prof. T. Leisinger (Swiss Federal Institute of Technology, ETH, Zürich, Switzerland) and Dr. T. Omori (University of Tokyo, Tokyo, Japan) for providing different chloroalkane-degrading strains.

References

Curragh H, Flynn O, Larkin MJ, Stafford TM, Hamilton JTG & Harper DB (1994) Haloalkane degradation and assimilation by *Rhodococcus rhodochrous* NCIMB13064. Microbiol. 140: 1433-1442

De Souza ML, Seffernick J, Martinez B, Sadowsky MJ & Wackett LP (1998) The atrazine catabolism genes *atzABC* are widespread and highly conserved. J. Bacteriol. 180: 1951-1954

Fetzner S & Lingens F (1994) Bacterial dehalogenases: biochemistry, genetics, and biotechnological applications. Microbiological Reviews 58: 641-685

Fulthorpe RR, McGowan C, Maltseva OV, Holben WE & Tiedje JM (1995) 2,4-Dichlorophenoxyacetic acid-degrading bacteria contain mosaics of catabolic genes. Appl. Environ. Microbiol. 61: 3274-3281

Herrick JB, Stuart-Keil KG, Ghiorse WC & Madsen EL (1997) Natural horizontal transfer of a naphthalene dioxygenase gene between bacteria native to a coal tar-contaminated field site. Appl. Environ. Microbiol. 63: 2330-2337

Janssen DB, Scheper A, Dijkhuizen L & Witholt B (1985) Degradation of halogenated aliphatic compounds by *Xanthobacter autotrophicus* GJ10. Appl. Environ. Microbiol 49: 673-677

Janssen DB, Jager D & Witholt B (1987) Degradation of n-haloalkanes and α,ω-dihaloalkanes by wild-type and mutants of *Acinetobacter* sp. strain GJ70. Appl. Environ. Microbiol. 53: 561-566

Janssen DB, Pries F, Van der Ploeg J, Kazemier B, Terpstra P & Witholt B (1989) Cloning of 1,2-dichloroethane degradation genes of *Xanthobacter autotrophicus* GJ10 and expression and sequencing of the *dhlA* gene. J. Bacteriol. 171: 6791-6799

Kulakova AN, Stafford TM, Larkin MJ & Kulakov LA (1995) Plasmid pRTL1 controlling 1-chloroalkane degradation by *Rhodococcus rhodochrous* NCIMB13064. Plasmid 33: 208-217

Kulakova AN, Larkin MJ & Kulakov LA (1997) The plasmid-located haloalkane dehalogenase gene from *Rhodococcus rhodochrous* NCIMB13064. Microbiol. 143: 109-115

Poelarends GJ, Wilkens M, Larkin MJ, Van Elsas JD & Janssen DB (1998) Degradation of 1,3-dichloropropene by *Pseudomonas cichorii* 170. Appl. Environ. Microbiol. 64: 2931-2936

Poelarends GJ, Van Hylckama Vlieg JET, Marchesi JR, Freitas Dos Santos LM & Janssen DB (1999) Degradation of 1,2-dibromoethane by *Mycobacterium* sp. strain GP1. J. Bacteriol. 181: 2050-2058

Sallis PJ, Armfield SJ, Bull AT & Hardman DJ (1990) Isolation and characterization of a haloalkane halidohydrolase from *Rhodococcus erythropolis* Y2. J. Gen. Microbiol. 136: 115-120

Sanger F, Nicklen S & Coulsen AR (1977) DNA sequencing with chain-terminating inhibitors. Proc. Natl. Acad. Sci. USA. 74: 5463-5467

Scholtz R, Schmuckle A, Cook AM & Leisinger T (1987) Degradation of eighteen 1-monohaloalkanes by *Arthrobacter* sp. strain HA1. J. Gen. Microbiol. 133: 267-274

Tardiff G, Greer CW, Labbé D & Lau PCK (1991) Involvement of a large plasmid in the degradation of 1,2-dichloroethane by *Xanthobacter autotrophicus* GJ10. Appl. Environ. Microbiol. 57: 1853-1857

Van den Wijngaard AJ, Van der Kamp KWHJ, Van der Ploeg J, Pries F, Kazemier B & Janssen DB (1992) Degradation of 1,2-dichloroethane by *Ancylobacter aquaticus* and other facultative methylotrophs. Appl. Environ. Microbiol. 58: 976-983

Van der Ploeg JR, Van Hall G & Janssen DB (1991) Characterization of the haloacid dehalogenase from *Xanthobacter autotrophicus* GJ10 and sequencing of the *dhlB* gene. J. Bacteriol. 173: 7925-7933

Verhagen C, Smit E, Janssen DB & Van Elsas JD (1995) Bacterial dichloropropene degradation in soil; screening of soils and involvement of plasmids carrying the *dhlA* gene. Soil Biol. Biochem. 27: 1547-1557

Wiegel JKW & Schlegel HG (1984) Genus *Xanthobacter*. In: Krieg NR & Holt G (Ed) Bergey's manual of systematic bacteriology, Vol 1 (pp325-333). The Williams & Wilkens Co., Baltimore

Yokota T, Fuse H, Omori T & Minoda Y (1986) Microbial dehalogenation of haloalkanes mediated by oxygenase or halidohydrolase. Agric. Biol. Chem. 50: 453-460

Yu F, Nakamura T, Mizunashi W & Watanabe I (1994) Cloning of two halohydrin hydrogen-halide lyase genes from *Corynebacterium* sp. strain N-1074 and structural comparison of the genes and gene products. Biosci. Biotechnol. Biochem. 58: 1451-1457

PART 3
BIODEGRADATION

MICROBIAL ASPECTS IN BIOREMEDIATION OF SOILS POLLUTED BY POLYAROMATIC HYDROCARBONS

PIERRE WATTIAU

Technical University of Munich, Institute for Water Quality Control and Waste Management, Am Coulombwall, D-85748 Garching, GERMANY Catholic University of Louvain, Bioengineering Unit, Place Croix du Sud 2/19, B-1348 Louvain-la-Neuve, Belgium email: wattiau@gebi.ucl.ac.be

Polyaromatic hydrocarbons (PAHs) are ubiquitous pollutants found in high concentrations at industrial sites associated with petroleum, coal tar, gas production and wood preservation industries. Due to their carcinogenic and mutagenic properties, PAHs are considered as environmental priority pollutants. They are stable and recalcitrant in soils as they are less easy to degrade than many other organic compounds. Though feasible, bioremediation of PAH-contaminated soils is seriously hampered by the low bioavailability of these compounds, making their removal a long and difficult process. Some PAH-degrading microorganisms seem however to be adapted to face the unfavourable physico chemical- properties of PAHs. Recent studies have shown that these peculiar organisms often belong to a discrete number of bacterial or fungal genera. In this review, both physico-chemical and metabolic aspects associated with the bacterial removal of PAHs from contaminated soils are discussed, as well as the perspectives for improving the associated biological technologies.

1. PAHs: toxic and recalcitrant compounds

PAHs are a class of organic compounds made up of two or more benzene rings fused in either linear, angular or cluster arrangement. Basically, they contain only C and H atoms, although S, N and O may substitute some carbon atoms in the aromatic rings to form the so-called heterocyclic subclass of polyaromatics. Examples of PAHs are illustrated in Fig. 1.

1.1. TOXICITY, SOURCES OF CONTAMINATION AND LEGAL DISPOSITIONS

PAHs are formed whenever organic matter is burnt but, as environmental pollutants, mainly issue from the processing, disposal and combustion of fossil fuels. They are constituents of fractionated oil products such as diesel and jet fuels, petroleum, lubricating oils, etc. PAH contamination at industrial sites is commonly associated with

S.N. Agathos and W. Reineke (eds.),
Biotechnology for the Environment: Strategy and Fundamentals, 69–89.
© 2002 *Kluwer Academic Publishers. Printed in the Netherlands.*

spills from storage tanks, transport, processing, use and disposal of PAH-containing fuels.

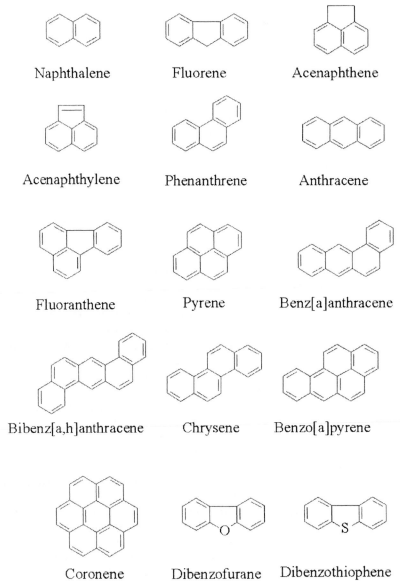

Figure 1. Structure of some common PAHs.

Heavily polluted industrial sites are often former gas plants, coking plants, petrochemical industries and wood-preserving products manufactures (reviewed in [1]).

As an indication, creosote and anthracene oil, which were widely used as wood-preservatives, contain up to 85% of PAHs by weight.

It is now well established that exposure to PAHs constitutes a risk for people living in industrialised areas [2,3]. Different carcinogenic, mutagenic and genotoxic activities were found associated with PAHs, ranging from inactive to highly potent [4]. Most non-substituted PAHs with two or three rings (e.g. naphthalene, fluorene, anthracene, phenanthrene and pyrene) are not carcinogenic in experimental animals. Benz[a]anthracene and chrysene are weakly so, whereas benzo[a]pyrene is a strong carcinogenic compound [5]. However, genotoxic tests based on mammalian DNA integrity (damage, mutation, and chromosomal abnormalities) [6] or on the Ames mutagenicity test in *Salmonella* [7] turn out to be positive for most PAHs. Therefore, the handling of material contaminated by these compounds requires care and safety precautions. The carcinogenic effects of PAHs on mammalian cells are a consequence of the metabolic activation to diol epoxides, which are highly reactive molecules that covalently bind to DNA. This activation occurs mainly in the microsomes of the endoplasmic reticulum and is catalysed by monooxygenase enzymes associated to cytochrome P-450 [8]. PAHs were also shown to affect the immune system of mammals [9].

PAHs are classified as priority pollutants by both the United States Environmental Protection Agency (EPA) and by the European Community [10,11]. A list of 16 compounds was published by the EPA and is used by US regulatory authorities to identify site contamination and specify monitoring parameters. In Europe, similar lists were proposed by different member states, the one of the Netherlands (now commonly referred as the "Dutch list") is the most widely accepted. These lists usually mention the limit PAH concentrations that should be considered as acceptable in soils and groundwater and supply "action" threshold values that may differ depending on the activities associated with the soil. Besides the governmental decisions regulating polluted soils, special dispositions have been taken to control the sources of contamination. A recent example is the drastic limitations regarding the marketing and use of creosotes that have been adopted by the EC countries (Directive 1999/833/EC). Nevertheless, the amount of PAH-polluted soil inherited from the industrial past is considerable and requires urgent treatment (Fig. 2).

1.2. POLYAROMATIC HYDROCARBONS AS LONG-TERM CONTAMINANTS

The persistence of PAHs in the environment is basically due to their low water solubility. PAHs accidentally introduced in soils rapidly escape from the water phase and become associated with sediments and soil particles, to which they strongly sorb [12,13]. From the microbial point of view, this dual segregation leads to a strongly reduced accessibility - *i.e.* a low bioavailability - of the PAHs as potential carbon sources.

Figure 2. Piles of PAH-contaminated soils at the soil remediation company SOILREM a.s., Kalundborg, Denmark (picture kindly provided by Susanne Schiøtz-Hansen).

As for genotoxicity, hydrophobicity and environmental persistence increase with the molecular size and the number of fused benzene rings [1,8]. The age of the contamination is also an important factor since the mean sorption distance of PAHs into soil particles is a time-dependent factor. Consequently, the reversal sorption of PAHs to the more accessible fraction of the soil is delayed in soils with long-term contamination history. Besides natural abiotic removal such as volatilisation, hydrolysis and leaching, it is widely accepted that the major natural process through which PAHs are removed from the environment is microbial degradation. In multiphase soil systems, biodegradation is accomplished by natural surface and subsurface microorganisms that can either mineralise PAHs into CO_2 and H_2O or partially transform them. The biodegradation of two- and three-ring PAHs has been shown to be extensive, whereas that of four- and more ring PAHs is considerably less significant [1,12,14].

2. PAHs-degrading microorganisms

A wide range of different microorganisms is able to partially metabolise PAHs. Fungi utilise cytochrome P-450 monooxygenases to form arene oxides that are further transformed to *trans*-dihydrodiols but might also rearrange into phenol derivatives, that are carcinogenic. Detoxification occurs via conjugation of the phenols with either sulphate or sugar derivatives. White rot fungi are lignolytic organisms that produce extracellular peroxidases with no or little substrate specificity that can convert PAHs to quinone derivatives [15]. A well-studied example is *Phanerochaete chrysosporium*, whose degrading capacities have been reviewed elsewhere [16]. PAH degradation by algae and cyanobacteria has also been reported [17,18].

2.1. BACTERIAL PAH-DEGRADATION PATHWAYS

Unlike fungal and mammalian cells, bacteria characteristically produce dioxygenases which incorporate two oxygen atoms into the substrate to form dioxethanes that are further oxidised to *cis*-dihydrodiols and dihydroxy products [1]. However, occasional monooxygenation or angular dioxygenation have been reported, for example in the bacterial metabolism of fluorene and dibenzofuran [19]. Catechol (1,2-dihydroxybenzene), gentisic acid (2,5-dihydroxybenzoic acid) and protocatechuic acid (3,4-dihydroxybenzoic acid) are the most common central intermediates in the bacterial PAH degradation pathways. These compounds are in turn metabolised to succinate, pyruvate, fumarate, acetate and acetaldehyde following different pathways that support protein synthesis and energy production and lead to the release of carbon dioxide and water. The intermediate compound produced depends on the position of the hydroxyl groups (*ortho* or *para*) in the initial diphenolic compounds. A number of different metabolic pathways have been established for the bacterial degradation of PAHs. These pathways are essentially based on the chemical analysis of metabolites. A compilation of the most common and best-documented pathways has been made available on the Internet [20]. PAHs are microbially degraded either as the sole carbon and energy source or by co-metabolism. As unique growth substrates, PAHs can be partially or completely mineralised, *i.e.* converted to carbon dioxide and water. As cometabolic substrates, they cannot support growth but are modified by enzymes produced in the course of the degradation of a growth-supporting, usually structurally related compound. Though little is known about the regulatory events controlling co-metabolism, it has been observed to contribute for a big extent to the degradation of high molecular weight PAHs and is believed to substantially initiate the removal of those recalcitrant compounds [21]. Anaerobic degradation of PAHs has also been observed. It is poorly documented so far but represents another source of valuable data for further research [22].

2.2. PAH-DEGRADING MICROORGANISMS

A large number of bacteria with PAH-degrading capabilities have been reported that were able to either completely assimilate a defined range of compounds, or to exhibit just partial metabolism and did so either as isolated organisms, or as part of consortia made up of several different organisms. This review is not intended to exhaustively report all these organisms, but rather to emphasise some microbiological aspects associated with PAH metabolism, *i.e.* (i) the relationship between the chemical structure of PAHs and the nature of the degrading organisms, (ii) the physiological advantages displayed by these organisms and (iii) what is known about the molecular mechanisms underlying these special adaptations.

Naphthalene is probably the simplest PAH. Naphthalene-degrading bacteria were used as illustrative examples to model the general principles of PAH metabolic pathways, enzymatic mechanisms and genetic regulation and has hence been abundantly studied [23-25]. The choice of naphthalene degradation as a model was explained by the relative ease of the handling and the genetic manipulation of the naphthalene-degrading bacteria, which are mainly pseudomonads. However, it is more and more accepted nowadays that data retrieved from these studies are far from being representative and in many aspects do not reflect the general characteristics of other PAH-degraders. Unlike most PAHs, naphthalene is fairly soluble in water (32 mg l^{-1} at 25°C, compared to 1.9 mg l^{-1} for fluorene, 1.0 mg l^{-1} for phenanthrene and much less than 1 mg l^{-1} for all the other PAHs). In addition, DNA hybridisation experiments have demonstrated that except for some phenanthrene degradation systems, no significant homology between PAH degradation genes can be observed with the naphthalene system of *Pseudomonas* when bacteria from different genera are considered [26-28]. 3-ring PAHs such as acenaphthene, acenaphthylene, fluorene and phenanthrene can be metabolised by a variety of different bacteria, including Gram-negative and Gram-positive representatives. Anthracene, although identical to phenanthrene in the number of aromatic rings, seems much more difficult to degrade probably as the consequence of its low (0.07 mg l^{-1}) solubility in water [12]. By contrast, data concerning the isolation and physiological description of microorganisms able to metabolise more complex PAHs are much scarcer. A short description of some PAH-degrading bacteria that are important for either historical reasons or for their peculiar degrading capacities is given below.

Early reports in the 1970s described the microbial oxidation of benzo[a]pyrene and benz[a]anthracene by a *Beijerinckia* sp. that had been chemically mutagenised [29]. Cometabolic biodegradation of fluoranthene and benzo[a]pyrene was also reported [30]. The first bacterium able to extensively metabolise a PAH with four aromatic rings was isolated in the late 1980s from oil field sediments. It turned out later to be a member of the *Mycobacterium* genus. It was described in 1988 as a Gram-positive rod that cometabolically degraded a number of PAHs including fluoranthene, pyrene and benzo[a]pyrene [31]. *Sphingomonas yanoikuyae* strain B1 (formerly described as a *Beijerinckia* sp.: see ref. [29]) was shown the same year to oxidise dibenz[a]anthracene [32]. In 1989, a bacterial community consisting of seven different strains was isolated from a soil contaminated with creosote that could use fluoranthene as sole source of carbon and energy [33]. In 1990, the first organisms able to utilise a four-ring PAH as sole source of carbon and energy were isolated: *Alcaligenes denitrificans* WW1 [34] and *Sphingomonas paucimobilis* EPA505 [35] were shown to grow on a mineral medium supplemented with fluoranthene. These organisms were also shown to cometabolically transform other PAHs composed of four fused aromatic rings. Additional fluoranthene-degraders of the *Sphingomonas* genus were isolated in the last decade [36]. *Mycobacterium* sp. strain PYR-1 was isolated during the same period and displayed a remarkable fluoranthene mineralisation rate [37]. This bacterium was also shown to metabolise other PAHs to various degrees, with the exception of chrysene. Since then, a number of interesting bacteria able to grow on three- and four-ring PAHs have been isolated. Some relevant examples are *Mycobacterium* sp. KR2 which grows on pyrene [38], *Mycobacterium* sp. BB1, which degrades fluoranthene and pyrene [39],

Rhodococcus sp. UW1 which degrades the same PAHs plus chrysene [40], *Mycobacterium* sp. RJGII-135 [41], another pyrene-degrader and *Mycobacterium* sp. CH1 which can grow on either pyrene or phenanthrene [26]. No bacteria have been isolated that could clearly grow with either benz[a]anthracene, dibenz[a,h]anthracene or benzo[a]pyrene as sole carbon and energy source. However, some bacterial and fungal co-cultures do so [42].

2.3. ISOLATION OF PAH-DEGRADING BACTERIA

Most of the isolated PAH-degrading bacteria were recovered following standard isolation methods of environmental microbiology. These methods make use of slurries consisting of aqueous soil suspensions supplemented with PAH crystals and assume that the desired microorganisms will at least partially colonise the aqueous phase after a defined incubation period, eventually followed by an enrichment procedure which is also based on diluted aqueous suspensions supplemented with PAH crystals [43,44]. In these procedures, microorganisms tightly adhering to the solid particles, *i.e.* those with highly hydrophobic cell surfaces or displaying for any reason strong binding capacities for the solid phase matrix, might not be recovered. In addition, these methods mostly select for fast-growing organisms. Using a biphasic aqueous-organic system consisting of dispersed silicone oil, Ascon-Cabrera and Lebeault [45] observed an increase in the degradation of xenobiotic compounds with low water solubility that was attributed to highly active microorganisms growing and spreading at the interfacial area, suggesting that non aqueous liquids might select and enrich for microbial populations specialised in the degradation of substrates poorly soluble in water. Based on these results and on the assumptions described above, a new method was introduced recently by Bastiaens *et al.* [46] for the isolation of PAH degrading bacteria from contaminated soils and sludges, and was compared with a common aqueous enrichment method. In this new method, PAHs are supplied sorbed to a solid phase in the form of a small filter made of synthetic polymers, and incubated for several days in an aqueous soil suspension so that PAH-degrading bacteria are enriched directly on the sorbing carrier. The method differs from the liquid biphasic aqueous-organic isolation system described by Ascon-Cabrera and Lebeault [45] in that PAHs are sorbed onto solid surfaces that mimic the static state of the solid matrix rather than dissolved in a non-aqueous liquid phase. A comparative study dealing with the same soil sample was undertaken and led to the isolation of different bacterial strains depending on the procedure. Classical liquid enrichments mainly led to *Sphingomonas* spp. while isolates enriched on membranes exclusively belonged to the *Mycobacterium* genus. Moreover, an anthracene-utilising *Mycobacterium* sp. (strain LB501T) was isolated from a Teflon membrane sorbed with anthracene, while no anthracene-utilising strains were isolated by the liquid enrichment procedure. The method is considered by its authors as a complementary system and not as a replacement of the liquid cell suspension method, since both methods seem to select for different kinds of bacteria. Finally, using different synthetic sorbing carriers, Grosser *et al.* have recently demonstrated that different phenanthrene-degrading bacteria are isolated depending on the bioavailability conditions created [47].

2.4. BACTERIA WITH HIGH MOLECULAR WEIGHT PAH-DEGRADING CAPACITIES

As described above, numerous members of the *Sphingomonas* and *Mycobacterium* genus have been isolated as degraders of PAHs and of other pollutants with low solubility in water. Interestingly, there seem to be a direct relationship between the complexity and recalcitrance of a defined PAH and the taxonomy of the corresponding degraders. Most bacteria selected on naphthalene belong to the fluorescent pseudomonads [48]. Many phenanthrene and anthracene utilising isolates are *Pseudomonas* spp. [49], *Sphingomonas* spp. [50] or Gram positive bacteria of the *Nocardia-Rhodococcus-Mycobacterium* group [26]. Above 3 aromatic rings, mainly Gram positive PAH-degrading bacteria have been reported, although some *Sphingomonas* degraders do exist [14]. A comparable observation has been made for the degradation of the most recalcitrant alkanes, as bacteria able to degrade long, polybranched alkanes belong to the same groups [51]. The genera *Sphingomonas* and *Mycobacterium* therefore seem to be specialised in degrading such less bioavailable compounds. Whether members of these taxons possess innate dispositions rendering them particularly well-adapted for the degradation of PAHs is an exciting hypothesis and deserves further research. They both have a particular type of cell wall, which may be important for the interaction with or the uptake of hydrophobic compounds. The next two sections briefly summarise the peculiar cell wall composition, the taxonomic position and some recent advances in the molecular biology associated with these two interesting bacterial genera.

2.4.1. Sphingomonas

Sphingomonas is a genus widespread in nature that has been found in very different ecological niches. It is part of the α-group of proteobacteria in the Gram-negative cluster [52]. Members of the *Sphingomonas* genus owe their name to the presence in the cell membrane of glycosphingolipids, which are unique compounds in the bacterial world. Glycosphingolipids are composed of sphingosine, a monounsaturated C-18 amino alcohol, esterified at the amino group with a fatty acid and condensed at the hydroxyl group with a sugar, resulting in an amide compound containing both hydrophilic and hydrophobic moieties (Fig. 3).

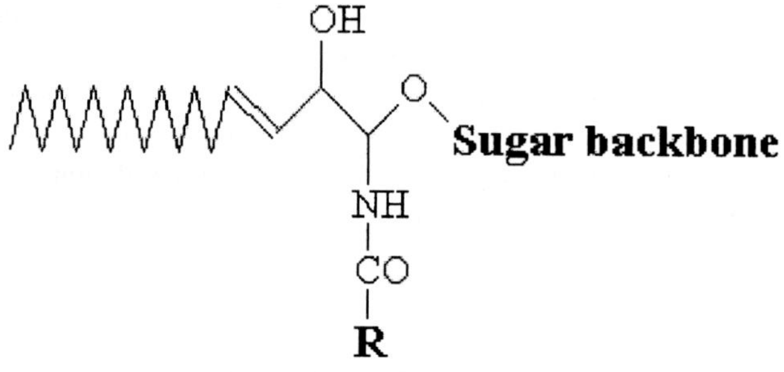

Figure 3. Structure of a typical sphingolipid. R is a fatty acid chain.

The structure of some new and typical sphingolipids from different *Sphingomonas* spp. have been reported recently [53,54]. Although the prevalence of *Sphingomonas* spp. in PAH-contaminated environments is well-documented, there is no direct evidence for the advantage that sphingolipids might confer to these bacteria regarding the uptake and/or metabolism of PAHs. It is clear however that the cell wall of these bacteria has a quite unique composition with peculiar physico-chemical properties. A large number of *Sphingomonas* strains have been isolated that displayed metabolic and mineralisation potentials towards various recalcitrant compounds ranging from natural biopolymers such as lignin degradation products to xenobiotics including solvents, pesticides, herbicides and PAHs [35,36,55-57]. PAH-degrading *Sphingomonas* were even found in deep terrestrial subsurface sediments, confirming the hypothesis that microbial degradation of PAHs is not a recently acquired metabolism [55]. In addition, the taxonomy of this genus has been revised and now includes a number of species that were initially misidentified, mainly as pseudomonads [35,52,56]. The development of molecular biology techniques and tools for *Sphingomonas* spp. has been stimulated during the last five years by the interesting metabolic potential displayed by members of this taxon. A number of mutagenesis and cloning tools have been adapted or constructed for use in these organisms. Tn5- and Tn7-based transpositions have been reported to be successful in a number of different *Sphingomonas* isolates as well as the use of wide-host range vectors [58,59]. ColE1-based replicons are not functional and can hence be used as vectors for the purposes of transposon delivery, integrative plasmid mutagenesis and allelic exchange[60]. A number of genes coding for enzymes participating in the metabolism of several xenobiotic substrates have been cloned and characterised [61-64]. Recently, the complete sequence of a 180-kb catabolic plasmid from the deep-subsurface organism *Sphingomonas aromaticivorans* F199 has been published [65]. This plasmid contains genes participating in the metabolism of several aromatic compounds, including toluene, xylene, *p*-cresol, naphthalene, dibenzothiophene and fluorene. No functional analysis of the individual catabolic genes is available yet. Paradoxically, the classical genetic analysis based on the phenotypic selection of mutants with altered PAH-degrading capacities is complicated by the fact that several functional pathways are often present in these organisms. In addition, several authors have reported that the PAH-degrading phenotype is easily lost if no selective pressure is maintained. These properties often lead to the isolation of non-specific mutants. Organisms with narrow substrate degradation capabilities may therefore be more convenient for such analysis. Indeed, mutants of *Sphingomonas* sp. LB126, a strain able to degrade fluorene but no other classical, including closely related, PAHs [46], can be obtained that are specifically affected in fluorene assimilation [66]. The mapping of such mini-Tn5 mutants engineered in LB126 has so far allowed the identification of a protocatechuic acid degradation pathway directly associated with fluorene mineralisation [60].

2.4.2. Mycobacterium

The genus *Mycobacterium* is a group consisting of Gram-positive bacteria with GC-rich DNA content that is part of a larger group known as the *Corynebacterium-Mycobacterium-Nocardia* group. These bacteria produce cell walls of unique structure containing *meso*-diaminopimelic acid as the diamino acid in the peptidoglycan. In addition to the classical envelope components, other important features are present in the mycobacterial cell wall. Arabinogalactan is a unique polysaccharide substituted with long-chain fatty acids known as mycolic acids (reviewed in [67]). Mycolic acids are high-molecular-weight α-alkyl, ß-hydroxy fatty acids. They are found mostly covalently attached to arabinogalactan by ester bonds, forming primarily tetramycolylpentarabinosyl clusters (Fig. 4). Compared to other mycolic acid-containing organisms, mycobacterial mycolic acids are long (C_{70} to C_{90}), have the largest α-chain (C_{20} to C_{25}), contain one or two typical double bonds or cyclopropane groups in the long chain moiety as well as some methyl or *ortho*-methyl substitutents. Mutant mycobacteria affected in the synthesis of mycolic acids were shown to be hypersensitive to hydrophobic antibiotics and permeable to hydrophobic dyes [68]. In addition to mycolic acids, the cell wall of mycobacteria contains typical glycolipids. These are either found covalently attached to the cell wall or free (extractable). Lipomannan and lipoarabinomannan are polysaccharides made up of a poly-D-mannose backbone, eventually branched with short poly-D-arabinose side chains (lipoarabinomannan) and both substituted with glycerol-esterified fatty acids (tuberculostearate and palmitate) via a phosphatidyl*myo*inositol link. The three major classes of extractable glycolipids are lipooligosaccharides, phenolic glycolipids and glycopeptidolipids. Lipooligosaccharides are composed of variable sugar residues linked to a tetraglucose core, which terminal residue is usually acylated at position 3, 4 and 6 by 2,4-dimethyltetradecanoic acid residues. Phenolic glycolipids owe their name to a C_{36} phenolic diol substituted by two molecules of a C_{34} fatty acid on one side and by an oligosaccharide made up of one to four *ortho*-methylated deoxy sugar residues at the other side. Glycopeptidolipids, the third class of mycobacterial glycolipids, are made up of a short peptide substituted with various oligosaccharides and with fatty acids. Finally, the presence of long chain diols (generally termed "waxes"), acylated trehaloses and sulpholipids has been reported in the mycobacterial envelope.

From the taxonomic point of view, the *Mycobacterium* genus contains today about one hundred established species to which at least 60 unidentified strains, including many PAH-degraders, should be merged. Mycobacteria used to be basically divided into the so-called "slow growers" and "fast growers" taxonomic groups. The classical differentiating characteristics are based on temperature-dependent growth rates, pigmentation, morphology and biochemical reactions such as nitrate reduction, catalase, Tween hydrolysis, arylsulfatase, urease and pyrazinamidase. Other methods based on lipid analysis, serotyping and DNA sequencing were later introduced and found invaluable to speed up and facilitate the identification process in clinical investigations [69]. Molecular taxonomy based on 16S rRNA has so far not invalidated the historical separation based on growth rate. Slow-growing mycobacteria typically contain the pathogenic species, whereas environmental isolates with PAH-degrading capacities cluster in the fast growing group.

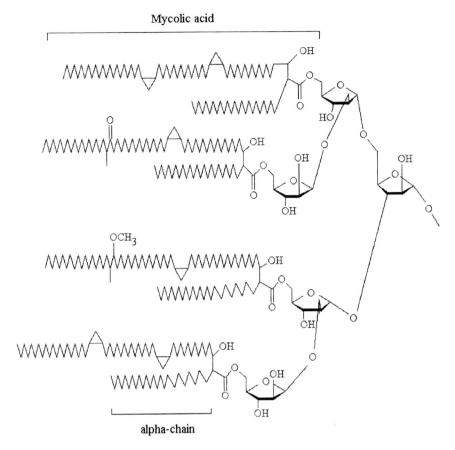

Figure 4. Structure of mycolic acids esterified with the sugar backbone as found in mycobacterial cell walls.

The molecular biology of mycobacteria faces a number of difficulties basically associated with their handling, *i.e.* slow growth rate, clumping, extremely resistant cell wall. In addition, the presence of efficient restriction/modification systems has been reported, which seriously limit the introduction of recombinant DNA. Genetics in mycobacteria is therefore very difficult and has not undergone the recent developments observed for example in *Sphingomonas* spp. Thanks to clinical microbiology research, some basic techniques and tools have been made available (reviewed in [70]). A few plasmid vectors have been shown to be functional in several slow- and fast-growers, the most popular being the pAL5000 plasmid originally isolated from *M. fortuitum* [71]. A transposon mutagenesis system based on a thermo-sensitive delivery plasmid has been constructed and shown to be functional in *M. fortuitum* and *M. smegmatis* [72]. However, its use requires the ability of the target strain to replicate at 42°C, a property that is not displayed by the most studied PAH-degrading mycobacteria [73]. To date, no information regarding the genetics associated with the uptake or metabolism of PAHs, or regarding any successful genetic experiment in a PAH-degrading *Mycobacterium* species of environmental origin has been reported. Preliminary experiments however showed that, using modified protocols, electroporation and stable replication of pAL5000-derivatives in at least some PAH-degrading mycobacteria is possible (unpublished data).

3. PAH bioremediation

Bioremediation technologies are aimed at utilising microorganisms, either native or sometimes from external inocula, for the breakdown of hazardous pollutants found in soil or groundwater. These technologies are cost-effective and, compared to traditional thermal or physico-chemical techniques, have the advantage to preserve the ecological parameters associated with the soil as a living biological system. Bioremediation technologies have undergone an exponential development since 1980 in the US, becoming very popular and gaining progressively the European market after the focus of the media on the of the clean-up of the Exxon Valdez oil spill in 1989 [74]. Soil bioremediation can be engineered either *in situ* or *ex situ*, depending on whether the contaminated soil is excavated or not. The great advantage of *in situ* technologies is their non-invasive character. The soil of the treated site (often of industrial type) is left in place, allowing human activities to be maintained at the surface. However, the degree of contaminant removal is often limited and the time required may exceed reasonable limits. The geological nature of the soil is also a very important parameter that will drastically affect the oxygen and nutrient flux engineered in the soil to stimulate the metabolism of the microorganisms. *Ex-situ* or "on-site" bioremediation technologies achieve higher clean-up levels in a shorter time. They require excavation and sometimes transport to specialised installations that increase the treatment costs, though generally to acceptable limits. Finally, treatments in bioreactors are usually very efficient since optimum-degradation conditions are easily and tightly maintained. However, though these bioreactors can be used on-site, running costs are usually higher than for *in situ* and other on-site treatments.

3.1. EVALUATION OF THE DIFFERENT TECHNOLOGIES

Despite the massive interest of the governments and of the private industries, and despite the success obtained with some more easily degradable or volatile compounds, bioremediation of recalcitrant pollutants such as PAHs remains in its infancy as the technology is not well established, the processes are time-consuming, and the clean-up level often limited [1,75]. Some successes were however obtained, mainly on soils with long-term PAH contamination history and for which no toxicity associated with additional pollutants (for example metals) waspresent. *In situ* treatments have been reviewed elsewhere [76]. Such treatments generally remove low molecular weight PAHs, leaving almost unchanged the content of high molecular weight compounds in the treated soil after reasonable time periods [1]. In addition to the difficulties inherent to the biological clean-up of hydrophobic, high molecular weight contaminants, *in situ* technologies for PAH bioremediation are strongly affected by the soil type and by the temperature. On-site bioremediation of soils contaminated with medium-distillate fuels has proven to be significant. These technologies allow pH and moisture control, fertilisation and tilling. In some instances, bioaugmentation with indigenous microorganisms has been applied [77]. At best, residual concentrations of barely less than 100 mg kg^{-1} total PAHs were obtained [1]. By contrast, the success of on-site treatment of soils contaminated with high molecular weight PAHs (mainly from gas work sites) has been so far only limited and the remaining total PAH content was usually of several hundreds of mg kg^{-1} (according to the drastic "Dutch list", the total PAH content requiring treatment has been fixed to 40 mg kg^{-1}) [1]. The use of electro-osmosis to increase the desorption rate has been experimented with some success [78]. Comparable clean-up results, though achieved in shorter time periods, were obtained with bioreactors. Abiotic losses of high molecular weight PAHs, which apparently desorbed from the soil particles and moved to the reactor slurry, were noticed [79].

PAH bioremediation has thus been proven feasible in some instances, especially for low molecular weight PAHs. Interestingly, soils with old contamination were found more suitable for bioremediation, as the probable consequence of microbial adaptation [1]. However, none of the technologies were so far able to significantly and reproducibly remove the most toxic, high molecular weight PAHs to acceptable levels. The following chapters emphasise some physico-chemical considerations and recent observations that might be important for the success of future PAHs bioremediation.

3.2. BIOAVAILABILITY: THE BOTTLENECK OF PAH BIOREMEDIATION

As discussed in the previous sections, PAH degradation is strongly affected by the low bioavailability of these compounds, as they are barely soluble in water and tend to sorb strongly to particles of organic matter. On the other hand, recent studies suggest that specific physiological properties of the PAH-degrading microorganisms might overcome the low availability of hydrophobic substrates. The general mechanisms promoting the biological availability of compounds with low water solubility include (i) production of biosurfactants or use of specific cell surface components with emulsifying properties, (ii) uptake systems with high substrate affinity, which efficiently reduce the substrate concentration close to the cell surface thereby increasing the diffusive substrate flux, and (iii) reduction of the distance between cells and substrate by means of cell surface adhesion. Thus, the natural bioremediation of such compounds in soil environments depends mainly on the physico-chemical characteristics of the PAHs, on the granulometric characteristics and chemical nature of the soil particles as well as on microorganism-specific properties.

3.2.1. Bacterial adhesion

Until recently, it was generally thought that PAHs are only utilised by microorganisms as substrates dissolved in the water phase. Therefore, in simple systems where desorption is not a limiting factor, PAH mineralisation was considered to be limited mainly by the dissolution rate. Desorption from soil particles, solubilisation in the water phase and diffusion to the microbial cell surface were believed to be the three crucial steps required before uptake and degradation of PAHs can take place. The surface area of the solid PAH was later noticed to be another crucial parameter that accounted not only for faster dissolution rates but also allowed another important mechanism to occur, *i.e.* direct uptake from sorbed/crystalline PAHs [80]. This phenomenon suggests that at least some microorganisms may utilise PAHs directly from the solid phase, and hence possess some special cell wall and/or binding properties. Bacteria specialised in adhesion to hydrophobic surfaces in the soil might have a selective advantage once they contain the necessary catabolic enzymes to metabolise the compound. From a physico-chemical point of view, bacteria in close contact with surfaces containing sorbed PAHs experience a micro-environment that is different from the surrounding bulk liquid. A higher PAHs concentration near the sorbent surface may render the substrates more readily available for adhered bacteria than for bacteria present in the aqueous phase of the soil. Therefore, it was postulated that PAH-degrading bacteria with highly efficient adhesion capacities might bypass the problems of low aqueous solubility and slow dissolution rates associated with some PAHs. Such bacteria (mainly mycobacteria) were indeed isolated by either classical or special techniques and shown to display strong adherence to hydrophobic surfaces, strong clumping/aggregating tendency and poor growth in aqueous suspensions [46,81]. Preliminary results showed that the growth rate of the anthracene-degrader *Mycobacterium* sp. LB501T [46] in liquid cultures is limited by the total substrate surface and hence by the amount of anthracene crystals in the suspension [82]. Accordingly, this bacterium was found able to form thick biofilms at the surface of the anthracene crystals (see Fig. 5). Comparable results were obtained with *Sphingomonas* strains grown in flow cells with PAHs as sole carbon source: some strains can be fed only with water-dissolved substrates, whereas other strains need to be tightly associated with the PAH surface, demonstrating a direct uptake from the crystal (A. Schnell and S. Wuertz, personal communication). Surprisingly, the uptake mechanism is not always related to the water solubility of the PAH compound, as the most soluble PAHs are not necessarily the ones taken up from the water-dissolved fraction.

According to the so-called colloid stability theory [83], in which electrostatic repulsion, van der Waals attraction and acid-base (hydrophobic) interactions are considered, the adhesion capacity is inversely correlated with a more negative surface charge and favoured by the hydrophobicity of the bacterial cells. However, recent comparative studies devoted to *Mycobacterium* and *Sphingomonas* isolates have shown that the negative charge and the hydrophobicity of the cell surface cannot account completely for the strong binding of some efficient PAH-degraders [46]. Other specific binding mechanisms, probably associated with peculiar cell wall polymers, are likely to occur in these strains. Based on the analysis of mutants, it has been shown that lipopolysaccharides and flagella are important adhesion-promoting factors in soil pseudomonads [84-86]. Whether the peculiar biopolymers exposed at the surface of *Sphingomonas* and *Mycobacterium* cells (see section 2.4) participate in binding to PAHs is unknown.

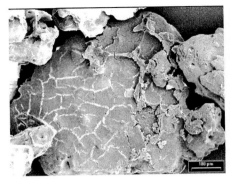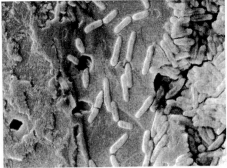

Figure 5. Scanning Electron Microscope pictures of anthracene crystals covered by a Mycobacterium *sp. LB501T biofilm. Left, general view (enlargement 100X). Right, detailed view of the biofilm (enlargement 5000X) (Pictures kindly provided by Lukas Y. Wick and Hauke Harms from the Swiss Federal Institute of Technology, Lausanne, Switzerland).*

3.2.2. Emulsifier- and surfactant-enhanced biodegradation

Surfactants are small amphipathic molecules that decrease the surface tension of aqueous solutions. They increase the aqueous concentration of PAHs by increasing their apparent solubility. In the absence of any toxic effect and if only uptake from the water-phase is considered, these molecules should theoretically increase the bacterial PAH-metabolism. Indeed, early studies have shown that phenanthrene-degradation by a *Mycobacterium* sp. was increased in the presence of Tween [87]. A similar effect was observed for the synthetic surfactant Biosoft HR33 on the degradation of phenanthrene in a biphasic liquid system, as the probable consequence of an increased partitioning rate [88]. However, more accurate studies on similar systems involving Triton X-100 later demonstrated a negative effect on cell adhesion at the interface [89]. Such an inhibitory effect of surfactants on adhesion and metabolism was also reported recently for a *Mycobacterium* strain and a *Pseudomonas* strain [81]. In real soil environments, similar antagonistic effects of surfactants have been reported. On one hand, sorption of PAHs to soil particles is increased as the consequence of sorption of the surfactant itself and, on the other hand, the aqueous concentration of PAHs increases via the micelles, the extent of both effects being dependent on the structure and concentration of the surfactant and on the nature of the soil. A positive effect of Witconol SN70 was observed for pyrene degradation by *Mycobacterium* sp. K-12 and B-24 in soils, but a negative effect was observed in soil slurries [90]. Non-ionic, slowly degradable surfactants were also found efficient in another study on pyrene removal from soil [91]. Initial studies with biosurfactants and bioemulsifiers were rather discouraging, as they did not significantly increase the rate of PAH metabolism. Bioemulsifiers and biosurfactants are high molecular weight molecules synthesised by various microorganisms and have gained increasing attention due to their diversity and environment-friendly nature. An extensive screening of strains isolated from PAH-contaminated soils for the production of biosurfactants was undertaken by Willumsen and Karlson [92] and revealed that most strains isolated on fluoranthene, phenanthrene and pyrene excreted molecules with emulsifying properties while strains isolated on anthracene rather produced molecules decreasing the surface tension. However, the fraction of the strains with true PAH-mineralisation capacities failed to produce any active molecule. Recently, alasan from *Acinetobacter radioresistens* KA53 was shown to increase the apparent solubility in water of several PAHs and to increase the mineralisation rate of fluoranthene and phenanthrene by *Sphingomonas paucimobilis* EPA505 by more than 2 times. However, this effect required an alasan concentration of 500 µg ml^{-1} [93]. The impact of surfactants on PAH bioavailability is thus very ambiguous. Negative effects may occur due to toxicity, lower availability of the contaminants in the micelles, substrate competition between PAHs and the surfactants themselves, increased mobility of the dissolved PAHs and prevention of adhesion of the PAH-degraders.

3.2.3. PAH uptake mechanisms

It is assumed that the internalisation of small hydrophobic molecules inside the bacterial cell is a passive, diffusive process. This has been indeed observed for naphthalene uptake by *Pseudomonas putida* [25]. Whether active, specific processes are required or complementary to the passive diffusion of PAHs is presently unknown. Such mechanisms have been observed for other pollutants and shown to increase the diffusive substrate flux [94]. Recently, DNA sequencing of the large catabolic pNL1 plasmid from the PAH-degrading *Sphingomonas aromaticivorans* revealed the existence of genes encoding large membrane proteins related to components of well-known transport systems [65]. Since these genes were found in close association with the catabolic genes likely participating in PAH metabolism, it is tempting to speculate a function in PAH uptake, although this has not been demonstrated yet.

4. Conclusion

Bioremediation of PAH-contaminated soils presents several difficulties that are typically associated with the nature of these pollutants. The main problem is the low bioavailability of PAHs that results from the unfavourable physico-chemical characteristics of these compounds. From the microbial point of view, potential metabolism exists and has been demonstrated for all classes of PAHs, including for molecules with high molecular weight. Although it has been shown that some fungal enzymes may attack virtually any PAH molecule, bacteria however seem to be limited to the metabolism of PAHs with up to four, possibly five rings. Extensive research during these last years has lead world-wide to the isolation of bacteria with exceptional, yet unreported metabolic capacities. These bacteria seem to overcome the low bioavailability of PAHs by developing highly adapted strategies such as hydrophobic, strongly adhering cell walls, peculiar membrane lipids and, possibly, efficient uptake systems. The exact nature of these peculiar adaptations and the molecular mechanisms that regulate their expression are so far totally unknown. Thanks to the tremendous progress of molecular microbiology initially stimulated by clinical research, the way is now open for a molecular study of these organisms. The biotechnological developments that are likely to arise from this research will certainly contribute to the improvement of the technologies associated with PAH bioremediation in the near future.

Acknowledgements

The author wishes to thank Dirk Springael for his remarks and improvements to the writing of the present review. Many thanks also to my colleagues L.Y. Wick, H. Harms, J.J. Ortega, L. Bastiaens, J. Parsons, R. Van Herwijnen, A. Schnell, S. Wuertz, P. Willumsen, U. Karlson, A. Johnsen and S. Schiøtz-Hansen for having allowed me to mention some of their unpublished results and for kindly providing me with illustrations. Research was supported by grants from the European Community (contracts n° BIO4-CT97-2015 and QLRT-1999-00326).

References

1. Wilson, S.C. and Jones, K.C. (1993) Bioremediation of soil contaminated with polynuclear aromatic hydrocarbons (PAHs): a review. Environ. Pollut. 81, 229-249.
2. Enzminger, J.D. and Ahlert, R.C. (1987) Environmental fate of polynuclear aromatic hydrocarbons in coal tar. Environ. Techn. Letters. 8, 269-278.
3. Menzie, C.A., Potocki, B.B. and Santodonato, J. (1992) Exposure to carcinogenic PAHs in the environment. Environ. Sci. Technol. 26, 1278-1284.
4. NIH (1985) Survey of compounds which have been tested for carcinogenic activity, NIH publication, NIH Press, Washington D.C.
5. Harvey, R.G. and Dunne, F.B. (1978) Multiple regions of metabolic activation of carcinogenic hydrocarbons. Nature 273, 566-568.
6. IARC (1984) International Agency for Research on Cancer, Monographs on the evaluation of the carcinogenic risk of chemical to humans, IARC Press, Lyon.
7. Ames, B.N., Mclann, J. and Yamashaki, E. (1975) Methods for detecting carcinogens and mutagens with the Salmonella/mammalian-microsome mutagenicity test. Mutat. Res. 31, 347-364.
8. Harvey, R. (1991) Polycyclic aromatic hydrocarbons: chemistry and carcinogenicity, in: M.M. Coombs, J. Ashby, M. Hicks and H. Baxter (eds.) Cambridge Monographs on Cancer Research, Cambridge University Press, Cambridge, pp. 396.
9. White, K.L.J. (1986) An overview of immunotoxicology and carcinogenic polycyclic aromatic hydrocarbons. Environ. Carcin. Revs. C4, 163-202.
10. Keith, L.H. and Telliard, W.A. (1979) Priority pollutants. I. A perspective view. Environ. Sci. Technol. 13, 416-423.
11. Murillo Mantilla, E. (1990) European Community policy with respect to soil contamination, Contaminated Land Workshop, (eds.),, Sevenage, UK.
12. Cerniglia, C.E. (1992) Biodegradation of polycyclic aromatic hydrocarbons. Biodegradation 3, 351-368.
13. Dzombak, D.A. and Luthy, R.G. (1984) Estimating adsorption of polycyclic aromatic hydrocarbons on soil. Soil Science 137, 292-308.
14. Kanaly, R.A. and Harayama, S. (2000) Biodegradation of high-molecular-weight polycyclic aromatic hydrocarbons by bacteria. J. Bacteriol. 182, 2059-2067.
15. Andersson, B.E. and Henrysson, T. (1996) Accumulation and degradation of dead-end metabolites during treatment of soil contaminated with polycyclic aromatic hydrocarbons with five strains of white-rot fungi. Appl. Microbiol. Biotechnol. 46, 647-652.
16. Reddy, C.A. and D'Souza, T.M. (1994) Physiology and molecular biology of the lignin peroxidases of Phanerochaete chrysosporium. FEMS Microbiol. Rev. 13, 137-152.
17. Cerniglia, C.E. (1981) Aromatic hydrocarbons: metabolism by bacteria, fungi, algae, Reviews in biochemical toxicology, Elsevier, Amsterdam.
18. Narro, M.L., Cerniglia, C.E., Van Baalen, C. and Gibson, D.T. (1992) Metabolism of phenanthrene by the marine cyanobacterium Agmenellum quadruplicatum PR-6. Appl. Environ. Microbiol. 58, 1351-1359.

19. Engesser, K.H., Strubel, V., Christoglou, K., Fischer, P. and Rast, H.G. (1989) Dioxygenolytic cleavage of aryl ether bonds: 1,10-dihydro-1,10-dihydroxyfluoren-9-one, a novel arenedihydrodiol as evidence for angular dioxygenation of dibenzofuran. FEMS Microbiol. Lett. 53, 205-209.

20. Ellis, L.B., Hershberger, C.D. and Wackett, L.P. (2000) The University of Minnesota Biocatalysis/Biodegradation database: microorganisms, genomics and prediction. Nucleic Acids Res. 28, 377-9.

21. Bouchez, M., Blanchet, D. and Vandecasteele, J.P. (1995) Degradation of Polycyclic Aromatic-Hydrocarbons by Pure Strains and by Defined Strain Associations - Inhibition Phenomena and Cometabolism. Appl. Microbiol. Biotechnol. 43, 156-164.

22. Coates, J.D., Woodward, J., Allen, J., Philp, P. and Lovley, D.R. (1997) Anaerobic degradation of polycyclic aromatic hydrocarbons and alkanes in petroleum-contaminated marine harbor sediments. Appl. Environ. Microbiol. 63, 3589-3593.

23. Barnsley, E.A. (1976) Role and regulation of the ortho and meta pathways of catechol metabolism in pseudomonads metabolizing naphthalene and salicylate. J. Bacteriol. 125, 404-408.

24. Barnsley, E.A. (1983) Bacterial oxidation of naphthalene and phenanthrene. J. Bacteriol. 153, 1069-1071.

25. Bateman, J.N., Speer, B., Feduik, L. and Hartline, R.A. (1986) Naphthalene association and uptake in Pseudomonas putida. J. Bacteriol. 166, 155-161.

26. Churchill, S.A., Harper, J.P. and Churchill, P.F. (1999) Isolation and characterization of a Mycobacterium species capable of degrading three- and four-ring aromatic and aliphatic hydrocarbons. Appl. Environ. Microbiol. 65, 549-552.

27. Goyal, A.K. and Zylstra, G.J. (1996) Molecular cloning of novel genes for polycyclic aromatic hydrocarbon degradation from Comamonas testosteroni GZ39. Appl. Environ. Microbiol. 62, 230-236.

28. Zylstra, G.J., Wang, X.P., Kim, E. and Didolkar, V.A. (1994) Cloning and analysis of the genes for polycyclic aromatic hydrocarbon degradation. Ann. N. Y. Acad. Sci. 721, 386-398.

29. Gibson, D.T., Mahadevan, V., Jerina, D.M., Yogi, H. and Yeh, H.J. (1975) Oxidation of the carcinogens benzo[a]pyrene and benzo[a]anthracene to dihydrodiols by a bacterium. Science 189, 295-297.

30. Barnsley, E.A. (1975) The bacterial degradation of fuoranthene and benzo(a)pyrene. Can. J. Microbiol. 21, 1004-1008.

31. Heitkamp, M.A. and Cerniglia, C.E. (1988) Mineralization of polycylic aromatic hydrocarbons by a bacterium isolated from sediment below an oil field. Appl. Environ. Microbiol. 54, 1612-1614.

32. Mahaffey, W.R., Gibson, D.T. and Cerniglia, C.E. (1988) Bacterial oxidation of chemical carcinogens: formation of polycyclic aromatic acids from benz[a]anthracene. Appl. Environ. Microbiol. 54, 2415-2423.

33. Mueller, J.G., Chapman, P.J. and Pritchard, P.H. (1989) Action of a fluoranthene-utilizing bacterial community on polycyclic hydrocarbon components of creosote. Appl. Environ. Microbiol. 55, 3085-3090.

34. Weissenfels, W.D., Beyer, M. and Klein, J. (1990) Degradation of phenanthrene, fluorene and fluoranthene by pure bacterial cultures. Appl. Microbiol. Biotechnol. 32, 479-484.

35. Mueller, J.G., Chapman, P.J., Blattmann, B.O. and Pritchard, P.H. (1990) Isolation and characterization of a fluoranthene-utilizing strain of Pseudomonas paucimobilis. Appl. Environ. Microbiol. 56, 1079-1086.

36. Shuttleworth, K.L., Sung, J.H., Kim, E. and Cerniglia, C.E. (2000) Physiological and genetic comparison of two aromatic hydrocarbon-degrading Sphingomonas strains. Mol. Cells 10, 199-205.

37. Kelley, I., Freeman, J.P., Evans, F.E. and Cerniglia, C.E. (1993) Identification of metabolites from the degradation of fluoranthene by Mycobacterium sp. strain PYR-1. Appl. Environ. Microbiol. 59, 800-806.

38. Rehmann, K., Noll, H.P., Steinberg, C.E.W. and Kettrup, A.A. (1998) Pyrene degradation by Mycobacterium sp. strain KR2. Chemosphere 36, 2977-2992.

39. Boldrin, B., Thiem, A. and Fritzsche, C. (1993) Degradation of phenanthrene, fluorene, fluoranthene, and pyrene by Mycobacterium sp. Appl. Environ. Microbiol. 59, 1927-1930.

40. Walter, U., Beyer, M., Klein, J. and Rehn, H.J. (1991) Degradation of pyrene by Rhodococcus sp. UW1. Appl. Microbiol. Biotechnol. 34, 671-676.

41. Schneider, J., Grosser, R., Jayasimhulu, K., Xue, W.L. and Warshawsky, D. (1996) Degradation of pyrene, benz[a]anthracene, and benzo[a]pyrene by Mycobacterium sp. strain RJGII-135, isolated from a former coal gasification site (vol 62, pg 14, 1996). Appl. Environ. Microbiol. 62, 1491-1491.

42. Boonchan, S., Britz, M.L. and Stanley, G.A. (2000) Degradation and mineralization of high-molecular-weight polycyclic aromatic hydrocarbons by defined fungal-bacterial cocultures. Appl. Environ. Microbiol. 66, 1007-1019.

43. Kiyohara, H., Nagao, K. and Yana, K. (1982) Rapid screen for bacteria degrading water-insoluble, solid hydrocarbons on agar plates. Appl. Environ. Microbiol. 43, 454-457.
44. Stieber, M., Haeseler, P., Werner, P. and Frimmel, F.H. (1994) A rapid screening method for microorganisms degrading PAHs in microplates. Appl. Microbiol. Biotechnol. 40, 753-755.
45. Ascon-Cabrera, M. and Lebeault, J. (1993) Selection of xenobiotic-degrading microorganisms in a biphasic aqueous-organic system. Appl. Environ. Microbiol. 59, 1717-1724.
46. Bastiaens, L., Springael, D., Wattiau, P., Harms, H., de Wachter, R., Verachtert, H. and Diels, L. (2000) Isolation of adherent polycyclic aromatic hydrocarbon (PAH)-degrading bacteria using PAH-sorbing carriers. Appl. Environ. Microbiol. 66, 1834-1843.
47. Grosser, R.J., Friedrich, M., Ward, D.M. and Inskeep, W.P. (2000) Effect of model sorptive phases on phenanthrene biodegradation: different enrichment conditions influence bioavailability and selection of phenanthrene-degrading isolates. Appl. Environ. Microbiol. 66, 2695-2702.
48. Dunn, N.W. and Gunsalus, I.C. (1973) Transmissible plasmid coding early enzymes of naphthalene oxidation in Pseudomonas putida. J. Bacteriol. 114, 974-979.
49. Menn, F.M., Applegate, B.M. and Sayler, G.S. (1993) NAH plasmid-mediated catabolism of anthracene and phenanthrene to naphthoic acids. Appl. Environ. Microbiol. 59, 1938-1942.
50. Kästner, M., Breuer-Jammali, M. and Mahro, B. (1994) Enumeration and characterization of the soil microflora from hydrocarbon-contaminated soil sites able to mineralize polycyclic aromatic hydrocarbons. Appl. Microbiol. Biotechnol. 41, 267-273.
51. Solano-Serena, F., Marchal, R., Casaregola, S., Vasnier, C., Lebeault, J.M. and Vandecasteele, J.P. (2000) A Mycobacterium strain with extended capacities for degradation of gasoline hydrocarbons. Appl. Environ. Microbiol. 66, 2392-2399.
52. Takeuchi, M., Sawada, H., Oyaizu, H. and Yokota, A. (1994) Phylogenetic evidence for Sphingomonas and Rhizomonas as nonphotosynthetic members of the alpha-4 subclass of the Proteobacteria. Int. J. Syst. Bacteriol. 44, 308-314.
53. Kawahara, K., Moll, H., Knirel, Y.A., Seydel, U. and Zahringer, U. (2000) Structural analysis of two glycosphingolipids from the lipopolysaccharide-lacking bacterium Sphingomonas capsulata. Eur. J. Biochem. 267, 1837-1846.
54. Naka, T., Fujiwara, N., Yabuuchi, E., Doe, M., Kobayashi, K., Kato, Y. and Yano, I. (2000) A novel sphingoglycolipid containing galacturonic acid and 2-hydroxy fatty acid in cellular lipids of Sphingomonas yanoikuyae. J. Bacteriol. 182, 2660-2663.
55. Balkwill, D.L., Drake, G.R., Reeves, R.H., Fredrickson, J.K., White, D.C., Ringelberg, D.B., Chandler, D.P., Romine, M.F., Kennedy, D.W. and Spadoni, C.M. (1997) Taxonomic study of aromatic-degrading bacteria from deep-terrestrial-subsurface sediments and description of Sphingomonas aromaticivorans sp. nov., Sphingomonas subterranea sp. nov., and Sphingomonas stygia sp. nov. Int. J. Syst. Bacteriol. 47, 191-201.
56. Masai, E., Katayama, Y., Nishikawa, S., Yamasaki, M., Morohoshi, N. and Haraguchi, T. (1989) Detection and localization of a new enzyme catalyzing the beta-aryl ether cleavage in the soil bacterium Pseudomonas paucimobilis SYK-6. FEBS Lett. 249, 348-352.
57. Nohynek, L.J., Suhonen, E.L., NurmiahoLassila, E.L., Hantula, J. and SalkinojaSalonen, M. (1996) Description of four pentachlorophenol-degrading bacterial strains as Sphingomonas chlorophenolica sp. nov. Syst. Appl. Microbiol. 18, 527-538.
58. Nagata, Y., Miyauchi, K., Suh, S.K., Futamura, A. and Takagi, M. (1996) Isolation and characterization of Tn5-induced mutants of Sphingomonas paucimobilis defective in 2,5-dichlorohydoquinone degradation. Biosci. Biotechnol. Biochem. 60, 689-691.
59. Wang, Y. and Lau, P.C.K. (1996) Sequence and expression of an isocitrate dehydrogenase-encoding gene from a polycyclic aromatic hydrocarbon oxidizer, Sphingomonas yanoikuyae B1. Gene 168, 15-21.
60. Wattiau, P., Bastiaens, L., van Herwijnen, R., Daal, L., Parsons, J.R., Renard, M.-E., Springael, D. and Cornelis, G.R. Fluorene degradation by Sphingomonas sp. LB126 proceeds through protocatechuic acid: a genetic analysis. Res Microbiol, in press.
61. Andujar, E., Hernaez, M.J., Kaschabek, S.R., Reineke, W. and Santero, E. (2000) Identification of an extradiol dioxygenase involved in tetralin biodegradation: Gene sequence analysis and purification and characterization of the gene product. J. Bacteriol. 182, 789-795.
62. Miyauchi, K., Adachi, Y., Nagata, Y. and Takagi, M. (1999) Cloning and sequencing of a novel meta-cleavage dioxygenase gene whose product is involved in degradation of gamma-hexachlorocyclohexane in Sphingomonas paucimobilis. J. Bacteriol. 181, 6712-6719.

63. Videira, P.A., Cortes, L.L., Fialho, A.M. and Sa-Correia, I. (2000) Identification of the pgmG gene, encoding a bifunctional protein with phosphoglucomutase and phosphomannomutase activities, in the gellan gum-producing strain Sphingomonas paucimobilis ATCC 31461. Appl. Environ. Microbiol. 66, 2252-2258.

64. Xun, L.Y., Bohuslavek, J. and Cai, M.A. (1999) Characterization of 2,6-dichloro-p-hydroquinone 1,2-dioxygenase (PcpA) of Sphingomonas chlorophenolica ATCC 39723. Biochem. Biophys. Res. Commun. 266, 322-325.

65. Romine, M.F., Stillwell, L.C., Wong, K.K., Thurston, S.J., Sisk, E.C., Sensen, C., Gaasterland, T., Fredrickson, J.K. and Saffer, J.D. (1999) Complete sequence of a 184-kilobase catabolic plasmid from Sphingomonas aromaticivorans F199. J. Bacteriol. 181, 1585-1602.

66. Bastiaens, L., Springael, D., Dejonghe, W., Wattiau, P., Verachtert, H. and Diels, L. A transcriptional luxAB reporter fusion responding to fluorene in Sphingomonas sp. LB126 and its initial characterisation for whole-cell bioreporter purposes. Res Microbiol, in press.

67. Brennan, P.J. and Nikaido, H. (1995) The Envelope of Mycobacteria. Annu. Rev. Biochem. 64, 29-63.

68. Liu, J. and Nikaido, H. (1999) A mutant of Mycobacterium smegmatis defective in the biosynthesis of mycolic acids accumulates meromycolates. Proc. Natl. Acad. Sci. USA 96, 4011-4016.

69. Luquin, M., Ausina, V., Lopez Calahorra, F., Belda, F., Garcia Barcelo, M., Celma, C. and Prats, G. (1991) Evaluation of practical chromatographic procedures for identification of clinical isolates of mycobacteria. J. Clin. Microbiol. 29, 120-130.

70. Parish, T. and Stoker, N.G. (1998) Mycobacteria protocols, Methods in Molecular Biology, Humana Press, Totowa, New Jersey.

71. Rauzier, J., Moniz-Pereira, J. and Gicquel-Sanzey, B. (1988) Complete nucleotide sequence of pAL5000, a plasmid from Mycobacterium fortuitum. Gene 71, 315-321.

72. Guilhot, C., Otal, I., Van Rompaey, I., Martin, C. and Gicquel, B. (1994) Efficient transposition in mycobacteria: construction of Mycobacterium smegmatis insertional mutant libraries. J. Bacteriol. 176, 535-539.

73. Govindaswami, M., Feldhake, D.J., Kinkle, B.K., Mindell, D.P. and Loper, J.C. (1995) Phylogenetic comparison of two polycyclic aromatic hydrocarbon-degrading mycobacteria. Appl. Environ. Microbiol. 61, 3221-3226.

74. Bragg, J.R., Prince, R.C., Harner, E.J. and Atlas, R.M. (1994) Effectiveness of bioremediation for the Exxon Valdez oil spill. Nature 368, 413-418.

75. Wang, X., Yu, X. and Bartha, R. (1990) Effect of bioremediation on polycyclic aromatic hydrocarbon residues in soil. Environ. Sci. Technol. 24, 1086-1089.

76. Madsen, E.L. (1991) Determining in situ bioremediation: Facts and challenges. Environ. Sci. Technol. 25, 1663-1673.

77. Kästner, M., Breuer-Jammali, M. and Mahro, B. (1998) Impact of inoculation protocols, salinity, and pH on the degradation of polycyclic aromatic hydrocarbons (PAHs) and survival of PAH-degrading bacteria introduced into soil. Appl. Environ. Microbiol. 64, 359-362.

78. Pamukcu, S., Filipova, I. and Wittle, J.K. (1995) The Role of Electroosmosis in Transporting PAH Compounds in Contaminated Soils, in: E.W. Brooman and J.M. Fenton (eds.) Proc. of the Symp. on Electrochemical Technology Applied to Environmental Problems, PV 95-12, The Electrochemical Society, pp. 252-266.

79. Mueller, G.J., Lantz, S.E., Blattmann, B.O. and Chapman, P.J. (1991) Bench-scale evaluation of alternative biological treatment processes for the remediation of pentachlorophenol- and creosote-contaminated materials: slurry phase bioremediation. Environ. Sci. Technol. 25, 1045-1055.

80. Thomas, J.M., Yordy, J.R., Amador, J.A. and Alexander, M. (1986) Rates of dissolution and biodegradation of water-insoluble organic compounds. Appl. Environ. Microbiol. 52, 290-296.

81. Stelmack, P.L., Gray, M.R. and Pickard, M.A. (1999) Bacterial adhesion to soil contaminants in the presence of surfactants. Appl. Environ. Microbiol. 65, 163-168.

82. Wick, L.Y., Colangelo, T. and Harms, H. (2001) Kinetics of mass transfer-limited bacterial growth on solid PAHs. Environ Sci Technol 35, 354-61.

83. Van Oss, C.J. (1994) Interfacial forces in aqueous media, New York.

84. DeFlaun, M.F., Tanzer, A.S., McAteer, A.L., Marshall, B. and Levy, S.B. (1990) Development of an adhesion assay and characterization of an adhesion-deficient mutant of Pseudomonas fluorescens. Appl. Environ. Microbiol. 56, 112-119.

85. DeFlaun, M., Marshall, B., Kulle, E. and Levy, S. (1994) Tn5 insertion mutants of Pseudomonas fluorescens defective in adhesion to soil and seeds. Appl. Environ. Microbiol. 60, 2637-2642.

86. DeFlaun, M.F., Oppenheimer, S.R., Streger, S., Condee, C.W. and Fletcher, M. (1999) Alterations in adhesion, transport, and membrane characteristics in an adhesion-deficient pseudomonad. Appl. Environ. Microbiol. 65, 759-765.
87. Tiehm, A. (1994) Degradation of polycyclic aromatic-hydrocarbons in the presence of synthetic surfactants. Appl. Environ. Microbiol. 60, 258-263.
88. Köhler, A., Schüttoff, M., Bryniok, D. and Knackmuss, H.J. (1994) Enhanced biodegradation of phenanthrene in a biphasic culture system. Biodegradation 45, 93-103.
89. Ortega-Calvo, J.J. and Alexander, M. (1994) Roles of bacterial attachment and spontaneous partitioning in the biodegradation of naphtalene initially present in nonaqueous-phase liquids. Appl. Environ. Microbiol. 60, 2643-2646.
90. Thibault, S.L., Anderson, M. and Frankenberger, W.T. (1996) Influence of surfactants on pyrene desorption and degradation in soils. Appl. Environ. Microbiol. 62, 283-287.
91. Madsen, E.L., Thomas, C.T., Wilson, M.S., Sandoli, R.L. and Bilotta, S.E. (1996) In situ dynamics of aromatic hydrocarbons and bacteria capable of AH metabolism in a coal tar waste-contaminated field site. Environ. Sci. Technol. 30, 2412-2416.
92. Willumsen, P.A. and Karlson, U. (1997) Screening of bacteria, isolated from PAH-contaminated soils, for production of biosurfactants and bioemulsifiers. Biodegradation 7, 415-423.
93. Barkay, T., Navon-Venezia, S., Ron, E.Z. and Rosenberg, E. (1999) Enhancement of solubilization and biodegradation of polyaromatic hydrocarbons by the bioemulsifier alasan. Appl. Environ. Microbiol. 65, 2697-2702.
94. Harms, H. and Zehnder, A.J. (1994) Influence of substrate diffusion on degradation of dibenzofuran and 3-chlorodibenzofuran by attached and suspended bacteria. Appl. Environ. Microbiol. 60, 2736-2745.

TRANSFER OF CATABOLIC PLASMIDS IN SOIL AND ACTIVATED SLUDGE: A FEASIBLE BIOAUGMENTATION STRATEGY?

EVA M. TOP
Laboratory of Microbial Ecology and Technology, Ghent University, Coupure Links 653, B-9000 Ghent, Belgium. Present address: University of Idaho, Department of Biological Sciences, Life Sciences Building, Moscow, ID 83844-3051 phone: 001/208-885-6185 or -4382 fax 001/208-885-7905 email: evatop@uidaho.edu

The list of known plasmids that carry genes, which code for the degradation of naturally occurring and xenobiotic compounds, is growing continuously. There is also more and more evidence that horizontal exchange of catabolic genes among bacteria in microbial communities plays an important role in the evolution of catabolic pathways. Less than ten years ago the suggestion was made to accelerate this natural gene exchange and pathway construction by introducing and subsequently spreading degradative genes, natural or modified, into well established, competitive indigenous microbial populations as a means of bioaugmentation of polluted sites. During the last decade this approach has only been investigated by a few groups. This paper will briefly review these studies, with emphasis on those that have clearly shown a direct effect of gene transfer on accelerated biodegradation. A few successful cases in soil indicate that the strategy could indeed work under certain conditions. In spite of the absence of specific data on this approach in activated sludge, the dissemination of introduced catabolic genes in such bioreactor communities could also be a promising bioaugmentation strategy. Further investigations in this area are obviously needed to improve our current knowledge on the efficiency of gene dissemination as a tool in bioremediation.

1. Catabolic plasmids and their role in bacterial adaptation to pollutants

During the last century industrial and agricultural activities have resulted in the release of large quantities of synthetic, sometimes harmful, compounds into the ecosphere. In soils contaminated with such toxic xenobiotic (=manmade) organic compounds there is a strong selective pressure for the development of bacterial strains with novel or improved biodegradative capabilities. Indeed, under favourable conditions indigenous microbial populations have shown a remarkable ability to degrade a range of such pollutants, previously considered recalcitrant (Top *et al.*, 2000). A growing interest in the use of these microbial degradation capacities to clean up polluted sites has led to an extensive isolation study of biodegradative strains. In these strains, specific catabolic pathways are often encoded on plasmids (Sayler *et al.*, 1990), of which several are self-

S.N. Agathos and W. Reineke (eds.),
Biotechnology for the Environment: Strategy and Fundamentals, 91–103.
© 2002 *Kluwer Academic Publishers. Printed in the Netherlands.*

transmissible and have a broad host range (Top *et al.*, 2000). This suggests that there is a horizontal gene pool in microbial communities of environmental habitats, which consists of these mobile genetic elements, and which allows bacterial populations to exchange useful catabolic functions across taxonomic boundaries, and thus acquire and construct new catabolic pathways (van der Meer, 1994). There are three distinct mechanisms by which plasmid encoded genes could be disseminated between bacteria: conjugation, transformation and transduction. Conjugation, which is a cell-contact-dependent parasexual process, whereby plasmids or transposons transfer from donor to recipient, is considered to play the most important role in the environment and has been studied most intensively. It can occur between different species, even between Gram-positive and Gram-negative species, and even between bacteria and yeasts and plants (Dröge *et al.*, 1999). Although natural transformation (uptake of naked DNA) and transduction (DNA transfer via bacteriophages) have not been studied so frequently in microcosms and in the field as conjugation, they are also considered to play an important role in the total gene flux (Dröge *et al.*, 1999). An important aspect of conjugative gene transfer is the host range of the plasmids: some plasmids have a narrow host range (NHR plasmids), while others can be transferred between bacteria of different genera (BHR plasmids, broad host range plasmids). The latter belong to incompatibility groups IncC, IncJ, IncN, IncP (=IncP1), IncQ and IncW (Top *et al.* 1992). These BHR plasmids may play a very important role in the dissemination of DNA between different groups of bacteria and thus also from introduced strains to a variety of strains in natural environments.

The list of known plasmids that carry genes which code for the degradation of naturally occurring and xenobiotic compounds is growing continuously as a result of the continuing efforts to isolate and characterise new mobile genetic elements. These plasmids are generally rather large - all are greater than 50 kb, while many are larger than 100 kb (Mergeay *et al.*, 1990). Many are as yet unclassified, but several plasmids that have been classified tend to be related to plasmids known in *Pseudomonas,* where there are thirteen identified incompatibility groups (Boronin, 1992). While some genes that encode degradation of naturally occurring organic compounds have been found on plasmids of the *Pseudomonas* incompatibility group IncP2, the best studied group of such plasmids belong to IncP9, which includes the original TOL plasmid pWW0 (Williams & Murray, 1974). Such plasmids appear to be widespread, can be associated with either resistance or degradative functions, and have a reasonably broad host range. Interestingly, while the groups IncP2 and IncP9 contain plasmids carrying genes that encode degradation of naturally occurring compounds, the plasmids encoding degradation of xenobiotic, often chlorinated, compounds, seem to belong mainly to the IncP1 group. These IncP1 plasmids are the most promiscuous self-transmissible plasmids characterised to date. If this correlation is true, it could suggest that bacterial adaptation is promoted by plasmid promiscuity.

There is little direct information available on where and when genetic adaptation has taken place in natural environments. One example is a quite recent study on the natural horizontal transfer of naphthalene dioxygenase genes in a coal tar-contaminated field site. Strong evidence is provided that horizontal gene transfer had indeed occurred between members of the soil bacterial community, and that a naphthalene-catabolic plasmid may have played a role in adaptation of this community to the coal tar contamination (Herrick *et al.*, 1997; Stuart-Kiel *et al.*, 1998). A thorough investigation about the origin of chlorobenzene (CB) degrading bacteria, which could be isolated from CB-contaminated groundwater, and not from the uncontaminated region outside this area, shows strong evidence that bacterial adaptation to the pollution occurred due to genetic recombination among bacteria in the aquifer, leading to the formation of a novel pathway for chlorobenzene degradation. This *in situ* genetic recombination apparently resulted in enhanced removal of chlorobenzenes from that environment (van der Meer *et al.*, 1998).

2. Dissemination of catabolic genes in the environment as a bioaugmentation strategy

The study of the transfer of catabolic genes in nature is not only important to better understand bacterial adaptation to man-made compounds entering the environment. They may also have potential applications in bioremediation of polluted soils and waters. Bioaugmentation is the application of indigenous or allochthonous wildtype or genetically modified organisms to polluted hazardous waste sites or bioreactors in order to accelerate the removal of undesired compounds. Although this augmentation of natural bacterial populations with highly efficient laboratory strains, containing new or existing biodegradative pathways, is attractive, these laboratory strains however have to compete with indigenous species adapted to the local conditions. This competition often results in a replacement of the laboratory strain by wild-type populations. This has been observed for inoculation of soils (Goldstein *et al.*, 1985; Alexander, 1994; Akkermans, 1994; van Veen *et al.*, 1997), activated sludge, and anaerobic reactors, where disappearance of introduced strains was observed as often as retention (McClure *et al.* 1991; Ahring *et al.* 1992; Nüßlein *et al.* 1992; Christiansen *et al.*, 1995; Selvaratnam *et al.* 1997; Miguez *et al.*, 1999; Tartakovsky *et al.*, 1999).

To overcome the problem of poor inoculum survival and activity, an alternative strategy has been suggested and examined by a few groups. It involves the introduction and subsequent transfer of genes encoding biodegradative pathways, located on conjugative plasmids, to the indigenous microorganisms of an ecosystem. The first publications, in which this strategy was suggested, were by Fulthorpe and Wyndham (1991), Brokamp and Schmidt (1991), Hickey *et al.* (1993), De Rore *et al.* (1994). Several studies have shown that conjugative plasmid transfer occurs at detectable frequencies in soils and activated sludge under simulated natural conditions. However, most of these studies have been performed with plasmids that encode antibiotic or heavy metal resistance genes (for a recent review on gene transfer in soil, see: Dröge *et al.*, 1999), and only a few have dealt with the dissemination of plasmid borne catabolic genes to and among indigenous soil or sludge bacteria. The strategy of spreading degradative genes into well established, competitive indigenous microbial populations as a means of bioaugmentation has been investigated even less thoroughly. The rest of this paper will briefly review these studies, with emphasis on those that have clearly shown a direct effect of gene transfer on accelerated biodegradation, and a few conclusions about the feasibility of this approach will be drawn.

2.1. CASES OF CATABOLIC PLASMID TRANSFER AND BIOAUGMENTATION IN SOIL

As mentioned above, a large number of studies have reported the occurrence of conjugative gene transfer between bacteria in soil (Dröge *et al.*, 1999; Hill and Top, 1998), but there is only little information about transfer of catabolic plasmids as a means of bioaugmentation. A few studies have reported the transfer of catabolic plasmids in soil, but without investigating whether this resulted in accelerated degradation of the pollutant for which the plasmid encodes degradation. For example in 1991 transfer of the catabolic plasmid pWWO-EB62, encoding degradation of ethylbenzoate, was examined between introduced strains in sterilised soil with and without addition of ethylbenzoate. Transfer was only detected when the donor and recipient strains were of the same species. A positive effect of ethylbenzoate on the number of transconjugants was not observed. On the contrary, a negative effect was seen i.e. a lag phase in the formation of transconjugants, which was probably due to toxic effects of the chemical on the recipient strain. These results are however difficult to interpret because of the sterile soil conditions, which are very different than the complexity of a real soil environment (Ramos-Gonzales *et al.*, 1991). Kinkle *et al.* (1993) showed that the plasmid pJP4, encoding 2,4-D (2,4-dichlorophenoxyacetic acid) degradation and mercury resistance, was transferred from an introduced *Bradyrhizobium japonicum* strain to several introduced *Bradyrhizobium sp.* strains in non-sterile soil. They only looked at the effect of the presence of mercury in the soil on the transfer, and concluded that the addition of mercury (up to 50 mg/kg soil) had no apparent stimulatory effect on the number of transconjugants obtained. Neilson *et al.* (1994) also followed transfer of pJP4, but now from *Alcaligenes eutrophus* JMP134 to *Variovorax paradoxus* in both sterile and non-sterile soil. Although ca. 10^3 CFU/g soil of transconjugants were detected in sterile soil, transfer of pJP4 to *V. paradoxus* in non-sterile soils could only be detected when the soil was amended with 1000 mg/kg 2,4-D. Daane *et al.*(1996) investigated the role of earthworms in the genetic transfer of plasmid pJP4 in soil. Transfer was observed, but since the soil contained no 2,4-D, no relationship with 2,4-D degradation could be made. In a next study the same group showed that earthworm activity could positively influence the transfer of plasmid pJP4 between spatially separated donor and recipient strains in soil columns, but also here no link with 2,4-D degradation was made (Daane *et al.*,1997).

The effect of transfer of a degradative plasmid in soil on the degradation of the corresponding pollutant has only been examined in a few studies. Brokamp *et al.* (1991) were probably the first to suggest the possible role of catabolic plasmid transfer in enhanced pesticide degradation in soil, but there was no direct evidence yet. They showed that a catabolic plasmid, encoding 2,2-dichloropropionate (DCPA) degradation, present in an inoculated *Alcaligenes xylosoxidans* strain, was transferred to soil bacteria in DCPA treated soil. Enhanced DCPA degradation in the inoculated soils was observed, but since the inoculum itself survived very well and could degrade the herbicide, there was no direct evidence that this positive effect was due to plasmid transfer. A second report that linked enhanced mineralisation of contaminants (PCBs) in soil after inoculation of a bacterial strain (degrading 3-chlorobenzoate) with potential genetic exchange between the inoculant and the indigenous soil bacteria was from

Hickey *et al.* (1993). Their hypothesis that the inoculum had taken up biphenyl degradative genes (*bph*) from indigenous soil bacteria, was based on the increasing frequency of isolates able to utilise both biphenyl (which could not be degraded by the original inoculum) and 2,5-DCB (which the original inoculum could degrade). Real evidence for transfer of (Cl-)biphenyl degradative genes into the inoculum was however not given. A later study of the same group (Focht *et al.*, 1996), showed more evidence that *bph* genes were transferred from indigenous biphenyl degraders to the inoculant. They showed that two isolated potential hybrid strains, most effective in metabolising PCBs, had the same REP-PCR fingerprints as the inoculant, which is strong evidence that they were indeed derived from this strain after having acquired *bph* genes.

In our laboratory, De Rore *et al.* (1994) studied the transfer of a non-recombinant biphenyl degradative transposon located on RP4, from an inoculated donor strain to indigenous bacteria in sandy soil. In spite of the fast disappearance of the donor the plasmid was transferred to bacteria belonging to a number of different genera. The transconjugant numbers were higher in biphenyl spiked soil than in non-treated soil (5.9×10^4 g^{-1} soil compared to 4.1×10^2 g^{-1} after 26 days). Most importantly, a clear effect of the transfer of the *bph* genes to the indigenous soil bacteria on the degradation of biphenyl (1000 mg/kg) was shown. Since the donor could not degrade biphenyl, even though it harboured the plasmid carrying the *bph* genes, the enhanced removal of biphenyl must have been due to transconjugants that had received the plasmid and expressed the *bph* genes.

The first report on a possible positive effect of gene transfer on the degradation of 2,4-D was presented by diGiovanni *et al.* (1996). They examined the frequency of plasmid transfer from *R. eutropha* JMP134, which contains pJP4 and can degrade 2,4-D, to indigenous bacteria. Similar to what was observed by Neilson *et al.*(1994), they only detected transconjugants when high concentrations of 2,4-D (1000 mg/kg soil) were added, and not when 100 or 500 mg 2,4-D/kg soil was present. The added 2,4-D was more rapidly degraded in soil microcosms inoculated with the donor strain. Since the donor strain, which can degrade 2,4-D, died off rapidly in the soil with 1000 mg/kg 2,4-D, the enhanced degradation in this soil was thought to be due to plasmid transfer. At the lower 2,4-D concentrations no evidence of plasmid mediated bioaugmentation could be given. A more recent study showed transfer of plasmid pJP4 from an *E. coli* strain ,which cannot degrade 2,4-D, to different indigenous bacteria in some but not all tested soils, and again only in the presence of high 2,4-D concentrations (500 or 1000 mg/kg). The effect of transfer on accelerated 2,4-D degradation was very small and only shown in one of the four soils studied (Newby *et al.*, 2000).

We have also used 2,4-D as a model compound to further investigate the possibility of soil bioaugmentation by catabolic plasmid transfer. Transfer of two other 2,4-D degradative plasmids, pEMT1 and pEMT3, isolated from agricultural soil (Top *et al.*, 1995), from an *E. coli* donor to the indigenous soil bacteria has been monitored in soil microcosms with and without 2,4-D (100 mg/kg soil) (Top *et al.*, 1998). Transfer of plasmid pEMT1k (a kanamycin marked derivative of pEMT1) to the indigenous microbial populations was observed in non-amended and 2,4-D-amended soil, and 2,4-D amendment had a clearly positive effect on transconjugant numbers ($2 \times 10^{+2}$ per donor in 2,4-D treated soil compared to 3.9×10^{-3} per donor in non-treated soil after 5 days). Transfer of plasmid pEMT3k was only observed in 2,4-D treated soil (1 per donor after 5 days). In each case transconjugants were confirmed by plasmid extractions and restriction digests of randomly picked colonies. Enhanced degradation of 2,4-D was clearly seen in one of the two soils tested after the donor was inoculated and had transferred its plasmid to the soil bacteria. Since this *E. coli* strain, although it had the 2,4-D degradative genes, could not degrade 2,4-D, the improved 2,4-D degradation in soil was clearly due to plasmid transfer.

Since *E. coli* is not a suitable strain for bioremediation purposes, we extended our study in the same soil without nutrients, with *P. putida* UWC3 as donor of plasmid pEMT1k. Again, transfer of plasmid pEMT1k was observed and correlated with an enhanced degradation of 2,4-D (Top *et al.*, 1999). However, since the donor strain was still present when degradation started (10^4 CFU/g soil), and was recently shown to partially degrade 2,4-D in sterile soil, its involvement in the degradation of 2,4-D could not be totally excluded. The most recent study in our laboratory investigated the transfer of the 2,4-D degradative plasmids pEMT1 and pJP4, from the same donor strain *P. putida* UWC3, as mentioned above, to the indigenous bacteria of the A- and B-horizon of a 2,4-D contaminated sandy-loam soil (Dejonghe *et al.*, 2000). Compared to the soil from the A-horizon (0-30 cm), the soil of the B-horizon (30-60 cm) has a different texture, organic matter content and thus also a different microbial community, with a different metabolic activity. The impact of these differences on the 2,4-D degradation capacity of the soils, the plasmid transfer rates and the subsequent accelerated 2,4-D degradation was investigated. Briefly, bioaugmentation with both plasmids was most successful in soil from the B-horizon, where the indigenous microbial community could not degrade the herbicide (100 mg/kg 2,4-D), even up to 89 days. Addition of the donor and subsequent plasmid transfer resulted in complete removal of 2,4-D within 19 days (Fig. 1). In the case of plasmid pEMT1, this success must be due to the transfer of the plasmid to the indigenous bacteria since the donor was very unstable and could not be detected anymore when degradation started (Fig. 1.A). In addition, DGGE (denaturing gradient gel electrophoresis) of the soil 16S rRNA gene pool was used for the first time to show that the *in situ* formation and subsequent proliferation of high numbers of 2,4-D degrading transconjugants caused clear changes in the soil bacterial community structure. This confirms that the inoculation and spread of a catabolic plasmid enables specific indigenous bacterial species to acquire new useful catabolic genes, which allows them to degrade the contaminant and to proliferate and outcompete other members that are unable to use this compound.

Transfer of catabolic genes as a bioaugmentation strategy has also been suggested in the context of rhizoremediaton. Crowley *et al.* (1996) observed increased 2,5-dichlorobenzoate (2,5-DCB) degradation in soil planted with bean (*Phaseolus vulgaris*) and inoculated with a *P. fluorescens* strain containing a plasmid encoded catabolic pathway for 2,5-DCB degradation. The fact that new degraders could only be isolated from inoculated soil, and not from soils, planted or not, without inoculum, was an indication that they were transconjugants which had received the catabolic plasmid from the donor. The overall enhanced degradation of 2,5-DCB in the inoculated and planted soil may have been partially due to this plasmid transfer, but hard evidence was lacking. Sarand *et al.* (2000) inoculated an indigenous *P. fluorescens* strain, supplied with the TOL plasmid pWWO::Km in soil microcosms with and without pine seedlings, mycorrhized with *Suillus bovinus*. After 3 months of regular treatment with m-toluate (mTA) the catabolic plasmid was found to be transferred into indigenous bacteria that belonged to at least two different genera. The inoculation also seemed to protect the plant and fungus from mTA, and since the donor could not be detected anymore, this positive effect was attributed to the degradation of mTA by the TOL plasmid containing indigenous transconjugants. The removal of mTA from this soil was however not measured analytically

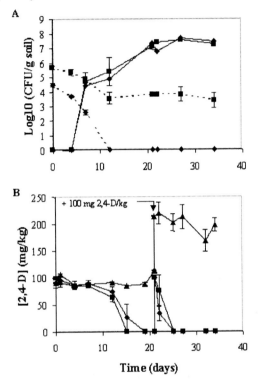

Fig. 1: Transfer of the plasmids pEMT1::lacZ and pJP4::lacZ in non-amended B-horizon soil with 2,4-D. Survival of donors (dotted lines) and formation of transconjugants (solid lines) (panel A). Effect of donor inoculation and subsequent plasmid transfer on the degradation of 2,4-D (panel B). The data points and error bars show the means and standard deviations based on data from duplicate microcosms. (▲) non-inoculated soil + 2,4-D; (◆) soil + 2,4-D + pEMT1::lacZ; (■) soil + 2,4-D + pJP4::lacZ. The arrow indicates the second amendment of 100 mg 2,4-D/kg soil on day 21 (adapted from Dejonghe et al., 2000)

These studies show first of all that a selective pressure in a soil habitat can strongly affect the extent of gene spread, even when the donor dies off quickly. It must be noted that the positive effect of the presence of the pollutant on the number of transconjugants is almost certainly due to the proliferation of the transconjugants, and is not a direct effect on the conjugation efficiency. Bioaugmentation after transfer of the introduced plasmid to indigenous soil bacteria was obtained in a few cases. The difficulties of finding a suitable system of a pollutant with a corresponding degradative mobile element, a donor strain which can be counterselected and which either dies off after a while or does not degrade the compound, and a sensitive method for detecting gene transfer, explains why the reported cases of plasmid mediated bioaugmentation are still limited. The still expanding knowledge on new mobile catabolic genes, and the improving methods to detect gene transfer *in situ* and to monitor low concentrations of pollutants, will certainly facilitate future studies in this area.

2.2. CASES OF SUCCESSFUL BIOAUGMENTATION IN ACTIVATED SLUDGE AND OTHER WASTEWATER TREATING BIOREACTORS AND THE ROLE OF CATABOLIC PLASMID TRANSFER

Only very few studies have evaluated the introduction of degradative bacteria with mobile catabolic genes into activated sludge in order to enhance biodegradation of xenobiotics. Nüßlein *et al.* (1992) augmented activated sludge with specific degradative genes by inoculating it with two genetically modified microorganisms (GMMs), a *Pseudomonas* sp. strain B13FR1(pFRC20P), designated FR120, and a *Pseudomonas putida* KT2440(pWWO-EB62), designated EB62. Strain FR120 contains an assembled ortho-cleavage route for simultaneous degradation of 3-chlorobenzoate (3CB) and 4-methylbenzoate (4MB). Its plasmid, pFRC20P harbours mobilisation but no transfer genes, and could thus only disseminate through conjugation when it is mobilised by a conjugative helper plasmid. The plasmid pWWO-EB62 is a derivative of the TOL plasmid, which can additionally degrade 4-ethylbenzoate (4EB), and is self-transmissible. Both GMMs, introduced into an activated sludge unit at initial densities of 10^6-10^7 CFU/ml, decreased to stable populations of 10^4-10^5 CFU/ml. Three days after inoculation of strain FR120 a drastic decrease in the concentrations of 3-CB and 4-MB (initially 1mM each) was observed. In the non-inoculated reactor 8 days of adaptation were needed before degradation of 4-MB occurred, while it took even more than 15 days to degrade all of the 3-CB. When 4 mM of 3-CB and 4-MB were present in the uninoculated microcosms, the indigenous microorganisms were killed, while a stable density of viable cells was maintained when FR120 was inoculated. The inoculation of EB62 did not significantly enhance the biodegradation of 4EB, because the indigenous microbial community was able to degrade this compound very easily. In filter matings the conjugative plasmid pWWO-EB62 was transferred to *Pseudomonas putida* UWC1 at a frequency of 10^{-1} transconjugants per donor cell, whereas the Tra⁻ Mob⁺ plasmid pFRC20C did not transfer (Nüßlein *et al.* 1992). Similarly transfer of plasmid pWWO-EB62 to UWC1 in the activated sludge unit was readily observed, whilst transfer of the non-conjugative plasmid pFRC20P was infrequent. The latter indicates however that mobilising plasmids, present in the activated sludge, must have been responsible for this mobilisation. Transfer of the plasmids to indigenous sludge bacteria was not investigated, and thus conclusions on its possible effect on the rate of biodegradation of the pollutants could not be drawn.

McClure *et al.* (1991) investigated the survival of a donor strain *P. putida* UWC1, and potential transfer of its 3-chlorobenzoate (3CB) degradative non-conjugative IncQ plasmid pD10 into indigenous activated sludge bacteria. Despite the long survival of this GMM in the sludge no breakdown of the 3CB was observed. However, an autochthonous sludge bacterium, which had taken up the plasmid from the *P. putida* donor strain in plate filter matings, reported in a previous study (McClure *et al.* 1989), could enhance the biodegradation after reinoculation in the activated sludge unit. These results indicate that activated sludge isolates, which are well adapted to the local conditions, can be very good host strains for catabolic genes, and thus good inoculants for bioaugmentation. The data did however not support the idea of spreading catabolic genes directly into an activated sludge community.

More recently Ravatn *et al.* (1998) monitored the transfer of the transposable element *clc*, coding for chlorocatechol degradation, from *Pseudomonas* sp. strain B13 to *Pseudomonas putida* F1 and to indigenous bacteria in a laboratory-scale activated-sludge microcosm. When strain F1 receives the *clc* element, it is able to degrade chlorobenzenes by complementation of two partial degradative pathways, while both the donor and recipient strain separately cannot degrade these compounds. Transfer of the *clc* element into the inoculated recipient was observed, but only when cell numbers of strain B13 were sufficiently high due to the addition of its substrate 3-chlorobenzoate, and when at the same time the selective substrate for the formed transconjugants, 1,4-dichlorobenzene (1,4-DCB) was added. These results show again the effect of the selective pressure on the number of transconjugants, and thus on the ability to detect them, due to the selective advantage that these transconjugants have, being the only ones able to use 1,4-DCB as carbon and energy source. The effect of this gene exchange on the degradation of this pollutant was however not investigated, since concentrations of 1,4-DCB were not measured.

It is clear that the number of studies about transfer of catabolic genes in activated sludge, and its effect on biodegradation of specific compounds is very limited. In order to understand if the introduction and dissemination of catabolic genes in such a bioreactor system is a feasible bioaugmentation strategy, more research is needed. We recently investigated the bioaugmentation of an activated sludge treating influent with 3-chloroaniline (3-CA). The inoculum, an indigenous *Comamonas testosteroni* strain, marked with the *gfp* gene, was able to remove all the 3-CA during ca. 2 weeks, but afterwards the removal efficiency decreased to ca. 50% (Boon *et al.*, 2000). Currently experiments are being performed to investigate if we can prolong the period of complete 3-CA removal by introducing another donor strain with plasmid encoded 3-CA degradative genes, which could be transferred to the sludge bacteria.

More knowledge on plasmid exchange in such activated sludge microbial communities would help to understand the conditions that promote or inhibit this process, and can thus help to improve the success of bioaugmentation by plasmid transfer. Transfer of conjugative plasmids and mobilisation of non-conjugative plasmids has been demonstrated in activated sludge reactors. In a sequencing batch reactor transfer of the IncW plasmid R388 to activated sludge bacteria was detected at a concentration of 4×10^3/ml as soon as one day after donor addition, but transconjugants could not be detected anymore after 20 days. In the fixed-bed reactor however the transconjugants remained detectable for a longer period. Mancini *et al.* (1987) observed mobilisation of the recombinant vector pHSV106 in a laboratory-scale waste treatment facility when mobilising strains were co-inoculated. The highest numbers of transconjugants were found at the bottom of the primary clarifier and the return-activated sludge from the secondary clarifier, where cell densities are the highest. Studies on the occurrence of natural plasmids in activated sludge and wastewater have shown that mobilising plasmids are indeed present in wastewater treatment plants and are actively mobilising non-conjugative (Tra⁻) plasmids (Bauda *et al.*, 1995; Fujita *et al.*, 1993; McPherson and Gealt, 1986; Top *et al.*, 1994). Normally, both the availability of nutrients and the presence of surfaces, which act as a support for the bacteria, have been shown to enhance the efficiency of conjugation by broad-host-range plasmids (Frank *et al.* 1996; Pukall *et al.* 1996). The ability of these surfaces to protect the conjugation pili has also been reported. Especially cells that contain a BHR plasmid form short rigid pili that may break more easily and are thus stabilised by the solid surfaces (Bradley, 1980; Willets, 1985; Libaron *et al.*, 1993). Because of their high cell density and the presence of biodegradable organic matter, sludge flocs, which contain the majority of the bacteria present in activated sludge, may form an ideal site for conjugation. However, because of the porous structure it is not clear whether these bacteria really have a greater chance to meet. Furthermore it is not investigated in depth whether exopolymeric material is not an obstacle for conjugation. In contrast with the collected knowledge on factors influencing conjugative gene transfer in soil, and on the importance of this gene exchange in bacterial adaptation in this habitat, there is very little known about the impact of gene transfer on the bacterial communities in activated sludge units.

In spite of the few results on horizontal transfer of catabolic genes in activated sludge, there is no reason not to believe that also in this ecosystem the catabolic potential could be widened by introducing and disseminating catabolic genes on mobile genetic elements.

3. Conclusions

This review shows that the cases where transfer of catabolic genes in soil microcosms or in activated sludge systems was observed and correlated with improved biodegradation of a certain organic pollutant, are still very limited. A few successes indicate that the approach can be feasible under certain conditions to accelerate the removal of pollutants in soil. It will certainly not be a magic solution to environmental pollution, since there are many more aspects to the problem of pollutant removal than the absence of the necessary genetic information (Romantschuk *et al.*, 2000). It is however interesting to see how in some cases supplying the polluted site with catabolic genes could enhance the biodegradation process. Further investigations in this area are obviously needed to improve our current knowledge on the efficiency of gene dissemination as a tool in bioremediation, and on the host-range of catabolic plasmids in the environment.

4. Acknowledgements

Parts of this review are adapted from other recent review papers (Top *et al.*, 2000; Van Limbergen *et al.*, 1998). Fig. 1 is part of a larger figure in Dejonghe *et al.* (2000), and this work was supported by the Fund for Scientific Research of Flanders (F.W.O.-Vlaanderen). E.M. Top is also indebted to the F.W.O.-Vlaanderen for a position as Research Associate (Onderzoeksleider). We thank W. Dejonghe for sharing literature references and papers.

References

Ahring, B.K., Christiansen, N., Mathrani, I., Hendriksen, H.V., Macario, A.J.L., and De Macario, E.C. (1992) Introduction of a de novo bioremediation ability, aryl reductive dechlorination, into anaerobic granular sludge by inoculation of sludge with Desulfomonile tiedjei, Appl. Environ. Microbiol. 58, 3677-3682.

Akkermans, A.D.L. (1994) Application of bacteria in soils: problems and pitfalls, FEMS Microbiol. Rev. 15, 185-194.

Alexander, M. (1994) Inoculation, in M. Alexander (ed.), Biodegradation and bioremediation, Academic Press, San Diego, California. pp. 226-247.

Bauda, P., Lallement, C., and Manem, J. (1995) Plasmid content evaluation of activated sludge, Wat. Res. 29, 371-374.

Boon, N., Goris, J., De Vos, P., Verstraete, W., and Top, E.M. (2000) Bioaugmentation of activated sludge by an indigenous 3-chloroaniline degrading Comamonas testosteroni strain I2gfp, Appl. Environ. Microbiol. 66, 2906-2913.

Boronin, A.M. (1992) Diversity and relationships of Pseudomonas plasmids. In E. Galli, S. Silver, S. and B. Witholt (eds.), Pseudomonas: Molecular Biology and Biotechnology, ASM Press, Washington. pp. 329-340.

Bradley, D.E. (1980) Morphological and serological relationships of conjugative plasmids, Plasmid 4, 155-169.

Brokamp, A., and Schmidt, F.R.J. (1991) Survival of Alcaligenes xylosoxidans degrading 2,2-dichloropropionate and horizontal transfer of its halidohydrolase gene in a soil microcosm, Curr. Microbiol. 22, 299-306.

Christiansen, N., Hendrikse, H.V., Jarvinen, T., and Ahring, B.K. (1995) Degradation of chlorinated aromatic compounds in UASB reactors, Water Sci. Technol. 3, 249-259.

Crowley, D.E., Brennerova, M.V., Irwin, C., Brenner, V., and Focht, D.D. (1996) Rhizosphere effects on biodegradation of 2,5-dichlorobenzoate by a luminescent strain of root-colonizing Pseudomonas fluorescens, FEMS Microbiol. Ecol. 20, 79-89.

Daane, L.L., Molina, J.A.E., Berry, E.C., and Sadowsky, M.J. (1996) Influence of earthworm activity on gene transfer from Pseudomonas fluorescens to indigenous soil bacteria, Appl. Environ. Microbiol. 62, 515-521.

Daane, L.L., Molina, J.A.E., and Sadowsky, M.J. (1997) Plasmid transfer between spatially separated donor and recipient bacteria in earthworm-containing soil microcosms, Appl. Environ. Microbiol. 63, 679-686.

Dejonghe, W., Goris, J., El Fantroussi, S., Höfte, M., De Vos, P., Verstraete, W., and Top, E.M. (2000). Effect of the dissemination of 2,4-dichlorophenoxyacetic acid (2,4-D) degradative plasmids on 2,4-D degradation and on the bacterial community structure in two different soil horizons, Appl. Environ. Microbiol. 66, 3297-3304.

De Rore, H., Demolder, K., De Wilde, K., Top, E., Houwen, F., and Verstraete, W. (1994) Transfer of the catabolic plasmid RP4::Tn4371 to indigenous soil bacteria and its effect on respiration and biphenyl breakdown, FEMS Microbiol. Ecol. 15, 71-84.

diGiovanni, G.D., Neilson, J.W., Pepper, I.L., and Sinclair, N.A. (1996) Gene transfer of Alcaligenes eutrophus JMP134 plasmid pJP4 to indigenous soil recipients, Appl. Environ. Microbiol. 62, 2521-2526.

Dröge, M., Pühler, A., and Selbitschka, W. (1999) Horizontal gene transfer among bacteria in terrestrial and aquatic habitats as assessed by microcosm and field studies, Biol. Fertil. Soils 29, 221-245.

Focht, D.D., Searles, D.B., and Koh, S.-C. (1996) Genetic exchange in soil between introduced chlorobenzoate degraders and indigenous biphenyl degraders, Appl. Environ. Microbiol. 62, 3910-3913.

Frank, N., Simao-Beaunoir, A.-M., Dollard, M.A., Bauda, P. (1996) Recombinant plasmid DNA mobilisation by activated sludge strains grown in fixed-bed or sequenced-batch reactors, FEMS Microbiol. Ecol. 21, 139-148.

Fujita, M., Ike, M., Suzuki, H. (1993) Screening of plasmids from wastewater bacteria. Water. Res. 27, 949-953.

Fulthorpe, R.R., and Wyndham, R.C. (1991) Transfer and expression of the catabolic plasmid pBRC60 in wild bacterial recipients in a freshwater ecosystem, Appl. Environ. Microbiol. 57, 1546-1553.

Goldstein, R.M., Mallory, L.M., and Alexander, M. (1985) Reasons for possible failure of inoculation to enhance biodegradation, Appl. Environ. Microbiol. 50, 977-983.

Hickey, W.J., Searles, D.B., and Focht, D.D. (1993) Enhanced mineralisation of polychlorinated biphenyls in soil inoculated with chlorobenzoate-degrading bacteria, Appl. Environ. Microbiol. 59, 1194-1200.

Hill, K. E., and Top, E.M. (1998) Gene transfer in soil systems using microcosms, FEMS Microbiol. Ecol. 25, 319-329.

Herrick, J.B., Stuart-Keil, K.G., Ghiorse, W.C., and Madsen, E.L. (1997) Natural horizontal transfer of a naphthalene dioxygenase gene between bacteria native to a coal tar-contaminated field site, Appl. Environ. Microbiol. 63, 2330-2337.

Kinkle, B.K., Sadowsky, M.J., Schmidt, E.L., and Koskinen, W.C. (1993) Plasmid pJP4 and R68.45 can be transferred between populations of *Bradyrhizobia* in nonsterile soil, Appl. Environ. Microbiol. 59, 1762-1766.

Libaron, P., Roux, V., Lett, M.C., Baleux, B. (1993) Effects of pili rigidity and energy availability on conjugative plasmid transfer in aquatic environments, Microbial Releases 2, 127-133.

Mancini, P., Fertels, S., Nave, D., and Gealt, M.A. (1987) Mobilization of plasmid pHSV106 *from Escherichia coli* HB101 on a laboratory-scale waste treatment facility, Appl. Environ. Microbiol. 53, 665-671.

McClure, N.C., Fry, J.C., and Weightman, A.J. (1991) Survival and catabolic activity of natural and genetically engineered bacteria in a laboratory-scale activated sludge unit, Appl. Environ. Microbiol. 57, 366-737.

McClure, N.C., Weightman, A.J., and Fry, J.C. (1989) Survival of *Pseudomonas putida* UWC1 containing cloned catabolic genes in a model activated sludge unit, Appl. Environ. Microbiol. 55, 2627-2634.

McPherson, P., Gealt, M.A. (1986) Isolation of indigenous wastewater bacterial strains capable of mobilizing plasmid pBR325, Appl. Environ. Microbiol. 51, 904-909.

Mergeay, M., Springael, D. and Top, E. (1990) Gene transfer in polluted soils, in J.C. Fry and M.J. Day (eds.), Bacterial Genetics in Natural Environments, Chapman & Hall, London, pp. 152-171.

Miguez, C.B., Shen, C.F., Bourque, D., Guiot, S.R., and Groleau, D. (1999). Monitoring methanotrophic bacteria in hybrid anaerobic-aerobic reactors with PCR and a catabolic gene probe, Appl. Environ. Microbiol. 65, 381-388.

Neilson, J.W., Josephson, K.L., Pepper, I.L., Arnold, R.B., diGiovanni, G.D., and Sinclair, N.A. (1994) Frequency of horizontal gene transfer of a large catabolic plasmid (pJP4) in soil, Appl. Environ. Microbiol. 60, 4053-4058.

Newby, D. T., Josephson, K. L., and Pepper, I. L. (2000) detection and characterization of plasmid pjp4 transfer to indigenous soil bacteria, Appl. Environ. Microbiol. 66, 290-296.

Nüsslein, K., Maris, D., Timmis, K., and Dwyer, D. (1992) Expression and transfer of engineered catabolic pathways harbored by *Pseudomonas spp.* introduced into activated sludge microcosms, Appl. Environ. Microbiol. 58, 3380-3386.

Pukall, R., Tschäpe, H., and Smalla, K. (1996) Monitoring the spread of broad host and narrow host range plasmids in soil microcosms, FEMS Microbiol. Ecol. 20, 53-66.

Ramos-Gonzalez, M., Duque, E., and Ramos, J.L. (1991) Conjugational transfer of recombinant DNA in cultures and in soils: host range of *Pseudomonas putida* TOL plasmids, Appl. Environ. Microbiol. 57, 3020-3027.

Ravatn, R., Zehnder, A.J.B., and van der Meer, J.R. (1998). Low-frequency horizontal transfer of an element containing the chlorocatechol degradation genes from *Pseudomonas putida* F1 and to indigenous bacteria in laboratory-scale activated-sludge microcosms, Appl. Environ. Microbiol. 64, 2126-2132.

Romantschuk, M., Sarand, I., Petänen, T., Peltola, R., Jonsson-Vihanne, M., Koivula, T., Yrjälä, K., and Jhaahtela, K. (2000) Means to improve the effect of in situ bioremediation of contaminated soil: an overview of novel approaches. Environmental Pollution 107, 179-185

Sarand, I., Haario, H., Jorgensen, K.S., and Romantschuk, M. (2000) Effect of inoculation of a TOL plasmid containing mycorrhizosphere bacterium on development of Scots pine seedlings, their mycorrhizosphere and the microbial flora in m-toluate-amended soil, FEMS Microbiol. Ecol. 31, 127-141.

Sayler, G.S., Hooper, G.S., Layton, A.C., and Henry King, J.M. (1990). Catabolic plasmids of environmental and ecological significance, Microb. Ecol. 19, 1-20.

Selvaratnam, C., Schoedel, B.A., McFarland, B.L., and Kulpa, C.F. (1997) Application of the polymerase chain reaction (PCR) and the reverse transcriptase/PCR for determining the fate of phenol-degrading *Pseudomonas putida* ATCC 11172 in a bioaugmented sequencing batch reactor, Appl. Microbiol. Biotechnol. 47, 236-240.

Stuart-Keil, K.G, Hohnstock, A.M., Drees, K.P., Herrick, J.B., and Madsen, E.L. (1998) Plasmids responsible for horizontal transfer of naphthalene catabolism genes between bacteria at a coal tar-contaminated site are homologous to pDTG1 from *Pseudomonas putida* NCIB 9816-4, Appl. Environ. Microbiol. 64, 3633-3640.

Tartakovsky, B., Levesque, M.-J., Dumortier, R., Beaudet, R., and Guiot, S.R. (1999) Biodegradation of pentachlorophenol in a continuous anaerobic reactor augmented with *Desulfitobacterium frappieri* PCP-1, Appl. Environ. Microbiol. 65, 4357-4362.

Top, E., De Smet, I., Mergeay, M., and Verstraete, W. (1994) Exogenous isolation of mobilizing plasmids from polluted soils and sludges, Appl. Environ. Microbiol. 60, 831-839.

Top, E.M., Holben, W.L., and Forney, L.J. (1995) Characterization of diverse 2,4-D degradative plasmids isolated from soil by complementation, Appl. Environ. Microbol. 61, 1691-1698.

Top, E. M., Maila, M.P., Clerinx, M., Goris, J., De Vos, P., and Verstraete, W. (1999) Methane oxidation as a method to evaluate the removal of 2,4-dichlorophenoxyacetic acid (2,4-D) from soil by plasmid mediated bioaugmentation, FEMS Microbiol. Ecol. 28, 203-213.

Top, E.M., Moënne-Loccoz, Y., Pembroke, T., Thomas, C.M. (2000) Phenotypic traits conferred by plasmids, in C.M. Thomas (ed.), The horizontal gene pool: bacterial plasmids and gene spread, Harwood Academic Publishers, The Netherlands, pp. 249-285.

Top, E. M., Van Daele, P., De Saeyer, N., and Forney, L.J. (1998) Enhancement of 2,4-dichlorophenoxyacetic acid (2,4-D) degradation in soil by dissemination of catabolic plasmids, Antonie van Leeuwenhoek. 73, 87-94.

Top, E., Vanrolleghem, P., Mergeay, M., and Verstraete, W. (1992) Determination of the mechanism of retro-transfer by mechanistic mathematical modeling, J. Bacteriol. 174, 5953-5960.

van der Meer, J.R. (1994) Genetic adaptation of bacteria to chlorinated aromatic compounds, FEMS Microbiology Reviews 15, 139-149.

van der Meer, J.R., Werlen, C., Nishino, S.F., and Spain, J.C. (1998) Evolution of a pathway for chlorobenzene metabolism leads to natural attenuation in contaminated groundwater, Appl. Environ. Microbiol. 64, 4185-4193.

Van Limbergen, H., Top, E.M., and Verstraete, W. (1998) Bioaugmentation in activated sludge: current features and future perspectives, Appl. Microbiol. Biotechnol. 50, 16-23.

van Veen, J.A., van Overbeek, L.S., and van Elsas, J.D. (1997) Fate and activity of microorganisms introduced into soil. Microbiol. Mol. Biol. Rev. 61, 121-135.

Willets, N. (1985) Plasmids, in J. Scaife, D. Leach and A. Galizzi (eds.), Genetics of bacteria, Academic Press, London, pp. 165-195.

Williams, P.A., and Murray, N.E. (1974) Metabolism of benzoate and the methylbenzoates by *Pseudomonas arvilla* mt-2: evidence for the existence of a TOL plasmid, J. Bacteriol. 120, 416-4

COPING WITH A HALOGENATED ONE-CARBON DIET: AEROBIC DICHLOROMETHANE-MINERALISING BACTERIA

STÉPHANE VUILLEUMIER

Institut für Mikrobiologie, ETH Zürich, ETH Zentrum/LFV,
Schmelzbergstr. 7, CH-8092 Zürich, Switzerland. Fax : +41 1 632 11 48,
email: vuilleus@micro.biol.ethz.ch

Abstract

The degradation by bacteria of man-made, often toxic halogenated chemicals present as contaminants in the environment is a subject that continues to fascinate scientists and the general public alike. This interest has stimulated research on the organisms capable of mineralising halogenated pollutants, and on the enzymes and genes involved in this metabolism. In the case of dichloromethane-mineralising aerobic bacteria, which are the subject of this review, dehalogenative metabolism is stripped down to its bare essentials: a single enzyme, dichloromethane dehalogenase, consisting of a single polypeptide, catalyses the cleavage of two carbon-halogen bonds from a single carbon atom, and allows growth of the bacterial host with dichloromethane as the sole carbon and energy source. The attractive simplicity of this system has made the mineralisation of dichloromethane by aerobic bacteria a well-studied and important paradigm for dehalogenative metabolism. However, recent evidence suggests that genes and proteins other than dichloromethane dehalogenase are also specifically required, and sometimes even essential, for growth of aerobic bacteria with dichloromethane. The characterisation of such accessory genes and proteins is a promising area of enquiry for the postgenomic age of bacterial biodegradation research.

1. The world of dichloromethane-degrading bacteria

Dichloromethane (DCM) is a solvent commonly used in a large variety of industrial applications [2], which is produced at an estimated level of 2-3 10^5 tons/year [2,66]. In contrast, evidence for the natural production of DCM is scarce: DCM formation during volcanic eruptions has been proposed [60], but no experimental data were provided to support this hypothesis. Plausible mechanisms for the synthesis of DCM by natural processes, involving, for example, the haloperoxidase catalysed halogenation of amino acids, have recently been reviewed [117]. A first, still uncertain estimate for the magnitude of natural production of DCM has been reported based on measurements of

S.N. Agathos and W. Reineke (eds.),
Biotechnology for the Environment: Strategy and Fundamentals, 105–130.

halogenated hydrocarbons in the troposphere, which suggests an uncharacterised oceanic source of DCM (Table 1). All things considered however, it seems plausible that human activity has significantly expanded the niches in which bacteria with the ability to degrade DCM can make a living. This strongly contrasts with the situation for other halogenated methanes (Table 1). In particular, chloromethane is produced naturally in very large amounts, and represents the most abundant halogenated carbon compound in the atmosphere.

Table 1. Tropospheric budgets of halogenated methanes [a]

	troposphere			fluxes		
	burden β	lifetime	conc.	industry	burning	natural
	(Tg Cl)	(yr)	(pptv)	(Tg Cl yr^{-1})	(Tg Cl yr^{-1})	(Tg Cl yr^{-1})
CH$_3$Cl	2.8	1.3	540	<0.01	0.64	0.57
CH$_2$Cl$_2$	0.25	0.4	17-40	0.49	0.05	0.16
CHCl$_3$	0.21	0.5	18	0.06	<0.01	0.50

[a] Data from [66]. β: amount of chlorine (1 Tg = 10^6 tons) present in the troposphere at any time, corresponding to the indicated concentration in parts per trillion (10^{-12}) volume (pptv). Emissions (fluxes) to the troposphere, arising from human activities (industry and burning) as well as from natural sources, are estimated from the observed or calculated sources and sinks for the different halomethanes.

Due to its low boiling point (40 °C; [85]) and high solubility in water (about 20 g/l, i.e. 235 mM; [85]), DCM easily escapes into the environment and is a frequently encountered pollutant, although both production [3] and emission [53] levels are now on the decrease compared to the 1980s. The contamination of groundwater and drinking water supplies by DCM became a source of serious concern in 1977, when DCM was included in the list of priority pollutants [67] because of evidence for genotoxic effects associated with DCM exposure (recently reviewed in [76]). This provided the incentive to investigate the degradation of DCM by bacteria, and to assess their potential for cleanup and prevention of enviromental contamination with DCM.

Carbon-halogen bonds are strong chemical linkages that are not easily cleaved abiotically [85,95,100], but biological degradation of halogenated methanes has been frequently observed in various soil and aqueous environments [85,118]. Both anaerobic and aerobic bacteria have been characterised that are able to degrade or mineralise DCM (Table 2). Many of these were isolated in the laboratory of Prof. T. Leisinger at ETH Zurich and, in the case of the aerobic DM strains (Table 2), characterised taxonomically in the laboratory of Prof. Y. Trotsenko in Pushchino.

Table 2. Dichloromethane degrading bacteria

Organism	Comments/Source	References
anaerobic		
Methanogenic enrichment culture	nonmethanogenic DCM degraders	[38]
Enrichment culture, digested sludge (Gram-positive DCM degraders)	fermentative culture (various electron end-acceptors)	[18]
Enrichment culture, DCM utilising fixed-bed reactor	fermentative culture (acetate, CO2, methane as products)	[109,130]
Defined mixed culture DM (Gram-positive strain DMA)	acetogenic culture derived from above (acetate, formate products)	[10]
Defined mixed culture DC (Gram-positive strain DMC)	culture derived from culture DM	[82]
Dehalobacterium formicoaceticum	strain derived from culture DC	[83,84]
aerobic		
Uncharacterised bacteria	activated sludge	[68,99]
Pseudomonas sp. strain LP	growth inhibited at > 0.1 mM	[77]
Methylosinus trichosporium OB3b	Co-metabolic oxidation by soluble methane monooxygenase	[52,94,119]
Strains DM20, DM21	trickle-bed bioreactor	[57]
Gram-negative methylotrophic strain MC8b		[58]
Acinetobacter sp. strain DCM	also grows well under anoxic nitrate-reducing conditions	[39]
Methylopila sp. strains DM1, DM3-DM9		[12,29,41,105]
Hyphomicrobium sp. strain DM2	also grows with nitrate as the terminal electron acceptor	[70,72,110]
Hyphomicrobium sp. strain GJ21		[25,61,71]
Methylobacterium dichloromethanicum DM4	best investigated strain	[43,45,74,75,105]
DCM-degrading sp. strain DM10	facultative autotroph	[30]
Methylophilus sp. strain DM11	fastest growing DCM degrader	[7,26,45,106]
Paracoccus methylutens DM12	facultative autotroph	[28]
Methylorhabdus multivorans DM13		[27]
DCM-degrading strain sp. DM14	facultative autotroph	[11]

Bioreactor studies performed with aerobic [42,57,63,111,134] and anaerobic (reviewed in [18]) DCM-degrading strains demonstrate that impressive maximal degradation rates of 12 kg DCM/m^3/d [111] and 1.25 kg DCM/m^3/d [18] can be achieved under aerobic and anaerobic conditions, respectively. However, biological strategies for the degradation of chlorinated chemicals based on bacteria appear to be insufficiently attractive to industry compared to, for example, waste incineration, reduction in the use of chlorinated solvents, and emission containment. Recent interest of research in the field of bacterial DCM dehalogenation has thus focussed to a large extent on the detailed characterisation of dehalogenation metabolism at the molecular level.

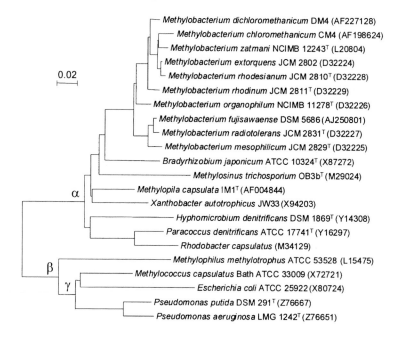

Figure 1. Unrooted phylogeny of 16S ribosomal DNA sequences from proteobacteria, with emphasis on bacterial genera able to grow aerobically with C1 compounds. The tree was generated from a sequence alignment generated with CLUSTALW [115] using the DNADIST and FITCH programs from the PHYLIP package [33]. The position of Escherichia coli, Pseudomonas aeruginosa *and* Pseudomonas putida *is shown for comparison.*

This work has yielded an overall picture of which enzymes are responsible for the dehalogenation of DCM by bacteria. Evidence is available that the anaerobic pathway of DCM degradation of the Gram-positive *Dehalobacterium formicoaceticum* is a multienzyme pathway involving corrinoid proteins, methyltransferases and tetrahydrofolate-based C1-interconverting enzymes [84]. Interestingly, similar combinations of enzymes are known for the mineralisation of chloromethane in anaerobic bacteria such as *Acetobacterium dehalogenans* (formerly strain MC) [87,131] and aerobic bacteria such as *Methylobacterium chloromethanicum* CM4 [112,121]. In contrast, aerobic pathways of DCM degradation involve other enzymatic mechanisms and are predominantly found in facultative methylotrophic Gram-negative α-proteobacteria (Figure 1) that are also capable of growing with C1 compounds such as methanol and methylamine. The obligate methanotroph *Methylosinus trichosporium* OB3b cometabolises DCM to carbon monoxide in a reaction catalysed by methane monooxygenase (Table 2), but is unable to use this metabolism to sustain growth, possibly because of the toxic products generated in this conversion [13,91]. In contrast, all other aerobic DCM-mineralising bacteria that have been investigated make use of the same enzyme, dichloromethane dehalogenase.

2. DCM dehalogenases: a particular brand of glutathione *S*-transferases

2.1. ENZYMES, GENES, AND REACTION MECHANISM

DCM dehalogenation activity was first noted to be highly induced during growth of different DCM-degrading strains with DCM. In cell-free extracts of *Hyphomicrobium* sp. strain DM2, DCM was found to be converted to formaldehyde and 2 mol of HCl in a reaction strictly dependent on glutathione [110] (Figure 2). This suggested that DCM dehalogenases were in fact glutathione *S*-transferases (GSTs), which at the time were well-characterised enzymes for the detoxification of electrophilic chemicals in mammals, but still essentially unheard of in bacteria (see [124] for a review). Dichloromethane dehalogenase was subsequently purified from this strain as a 34 kDa cytoplasmic protein whose expression is highly induced during growth with DCM [70]. Sequence analysis of *dcmA*, the gene encoding DCM dehalogenase, from *Methylobacterium dichloromethanicum* DM4 [74], and later from another DCM-degrading strain *Methylophilus* sp. strain DM11 [7], confirmed that DCM dehalogenases indeed belongs to the Theta class of the GST superfamily.

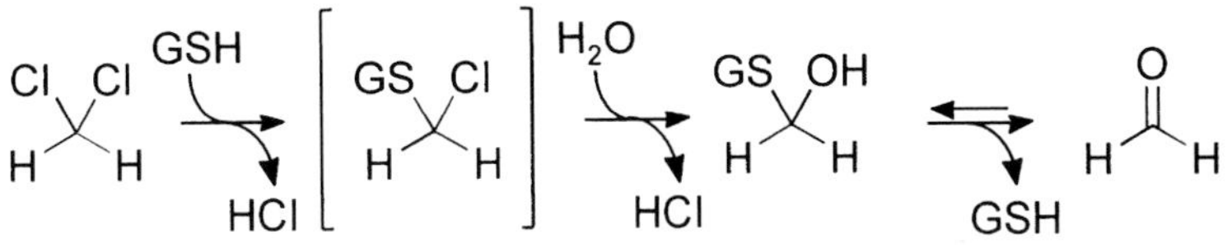

Figure 2. Schematic representation of the reaction mechanism of DCM dehalogenases. GSH, glutathione

The DCM dehalogenases from strain DM4 (287 residues) and DM11 (266 residues) are closest relatives in sequence databases, but only 56% identical at the protein level. They are only 15-25% identical to other GSTs, but no significant similarity to other families of proteins can be detected. DCM dehalogenases display a number of unusual properties that distinguish them from the bulk of GST enzymes [125,126]: (i) DCM dehalogenases appear essentially specific for halogenated methanes, achieving a 10^{10}-fold rate increase of DCM conversion over the abiotic reaction [8]; (ii) the glutathione cofactor is regenerated in its reduced form after completion of the reaction; (iii) the expression of DCM dehalogenases can be induced by their substrate to very high levels (up to about half of the total cytoplasmic protein under substrate limiting conditions; D. Gisi and S.V., unpublished observations). Most GSTs, in contrast, catalyse the conjugation of structurally diverse electrophilic compounds with glutathione, which is consumed in the reaction, and are usually expressed at low levels [124,126].

The details of the mechanism of the dehalogenation reaction still remain to be determined. In particular, the occurrence of the short-lived *S*-chloromethylglutathione as a likely intermediate in the reaction (Figure 2) remains to be demonstrated (see Section 3). Regarding the catalytic determinants of DCM conversion by DCM dehalogenases, it seems evident, given the quite different sequences of DM4 and DM11 enzymes, that the ability to degrade DCM can be maintained with only moderate sequence conservation. However, the GST framework is structurally strongly conserved throughout this enzyme family despite levels of sequence identity often below 20% [5,126]. The key question is therefore to understand how this structural framework has been adapted to achieve the observed specificity for DCM catabolism, and which residues are major determinants of DCM dehalogenase activity.

2.2. PROTEIN ENGINEERING STUDIES OF DCM DEHALOGENASE

Protein engineering studies have been carried out with the DCM dehalogenase from *Methylophilus* sp. strain DM11 which, unlike the enzyme from *Methylobacterium dichloromethanicum* DM4, can be expressed in active form in *E. coli* [7]. A serine residue (Ser12 in the DM11 enzyme) is conserved in several Theta class and bacterial GST, and is oriented towards the glutathione thiol in the X-ray structure of the insect GST from *Lucilia cuprina* [129]. This serine residue was found to be a major determinant of DCM dehalogenase function [123], rather than the conserved nearby tyrosine residue, which is involved in enhancing nucleophilicity of the glutathione thiol in mammalian enzymes [5]. Other residues which were investigated, such as Val18 [125] and Trp117 [14], affect the affinity for glutathione and DCM, but the underlying structural changes mediating these effects remain obscure. Considering the sequence plasticity of proteins of the GST superfamily, it seems likely that many other residues in DCM dehalogenases and other bacterial GSTs can modulate binding and reactivity of the glutathione cofactor as well as substrate binding. Indeed, in the case of the GST from *Proteus mirabilis*, for which a three-dimensional structure is available [102], all tyrosine, serine and cysteine residues near the N-terminus of the GST from *Proteus mirabilis* were exchanged with alanine without significant alteration of the measured kinetic parameters of the enzyme [15]. Thus, pinpointing which residues are important for catalysis may be problematic even when a detailed structure of the protein is available.

2.3. NATURAL VARIATION IN BACTERIAL DCM DEHALOGENASES

In an alternative approach towards defining the catalytic determinants of DCM dehalogenases, the kinetic properties of DCM dehalogenases of different strains closely related to strain DM4 were investigated. Such DCM dehalogenases were originally left aside because their size, specific activity in cell-free extracts, and crossreactivity with dehalogenase-specific antibodies are very similar to that of strain DM4 [71]. In addition, all DM strains, with the exception of strain DM11, display a 4 kb *Bam*HI DNA fragment that hybridises with a *dcmA* probe from strain DM4 [106]. The sequencing of a 10 kb DNA region containing the *dcmA* gene in *Methylobacterium dichloromethanicum* DM4 [105] later revealed the presence of a complex arrangement of insertion elements around the *dcmA* gene (Figure 3). This arrangement varies significantly in different DCM-degrading strains with a DCM dehalogenase closely related to that of strain DM4, and strongly suggests that horizontal transfer has been instrumental in the spread of DCM utilisation among bacteria [105]. Moreover, a sequence encoding a protein fragment with significant sequence identity (24% over 118 residues) to a GST from a schistosome worm was recently detected immediately downstream of the reported DCM dehalogenase DNA region (Figure 3). This observation of a typically eukaryotic GST sequence in a bacterial genome is unprecedented to our knowledge, and suggests that horizontal gene transfer events may have occurred on a larger scale than previously thought during spread of the DCM-utilising trait.

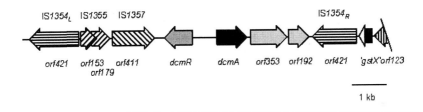

Figure 3. Current map of the dcm region in Methylobacterium dichloromethanicum *DM4. The* dcm *fragment common to DM-strains related to strain DM4 comprises* dcmR, *the gene involved in regulation of DCM dehalogenase expression [75], and genes* orf353 *and* orf192 *of unknown function in addition to* dcmA *[105]. Striped arrows indicate genes with strong similarity to transposases. The sequence labelled* 'gstX' *denotes a DNA stretch encoding a protein fragment most similar to schistosomal GST.*

This motivated us to check whether differences in the pattern of insertion sequences flanking the *dcmA* gene were associated with sequence variation in the DCM dehalogenase gene itself, and if so, if these changes influenced the kinetic properties of the corresponding enzyme [127]. With the exception of strain DM11, previously isolated DCM-degrading strains DM1-DM14 and GJ21 (Table 2) contained *dcmA* genes so similar to that of DM4 that they could be amplified using primers corresponding to sequences flanking the *dcmA* gene of strain DM4 (Figure 4). PCR products were also obtained with DNA extracted from sludge samples and from enrichment cultures growing with DCM, as well as from a prototype bioreactor for DCM removal from waste air [134,135].

Although some of the obtained PCR products that were cloned were identical to those of strain DM4, 8 different gene sequences encoding 7 different protein sequences were also obtained (Figure 4). The nucleotide exchanges observed display a striking bias towards base changes leading to differences in the protein sequence of the DCM dehalogenase. Phylogenetic analysis of DM4-related sequences do not provide a satisfactory model for a single evolutionary pathway connecting different DCM dehalogenases (data not shown). This constitutes additional evidence for horizontal transfer during spread of the *dcmA* gene. In addition, the strong divergence of the DM11 sequence prevents a reliable rooting of the phylogenetic tree of all DCM dehalogenases: at least 332 substitutions are required to obtain the DM4 sequence from the DM11 sequence [79], compared to the most divergent DM4-like sequence obtained, that of strain GJ21, with just 15 differences.

Stéphane Vuilleumier

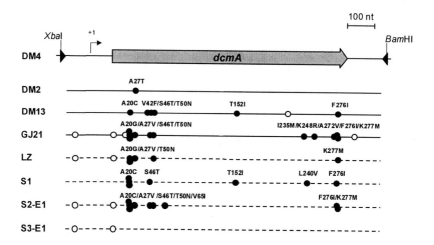

Figure 4. Overview of observed sequence variability in the PCR amplified dcmA *sequence region of DCM-degrading strains DM2, DM13, GJ21, of LZ, a trickling-filter bioreactor originally inoculated with strain DM4 [134], sludge from industrial wastewater treatment plant S1, and DCM-degrading enrichment cultures S2-E1 and S3-E1 obtained from sludge samples. The position of primers (triangles) used in PCR detection and cloning of* dcmA *genes is shown. The transcriptional start of the Methylobacterium dichloromethanicum DM4* dcmA *gene is indicated as "+1". Sequence variation in cloned PCR amplicons is indicated by circles (open circles, silent nucleotide changes; filled circles, nucleotide changes leading to an altered protein sequence relative to the DCM dehalogenase of strain DM4, as indicated).*

The consequences of sequence variation at the enzyme level were evaluated for the DCM dehalogenase from strains DM2, DM13 and GJ21 purified from their natural host [127]. The kinetic parameters k_{cat} and K_m of these enzymes very closely resemble those of the DM4 strain, but differ significantly from those of the DM11 enzyme. Their pH optima are closely similar, with the DM13 and GJ21 enzymes being more active at the low and high ends of the pH scale. Previously unremarked is the qualitative difference in pH profiles between DM4-like and DM11 enzymes, the latter showing a plateau extending up to pH 10 and not a pH optimum around pH 8.5.

Sequence variation may result in increased structural stability rather than in increased catalytic proficiency. Single amino acid changes can significantly affect both the stability and the kinetic properties of an enzyme [35]. The stability of DCM dehalogenases to urea denaturation is low but remarkably similar, although DM13 and GJ21 enzymes are slightly more stable than the DM2 and DM4 enzymes.

The lack of detectable functional consequences of sequence variation does not exclude that such variation confers a selective advantage under certain conditions *in vivo*, or that it affects properties of the enzyme that are not manifested in the *in vitro* assays [44]. It is unlikely that variation occurred only at functionally strongly constrained positions in DCM dehalogenases, since no significant differences in enzymatic properties of different DCM dehalogenases were found. On the other hand, base changes leading to synonymous codons were rarely observed (Figure 4). This would indicate that codon usage in the highly expressed *dcmA* gene is strongly constrained [1], but that the *dcmA* genes from different DCM degraders recovered in our analysis were acquired too recently for codon optimisation to already have taken place. In summary, one can only note that both the sequence and the catalytic properties of DCM dehalogenases are strongly conserved among different DCM-degrading strains, and that this has prevented further characterisation of the residues involved in enzyme catalysis.

2.4. RAT GST T1-1: DCM DEHALOGENATION ON THE LOOSE

Certain mammalian glutathione S-transferases of the Theta class (GST T1-1) are the most closely related enzymes to bacterial DCM dehalogenases (25% identity at the protein level). Although GST T1-1 enzymes also efficiently convert DCM [62,88,128], GST T1-1 mediated DCM turnover is associated with DCM tumourigenicity in mice and with mutagenicity in *Salmonella* Ames tester strains [113]. Whether GST T1-1 dependent toxicity of DCM also leads to cancer in humans is still intensely debated [2,23,50,76,80]. Oxidation of DCM by cytochrome P-450 is thought to be the major pathway for DCM metabolism at low levels of DCM exposure, whereas the GST-dependent pathway operates at high concentrations of DCM [50]. Analysis of the toxic effect of DCM is complicated by genetic polymorphism of the *GSTT1* gene in human populations (see [69]), as well as by variation in the induction of the enzyme expression by various chemicals [107] and in different organs and subcellular compartments [114].

These observations were intriguing considering that methylotrophic bacteria expressing DCM dehalogenase/GST grew with DCM without signs of adverse effects. We thus investigated whether mammalian DCM dehalogenases would support growth of *Methylobacterium* with DCM as the sole carbon source. On the basis of the higher reported specific activity of the rat GST T1-1 enzyme with DCM [88], we surmised that cultivation of a transconjugant of strain DM4-2cr expressing the rat *GSTT1* gene on a plasmid would select for mutants of this gene affording superior growth of the transconjugant strain with DCM as the carbon source. Much to our surprise, not only is a DM4-2cr/*rGSTT1* transconjugant unable to grow with DCM, but the presence of DCM is strongly detrimental to growth of the strain with methanol (Figure 5).

The growth-inhibiting effect of DCM conversion by rGST T1-1 was shown to have a genotoxic component: expression of rat GST T1-1 in the background of the *Salmonella typhimurium* TA1535 tester strain, a histidine auxotroph, yields high levels of prototrophic revertants in the presence of DCM, in contrast to expression of bacterial DCM dehalogenase from *Methylophilus* sp. strain DM11 in an otherwise identical strain [46]. This difference in mutagenicity beween rat and bacterial enzymes correlates with a previously overlooked thousand-fold difference in affinity for DCM: the rat enzyme has only negligible affinity for DCM ($K_m \sim 50$ mM) compared to bacterial enzymes ($K_m = 10$-60 µM) [46].

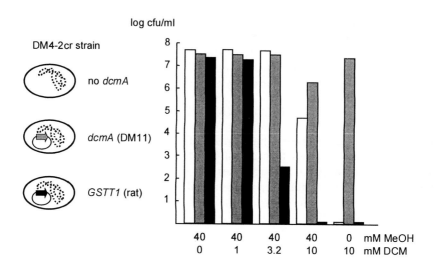

Figure 5. Viability, expressed as colony forming units per milliliter (cfu/ml), of Methylobacterium dichloromethanicum *DM4-2cr and transconjugants in the presence of different concentrations of dichloromethane.* Methylobacterium dichloromethanicum *DM4-2cr lacking the DCM dehalogenase gene (white bars) and the transconjugants DM4-2cr/dcmA (grey) and DM4-2cr/GSTT1 (black) with a plasmid for expression of bacterial DCM dehalogenase from* Methylophilus sp. *strain DM11 or rat liver GST T1, respectively, were plated on minimal medium agar and incubated in gas-tight containers with different concentrations of DCM and 40 mM methanol.*

We then investigated whether the *Methylobacterium dichloromethanicum* DM4-2cr/*rGSTT1* transconjugants that survived treatment with 3 mM DCM (Figure 5) in fact displayed reduced DCM dehalogenase activity [47]. Only two-thirds of the surviving transconjugants still contained the *rGSTT1* gene, and only about a quarter expressed the protein at significant levels judged by SDS-PAGE. Sequence analysis of the *GSTT1* gene in such transconjugants yielded 11 different sequences encoding GST T1 variants, with changes in the protein sequence ranging from single amino acid exchanges to insertions and deletions of 2 to 32 residues. Sequence variation appeared to cluster around the glutathione activation site (Figure 6) which, in a structural model of GST T1 [36], is shielded from the solvent by the C-terminal helix believed to cap the active site of the enzyme [36,37,103].

The GST T1 protein variants that could be expressed and purified as His-tagged proteins in *E. coli* [47] displayed a marked reduction of turnover number to a few percent of the value obtained for wild-type protein with DCM. Interestingly, the affinity of the wild-type rat GST T1-1 enzyme for ENPP was also decreased in the recovered variants, suggesting that the active site of the enzyme is the same for ENPP and DCM despite the extremely low affinity of the enzyme for DCM, and that GST T1 variants with unspecifically decreased catalytic properties were recovered after DCM exposure. Nevertheless, the relative catalytic efficiency of the enzyme variants with ENPP compared to DCM was increased in some cases.

Taken together, these experiments indicate that unlike mammalian enzymes, bacterial DCM dehalogenases are able to convert DCM in a way that minimises the genotoxic effects of DCM conversion. This constitutes an interesting example of two enzymes with related sequences which perform the same enzymatic reaction, but with widely different consequences for the host. The molecular basis for the apparently safer conversion of DCM by bacterial DCM dehalogenases is still unclear. One favoured hypothesis is that the high affinity of bacterial DCM dehalogenases for its substrate DCM also applies for binding of a toxic intermediate in the dehalogenation reaction, presumably *S*-chloromethylglutathione (Figure 2). As discussed in the next section, this compound is a prime suspect for the observed toxic effects of DCM. Alternatively, bacterial DCM dehalogenases may accelerate the decomposition of the toxic intermediate, thus reducing the time during which it can exert its toxic effects.

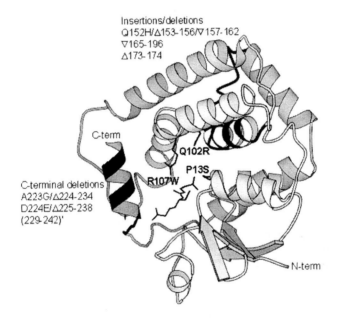

Figure 6. Molscript [73] representation of the structural model of the GST T1 monomer [36] indicating the positions of sequence differences in GST T1 variants. Side chains are indicated at positions where single amino acid exchanges were found, and the protein chain is highlighted in black where insertions and deletions were observed.

3. DNA alkylation, a major problem of glutathione-mediated DCM conversion

In the light of our results on the toxicity of rat GST T1-1 mediated DCM conversion in bacteria, it is clear that bacteria utilising DCM as the sole carbon source not only face the problem of catalysing a chemically unfavourable reaction, but also that of dealing with a potentially lethal transformation. This raises the question as to whether methylotrophic bacteria growing with DCM as the sole carbon source are more resistant to the toxic effects of DCM than other bacteria. It has been assumed by some authors that DCM can be transformed in bacteria such as *Salmonella typhimurium* by uncharacterised glutathione *S*-transferases that may resemble DCM dehalogenases from methylotrophic bacteria [50]. However, no *dcmA* gene could be detected by PCR or hybridisation analysis in the *Salmonella typhimurium* strains used for mutagenicity assessment, nor have such genes yet been detected in any fully sequenced bacterial genome (S.V., unpublished data).

Table 3. Representative studies on dichloromethane genotoxicity in S. typhimurium strains

Strain	GST	[DCM]	Revertants [a] (background)	Remarks	References
TA100	± S9 extract	8.4% v/v (~1.3 M)	1000 (100)	Independent of extracellular addition of S9 fraction from rat liver extract containing GST enzymes; formaldehyde is mutagenic	[49,93]
TA1535	GST T1-1 [b]	2 mM	50 (20)	DCM is mutagenic in a dose dependent manner; formaldehyde is not mutagenic	[113]
TA1535	GST T1-1 [b]	0.13 mM	150 (25)	DCM is only mutagenic in strain TA1535 in the presence of rGST T1-1 enzyme	[96]
TA1535	GST T1-1 [b]	400 ppm	140 (25)	Mutations involve GC/AT transitions only; different mutation spectrum from *S. ty.* TA100	[21]
NM5004	GST T1-1 [c]	20 mM	630 (100) [c]	70% viability at 20 mM DCM	[92]
TA1535	DcmA [d]	20 mM	20 (5)	Expression of bacterial DcmA/GST from *Methylophilus* strain DM11 is less genotoxic than that of rat GST T1	[46]
TA1535	GST T1-1 [d]	20 mM	90 (5)		

[a] His+ revertants per plate over spontaneous background reversion rate
[b] *Salmonella typhimurium* TA1535/rGSTT1 expresses the rat liver GST T1-1 enzyme intracellularly from a plasmid-encoded gene. The TA1535 strain, a *uvrB* mutant incapable of nucleotide excision repair and thus more sensitive to mutagenic lesions, contains the missense base substitution *hisG46* involving a GC base pair. Strain TA100 is identical to strain TA1535 except for the addition of the pKM101 plasmid, which encodes an error prone repair system containing the genes *mucA* and *mucB* [59]. These genes are functionally equivalent to the chromosomal genes *umuD* and *umuC* involved in the SOS response of *E. coli*.
[c] The SOS/*umu* system is based on the induction of *umuC* gene expression by genotoxic agents as part of the SOS response. Gene expression is quantified as β-galactosidase activity arising from the *umuC'-'lacZ* gene fusion. Strain NM 5004 carries the genes *rGSTT1* and *umuC'-'lacZ* on the same plasmid.
[d] The dcmA DCM dehalogenase gene from *Methylophilus* sp. strain DM11 and the rat *GSTT1* gene were provided on the IPTG-inducible plasmid pTrc99A.

A number of studies have been devoted to the mutagenicity of DCM in *Salmonella typhimurium* tester strains (Table 3). The sometimes conflicting results that have been reported have generated some confusion regarding the toxic agent associated with GST-mediated conversion of DCM. Both formaldehyde, the product of GST-mediated dehalogenation of DCM, as well as reactive intermediates resulting from activation of DCM by glutathione, are likely to give rise to DNA adducts and cause toxic and mutagenic effects by impairing DNA replication. Whereas it is usually assumed that DCM activation by glutathione to *S*-chloromethylglutathione (Figure 2) results in a monoalkylated DNA adduct [20,50], formaldehyde leads predominantly to protein-DNA and protein-RNA crosslinks [16]. However, both DCM and formaldehyde exposure may lead to single stranded DNA breaks [48]. One strong argument against formaldehyde being the main causative agent of GST-mediated toxicity of DCM is as follows. Bulky DNA lesions are usually repaired by the Uvr system [104]. Many *Salmonella typhimurium* Ames tester strains, including strains TA1535 and TA100 (Table 3), carry the same *uvrB* mutation which make them more sensitive to DNA alkylation. However, *Salmonella typhimurium* strain TA100 differs from strain TA1535 in that it is contains plasmid pMK101 carrying the *mucAB* genes involved in error-prone DNA excision repair [93]. Thus, strain TA100, unlike strain TA1535, is excision-repair proficient [59]. Because formaldehyde is known to require a proficient nucleotide repair machinery to unfold its mutagenic effects [133], formaldehyde is genotoxic in strain TA100 [93] but not in strain TA1535 (Table 3). Hence, the formaldehyde produced from DCM by GST-mediated conversion cannot be the cause of the observed mutagenic effects in *Salmonella typhimurium* TA1535. Indeed, comparison of the GST-mediated toxicity of DCM turnover by DCM-active GST T1 from rat and by bacterial DCM dehalogenase expressed in *Methylobacterium* or *Salmonella typhimurium* TA1535 showed that expression of the rat enzyme is more toxic and mutagenic to the host in the presence of DCM, despite a lower conversion rate of DCM to formaldehyde [46]. These studies provided a clear indication that a reaction intermediate, rather than DCM or formaldehyde, is the causative agent of glutathione-dependent DCM dehalogenase/GST mediated DCM toxicity. Evidence for this intermediate being *S*-chloromethylglutathione (Figure 2) is still patchy but suggestive. Transient formation of a compound with an ^{19}F-NMR signal compatible with *S*-fluoromethylglutathione was observed when DCM dehalogenase from *Methylophilus* sp. strain DM11 was incubated with glutathione and chlorofluoromethane [8]. *S*-fluoromethylglutathione, a lesser-reactive fluorinated homolog of *S*-chloromethylglutathione, is hydrolysed with a half-life of 5.8 min at room temperature in D_2O [20]. Chemically synthesised *S*-chloromethylglutathione was demonstrated to alkylate deoxyguanosine *in vitro* [20], and the product of this reaction was identified as *S*-[1-N^2-deoxyguanosinyl-methyl]glutathione [20,113]. These data all support the idea that *S*-chloromethylglutathione may exert a toxic effect by alkylating DNA bases, leading to impaired DNA replication and mutagenic lesions. We recently obtained experimental evidence for the formation of DNA adducts during GST-mediated conversion of DCM by purified DCM dehalogenase, as well as for the occurrence of DNA damage during growth of *Methylobacterium dichloromethanicum* DM4 with DCM [65]. Thus, conversion of DCM by DCM dehalogenases/GST most likely leads to DNA damage in the bacterial cell.

4. Searching for accessory genes and proteins of bacterial DCM metabolism

GST-mediated dehalogenation of DCM is likely to impose quite drastic requirements on bacterial metabolism. As discussed above, the product of dehalogenation, formaldehyde, certainly constitutes a toxic threat to the cell, although DNA alkylation by *S*-chloromethylglutathione, the probable intermediate in the dehalogenation of DCM, is likely to have more damaging consequences. However, other aspects of bacterial dehalogenation metabolism have been rather neglected until now and, for a large part, remain to be explored. In particular, high concentrations of protons and chloride ions are steadily generated in the cytoplasm during growth with DCM as the sole carbon source (Figure 7). The solutions adopted by dehalogenating bacteria to face with this situation are unknown. Available experimental data bearing on this topic are discussed in the following.

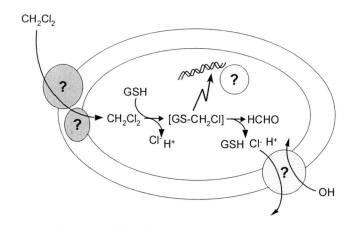

Figure 7. Schematic view of the integration of DCM degradation within cellular metabolism in a DCM-mineralising bacterium such as Methylobacterium dichloromethanicum DM4. DCM dehalogenase catalyses the nucleophilic attack of glutathione onto DCM, yielding the short-lived and toxic glutathione conjugate S-chloromethylglutathione and one equivalent of HCl. Subsequent hydrolysis results in the formation of formaldehyde and of a second molecule of HCl, as well as in the regeneration of glutathione. Other proteins than the DCM dehalogenase may be associated with the dehalogenation process, including enzymes involved in the response to genotoxic effects arising from GST-mediated turnover of DCM (white), in the excretion of chloride ions and the neutralisation of protons produced in the dehalogenation reaction (light grey), as well as membrane-associated proteins involved in the accumulation of growth substrate under growth-limiting conditions (indicated in dark grey).

4.1. BACTERIAL GROWTH WITH DICHLOROMETHANE: A COLLECTION OF OPEN QUESTIONS

The properties of bacterial DCM dehalogenases of course strongly influence bacterial growth with DCM. The DCM dehalogenase from strain DM11 displays a higher turnover rate and a lower affinity for DCM than the enzyme from strain DM4 (k_{cat} of 3.3 s^{-1} vs 0.6 s^{-1} and K_m of 59 µM vs 9 µM [123], respectively). Hence, a bacterium expressing the DM11 enzyme grew better with DCM under conditions of substrate excess than an otherwise identical strain expressing the DM4 enzyme. However, during growth under limiting conditions of DCM as the growth substrate in chemostat experiments, the reverse applied [45]. In addition, the growth rate sustained by strain DM4 under these conditions was dramatically higher than expected from the turnover number and expression level of DCM dehalogenase [45]. This allows one to consider the existence of an accumulation system for DCM operating under strongly limiting conditions of growth substrate. An active transport system would not be unprecedented in methylotrophic bacteria: a high-affinity transport system for short-chain amides and urea, consisting of an outer-membrane porin and a periplasmic amide-urea binding protein, is induced under limiting conditions of these growth substrates in *Methylophilus methylotrophus* [89,90]. As with *Methylobacterium dichloromethanicum* DM4 growing with DCM, the Monod constant K_s of this system (equivalent to the residual growth substrate concentration at half-maximal growth rate) is in the micromolar range, and is lower than the K_m of the catabolic enzyme by at least an order of magnitude.

Given their low turnover number, the expression level of DCM dehalogenases is also an important parameter for bacterial growth with DCM. In strain DM4, expression of the structural DCM dehalogenase gene *dcmA* is controlled by the adjacent, divergently transcribed gene *dcmR* (Figure 3) which, in the absence of DCM, acts as a repressor of gene expression at the level of transcription [75]. DCM dehalogenase expression is strongly induced to about 20% of the total cytoplasmic protein during growth with DCM in shaken batch cultures, and induced to even higher levels under DCM-limited conditions in a chemostat (D. Gisi and S.V., unpublished observations). How this is achieved in detail remains to be understood.

GST-mediated metabolism of DCM is likely to influence cellular metabolism through the many physiological functions of glutathione [6,32,97,122] (see also [108] for a recent selection of key references). Indeed a strain actively dehalogenating DCM by way of a highly expressed DCM dehalogenase probably devotes a significant proportion of its glutathione pool to dehalogenative metabolism: the glutathione content in cell-free extracts of strain DM4 is about 30 nmol/mg protein, while formaldehyde is produced at about 200 nmol/min/mg protein (M. Kayser and S.V., unpublished data). In addition, formaldehyde is present almost entirely as the corresponding glutathione conjugate in solution [116] (K_{diss} 1.5 mM). In *E. coli*, the glutathione conjugates of toxic electrophilic agents such as N-ethylmaleimide, methylglyoxal and 1-chloro-2,4-dinitrobenzene are known to activate KefB and KefC potassium efflux systems, causing acidification of the cytoplasm and affording protection against toxic electrophiles [34]. This emphasises another potential link of glutathione metabolism with DCM dehalogenation, i.e. the likely involvement of glutathione in responding to disturbances in pH and ion balance, recently illustrated by the protection of rhizobial bacteria against acid shock and osmotic stress offered by glutathione [98].

Investigation of pH stress in bacteria until now have mostly addressed the bacterial response to an extracellular pH shock. In the case of GST-mediated DCM conversion, however, acid is produced intracellularly at a high rate, potentially leading to a rapid drop of pH. Given that the buffering capacity of the bacterial cytoplasm is quite limited in the neutral pH range [9], significant adaptations to acid metabolism are likely to take place in DCM-degrading strains. Nothing is yet known about pH homeostasis in bacteria growing with DCM or indeed with any other halogenated carbon source. It is only clear that protons and chloride ions are rapidly excreted into the medium in bacteria performing dehalogenation of DCM. Lower carbon yields are obtained with DCM than with methanol [106], and this may reflect in part the burden faced by methylotrophic bacteria due to constant acidification of the cytoplasm during growth with DCM.

The intracellular production of chloride ions accompanying bacterial dehalogenation metabolism represents another area which remains to be investigated. Anion homeostasis is central to the regulation of turgor and of the potassium pool [86], but the regulation of intracellular chloride is still poorly characterised. A chloride channel has been detected in *E. coli* [81], but whether such a channel can participate in the efflux of this anion is unknown. With the exception of strain DM6, however, the majority of DCM-degrading strains were noted to possess a low tolerance to the presence of sodium chloride in the medium [29].

Knowledge of salt and pH homeostasis is most advanced in *E. coli*. We therefore investigated dehalogenative metabolism in an *E. coli* strain expressing DCM dehalogenase from *Methylophilus* sp. strain DM11, to probe the capacity of *E. coli* cells to cope with the multiple stresses posed by dehalogenation of chlorinated methanes [31]. Addition of 0.3 mM DCM to *E. coli* expressing DCM dehalogenase led to a very rapid lowering of pH_i from its normal value of pH 7.8 to pH 7.4-7.5. It is known that when the cytoplasmic pH falls below pH 7.4, the growth rate falls by 50% [101]. In contrast, growth was completely but transiently inhibited during DCM dehalogenation, and resumed at a lower rate identical to that observed in the presence of 0.3 mM formaldehyde. This demonstrated that the change in the cytoplasmic pH and the formaldehyde generated from DCM were not the sole causes of inhibition, and suggested that, in *E. coli* as well, a reaction intermediate contributes to the toxicity of GST-mediated conversion of DCM.

Evidently, *E. coli* is ill-equipped to cope with the formaldehyde generated from DCM. It is thus worth speculating on which variations of C1 metabolism may be of advantage in methylotrophic DCM-degrading strains. Methylotrophic bacteria have developed several metabolic strategies for the assimilation of carbon and the generation of energy from C1 compounds [4,24]. The DCM-degrading strains that have been isolated (Table 2) cover most of the known pathways for assimilation and mineralisation of C1 compounds. For example, *Methylobacterium dichloromethanicum* DM4 and *Hyphomicrobium* sp. strains DM2 and GJ21 employ an incomplete version of the serine cycle for carbon assimilation, and thus most likely possess two linear multistep pathways involving either folate or methanopterin-dependent C1 interconverting enzymes for mineralisation of C1 compounds from formaldehyde [17] (Figure 8). By contrast, facultatively autotrophic DCM-degrading strains DM10 (29) and DM12 (28) are α-proteobacteria utilising the ribulose-bisphosphate pathway of carbon assimilation, and the β-proteobacterium *Methylophilus* sp. strain DM11 utilises the ribulose monophosphate pathway (27). One might expect that strains with glutathione-dependent enzymes for formaldehyde conversion possess an advantage during GST-mediated growth with halogenated compounds. However, only some of the DCM-degrading strains that were isolated possess enzymes for a glutathione-dependent pathway for formaldehyde oxidation, which has been best characterised in *Paracoccus denitrificans* [120].

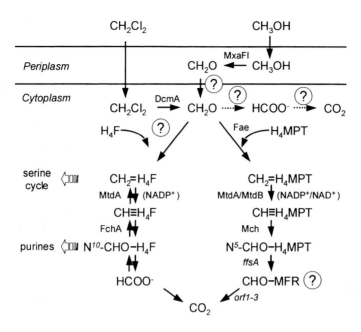

Figure 8. Current model of C1-metabolism in M. dichloromethanicum DM4 (adapted from the known pathways of Methylobacterium extorquens *AM1 ([17] and J. Vorholt, pers. comm.). H₄F, tetrahydrofolate; H₄MPT, dephosphotetrahydromethanopterin; MxaFI, methanol dehydrogenase; DcmA, dichloromethane dehalogenase; Fae, formaldehyde activating enzyme; MtdA, methylene tetrahydrofolate or tetrahydromethanopterin dehydrogenase; MtdB, methylene tetrahydromethanopterin dehydrogenase FchA, methenyl tetrahydrofolate cyclohydrolase; Mch, methenyl tetrahydromethanopterin cyclohydrolase; ffsA, formylmethanofuran tetrahydro-methanopterin formyltransferase; orf1-3, formylmethanofuran dehydrogenase; MFR, methanofuran cofactor, whose presence in* Methylobacterium *remains to be demonstrated. Dashed lines denote a pathway whose relevance for carbon mineralisation in* Methylobacterium *is unknown.*

Finally, it is worth noting that the physiological impact of formaldehyde production may differ depending on whether formaldehyde is produced from methanol or from DCM. In *Methylobacterium*, methanol dehydrogenase (MxaFI) is a periplasmic enzyme, whereas DCM dehalogenase (DcmA) is located in the cytoplasm (Figure 8; [78]). Crucially, a transconjugant of the type strain *Methylobacterium extorquens* AM1 expressing active DCM dehalogenase of *M. dichloromethanicum* DM4 from a plasmid appears unable to grow with DCM as the carbon source. In addition, growth of this AM1 transconjugant with methanol is strongly inhibited in the presence of DCM (M. Kayser and S.V., unpublished data). This is certainly suggestive that genes other than that of DCM dehalogenase are required for growth of methylotrophic bacteria with DCM.

4.2. MINITN5 MUTAGENESIS: SHOOTING IN THE DARK AND HITTING THE LIGHT SWITCH

We have used random mutagenesis to identify genes associated with DCM metabolism in *Methylobacterium dichloromethanicum* DM4 [64]. Plasmid pUTminiTn5*lacZ* encoding a minitransposon element bearing a kanamycin resistance gene [19] was conjugated into strain DM4. This plasmid is unable to replicate in *Methylobacterium* and kanamycin-resistant mutants were obtained upon transposition of the mobile element from the pUT plasmid into the chromosome of strain DM4. Screening of kanamycin resistant transconjugants for lack of growth with DCM as the carbon source yielded several mutants whose growth with DCM was impaired on solid medium, but which nevertheless displayed glutathione and DCM dehalogenase activity levels comparable to the wild-type strain.

Mutant DM4-1445 showed the most clearcut phenotype. This strain was unable to grow with DCM as the sole carbon source in liquid culture and was therefore characterised in detail. The gene disrupted by minitransposon insertion in this mutant is *polA*, encoding DNA polymerase I, an enzyme with well-known functions in DNA repair [40]. The PolA protein sequence from *Methylobacterium* features three consecutive domains corresponding to the 5'-3' exonuclease, proofreading 3'-5' exonuclease and DNA polymerase activities of DNA polymerase I from *E. coli* (the latter two domains constitute the so-called Klenow fragment). Growth of the *polA* mutant with DCM was restored by a plasmid-encoded intact copy of the disrupted gene. In contrast to extracts of wild-type and complemented mutant strains, no DNA polymerase activity was detected in cell-free extracts of the *polA* mutant in SDS-PAGE gels containing nicked DNA after incubation with radiolabelled deoxynucleotides. As observed previously with *polA* mutants of various bacteria (e.g. [51]), mutant DM4-1445 shows a markedly lower resistance to DNA crosslinking by UV light or mitomycin. More specifically, the *polA* mutant from strain DM4 is also sensitive to treatment with DCM (Figure 9). This effect is more pronounced in the presence of DCM during post-challenge incubation on solid medium, probably because of impaired recovery from the initial treatment with DCM during growth (Figure 9B).

The finding that DNA polymerase I is essential for DCM metabolism in *Methylobacterium* provided the first indication, at the physiological level, that GST-mediated conversion of DCM causes extensive DNA damage. It also suggested an important role of the DNA repair machinery in the degradation of halogenated, DNA-alkylating compounds by bacteria. The less pronounced growth defect of another DM4 mutant disrupted in the *uvrA* gene provides additional support for this notion. An obvious role for DNA polymerase I in DCM metabolism would be in nucleotide excision repair of bulky base adducts formed upon reaction of DNA with the probable DCM dehalogenase reaction intermediate *S*-chloromethylglutathione (see Section 3 above), and in preventing stalling of the UvrABC excision nuclease protein complex [104], which presumably recognises such lesions.

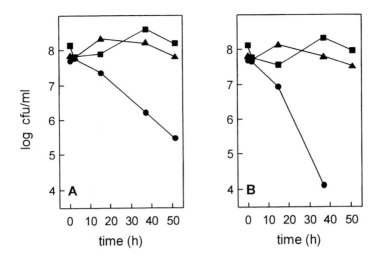

Figure 9. Effect of DCM on the viability of wild-type Methylobacterium dichloromethanicum DM4 and of the polA mutant. M. dichloromethanicum DM4 wild-type (squares), polA mutant DM4-1445 (circles) and mutant DM4-1445 complemented with a plasmid-encoded copy of the wild-type polA gene (triangles) were grown with methanol and treated with 10 mM DCM at $OD_{600}=0.05$. The viability of cultures treated with DCM relative to that of untreated cultures was determined after further growth to $OD_{600}=0.8$ and plating on minimal medium with either 40 mM methanol (A) or a mixture of 40 mM methanol / 10 mM DCM (B) as the carbon source.

More generally, this random mutagenesis approach has provided the first indications that other genes than the DCM dehalogenase gene itself are required for productive dehalogenative metabolism with DCM as the sole carbon source. As well as revealing genes involved in DNA repair, sequence analysis of the disrupted gene in other miniTn5 mutants suggests that anion homeostasis, for instance, is also likely to be associated with this metabolism. In particular, a miniTn5 mutant disrupted in a gene encoding a protein with strong similarity to various bacterial membrane proteins involved in ion transport and osmosensing [132] also shows altered growth properties when sodium chloride is added to the medium (S.V., unpublished data). The metabolic defects in this and other miniTn5 mutants of strain DM4 displaying a growth phenotype with DCM now needs to be characterised in detail.

5. Conclusions and perspectives

Investigations of DCM dehalogenation by methylotrophic bacteria go back more than twenty years. Significant information has been obtained on the bacteria and the corresponding dehalogenases involved. Nevertheless, the molecular basis by which the conserved structural framework of GST is adapted for the degradation of DCM remains elusive. In particular, the marked differences in the kinetic properties and toxicity of DCM dehalogenases/GST from bacteria and mammals are especially intriguing. Hence, crystallisation of bacterial DCM dehalogenases for structural studies is an important next target. On the other hand, the limited sequence variation among DCM dehalogenases from DCM-degrading strains, enrichment cultures and sewage sludge [127] has provided only limited insights into the workings of the enzyme, although this variation may provide a fine-tuning of dehalogenase activity *in vivo* that was undetectable *in vitro*. It is likely that methods for the experimental evolution of enzymes, which have already found applications in studies of GST function [22,54,55], will now be required to understand how subtle changes in the DCM dehalogenase sequence affect the relative fitness of DCM-degrading bacteria under strongly selective conditions.

Nevertheless, studies at the enzyme level have demonstrated that the dehalogenase enzyme itself is not the only factor that is important for productive dehalogenation, but that accessory genes are also involved in this metabolism. The genome sequence of the type strain *Methylobacterium extorquens* AM1, which is currently being determined, and the availability of high-throughput methods for analysis of gene expression [56], will be invaluable in defining which other genes are involved in dehalogenative metabolism, and in evaluating which adaptations have been specifically established by methylotrophic bacteria to grow with dichloromethane as the sole carbon source. Such endeavours should have fruitful implications for the field of bacterial dehalogenation as a whole, especially with respect to the development and application of dehalogenating bacteria for practical purposes.

Acknowledgments

I am grateful to Thomas Leisinger for his unfailing trust and support over the years, and to the Ph. D. and diploma students mentioned in the text who contributed to this work. I also thank Ian Booth and Michael Parker for collaborations, and Yuri Trotsenko, Michael Kertesz and Jan van der Ploeg for comments on the manuscript. Support by ETH Zürich and by grants 5002-037905 (Biotechnology Priority Program) and 3100-050602.97 of the Swiss National Science Foundation is gratefully acknowledged.

References

1. Akashi, H., and Eyre-Walker, A. (1998) Translational selection and molecular evolution, Curr. Op. Genet. Dev. 8, 688-693.
2. Alliance, H.S.I. (1998) White paper on methylene chloride, [online] http://www.hsia.org/white_papers/methchlor.htm (June 1998, last accessed 11-07-01).
3. Anonymous (1993) Facts and figures for the chemical industry, Chem. Eng. News 71, 38-83.
4. Anthony, C. (1982) The biochemistry of methylotrophs, Academic Press, London.
5. Armstrong, R.N. (1997) Structure, catalytic mechanism, and evolution of the glutathione transferases, Chem. Res. Toxicol. 10, 2-18.
6. Åslund, F., and Beckwith, J. (1999) Bridge over trouble waters: sensing stress by disulfide bond formation, Cell 96, 751-753.
7. Bader, R., and Leisinger, T. (1994) Isolation and characterization of the *Methylophilus* sp. strain DM11 gene encoding dichloromethane dehalogenase/glutathione *S*-transferase, J. Bacteriol. 176, 3466-3473.
8. Blocki, F.A., Logan, M.S.P., Baoli, C., and Wackett, L.P. (1994) Reaction of rat liver glutathione *S*-transferases and bacterial dichloromethane dehalogenase with dihalomethanes, J. Biol. Chem. 269, 8826-8830.
9. Booth, I.R. (1985) Regulation of cytoplasmic pH in bacteria, Microbiol. Rev. 49, 359-378.

10. Braus-Stromeyer, S.A., Hermann, R., Cook, A.M., and Leisinger, T. (1993) Dichloromethane as the sole carbon source for an acetogenic mixed culture and isolation of a fermentative, dichloromethane-degrading bacterium, Appl. Environ. Microbiol. 59, 3790-3797.
11. Braus-Stromeyer., S. (1993) Anaerobic degradation of chlorinated methanes. Ph. D. thesis No. 10324, ETH Zürich, Switzerland,.
12. Brunner, W., Staub, D., and Leisinger, T. (1980) Bacterial degradation of dichloromethane, Appl. Environ. Microbiol. 40, 950-958.
13. Byers, H.K., and Sly, L.I. (1993) Toxic effects of dichloromethane on the growth of methanotrophic bacteria, FEMS Microbiol. Ecol. 12, 35-38.
14. Cai, B., Vuilleumier, S., and Wackett, L.P. (1998) Purification and characterization of the mutant enzyme W117Y of the dichloromethane dehalogenase from *Methylophilus* sp. strain DM11, Ann. NY Acad. Sci. 864, 210-213.
15. Casalone, E., Allocati, N., Ceccarelli, I., Masulli, M., Rossjohn, J., Parker, M.W., and Di Ilio, C. (1998) Site-directed mutagenesis of the *Proteus mirabilis* glutathione transferase B1-1 G-site, FEBS Lett. 423, 122-124.
16. Casanova, M., Bell, D.A., and Heck, H.d.A. (1997) Dichloromethane metabolism to formaldehyde and reaction of formaldehyde with nucleic acids in hepatocytes of rodents and humans with and without glutathione *S*-transferase T1 and M1 genes, Fund. Appl. Toxicol. 37, 168-180.
17. Chistoserdova, L., Vorholt, J.A., Thauer, R.K., and Lidstrom, M.E. (1998) C-1 transfer enzymes and coenzymes linking methylotrophic bacteria and methanogenic Archaea, Science 281, 99-102.
18. De Best, J.H., Ultee, J., Hage, A., Doddema, H.J., Janssen, D.B., and Harder, W. (2000) Dichloromethane utilization in a packed-bed reactor in the presence of various electron acceptors, Water Res. 34, 566-574.
19. De Lorenzo, V., and Timmis, K.N. (1994) Analysis and construction of stable phenotypes in Gram-negative bacteria with Tn5- and Tn10-derived minitransposons, Methods Enzymol. 235, 386-405.
20. Dechert, S. (1995) Untersuchungen zum Wirkmechanismus der Mutagenität und Tumorigenität von Dichlormethan und seinen Metaboliten, Ph. D. Thesis, Universität Würzburg, Germany.
21. DeMarini, D.M., Shelton, M.L., Warren, S.H., Ross, T.M., Shim, J.-Y., Richard, A.M., and Pegram, R.A. (1997) Glutathione *S*-transferase-mediated induction of GC - AT transitions by halomethanes in *Salmonella*, Environ. Mol. Mutagen. 30, 440-447.
22. Demartis, S., Huber, A., Viti, F., Lozzi, L., Giovannoni, L., Neri, P., Winter, G., and Neri, D. (1999) A strategy for the isolation of catalytic activities from repertoires of enzymes displayed on phage, J. Mol. Biol. 286, 617-633.
23. Dhillon, S., and Von Burg, R. (1995) Methylene chloride, J. Appl. Toxicol. 15, 329-335.
24. Dijkhuizen, L., Levering, P.R., and de Vries, G.E. (1992) The physiology and biochemistry of aerobic methanol-utilizing Gram-negative and gram-positive bacteria, in J.C. Murrell and H. Dalton (ed.), Methane and methanol utilizers, vol. 5, Plenum Press, New York, pp. 149-181.
25. Diks, R.M., and Ottengraf, S.P. (1991) Verification studies of a simplified model for the removal of dichloromethane from waste gases using a biological trickling filter (part II), Bioprocess. Eng. 6, 131-140.
26. Doronina, N.V., and Trotsenko, Y.A. (1994) *Methylophilus leisingerii* sp. nov., a new species of restricted facultatively methylotrophic bacteria, Microbiology (Russian) 63, 298-302.
27. Doronina, N.V., Braus-Stromeyer, S.A., Leisinger, T., and Trotsenko, Y.A. (1995) Isolation and characterization of a new facultatively methylotrophic bacterium: description of *Methylorhabdus multivorans*, gen. nov., sp. nov., Syst. Appl. Microbiol. 18, 92-98.
28. Doronina, N.V., Trotsenko, Y.A., Krausova, V.I., and Suzina, N.E. (1998) *Paracoccus methylutens* sp. nov. - a new aerobic facultatively methylotrophic bacterium utilizing dichloromethane, Syst. Appl. Microbiol. 21, 230-236.
29. Doronina, N.V., Trotsenko, Y.A., Tourova, T.P., Kuznetzov, B.B., and Leisinger, T. (2000) *Methylopila helvetica* sp. nov. and *Methylobacterium dichloromethanicum* sp. nov. - novel aerobic facultatively methylotrophic bacteria utilizing dichloromethane, Syst. Appl. Microbiol. 23, 210-218.
30. Doronina, N.V., Trotsenko, Y.A., Tourova, T.P., Kuznetsov, B.B., and Leisinger, T. (2001) *Albobacter methylovorans* gen. nov. sp. nov., a novel aerobic, facultatively autotrophic and methylotrophic bacterium that utilizes dichloromethane, Int. J. Syst. Evol. Microbiol. 51, 1051-1058.
31. Evans, G., Ferguson, G.P., Booth, I.R., and Vuilleumier, S. (2000) Growth inhibition of *Escherichia coli* by dichloromethane in cells expressing dichloromethane dehalogenase/glutathione *S*-transferase, Microbiology 146, 2967-2975.
32. Fahey, R.C., and Sundquist, A.R. (1991) Evolution of glutathione metabolism, Adv. Enzymol. Rel. Areas Mol. Biol. 64, 1-53.

33. Felsenstein, J. (1993) PHYLIP (Phylogeny inference package) version 3.5c. Distributed by the author. Department of Genetics, University of Washington, Seattle.
34. Ferguson, G.P. (1999) Protective mechanisms against toxic electrophiles in *Escherichia coli*, Trends Microbiol. 7, 242-247.
35. Fersht, A.R. (1999) Structure and mechanism in protein science, W. H. Freeman, New York.
36. Flanagan, J.U., Rossjohn, J., Parker, M.W., Board, P.G., and Chelvanayagam, G. (1998) A homology model for the human Theta-class glutathione transferase T1-1, Proteins Struct. Funct. Genet. 33, 444-454.
37. Flanagan, J.U., Rossjohn, J., Parker, M.W., Board, P.G., and Chelvanayagam, G. (1999) Mutagenic analysis of conserved arginine residues in and around the novel sulfate binding pocket of the human Theta class glutathione transferase T2-2, Protein Science 8, 2205-2212.
38. Freedman, D.L., and Gossett, J.M. (1991) Biodegradation of dichloromethane and its utilization as a growth substrate under methanogenic conditions, Appl. Environ. Microbiol. 57, 2847-2857.
39. Freedman, D.L., Smith, C.R., and Noguera, D.R. (1997) Dichloromethane biodegradation under nitrate reducing conditions, Water Environ. Res. 69, 115-122.
40. Friedberg, E.C., Walker, G.C., and Siede, W. (1995) DNA repair and mutagenesis, ASM Press, Washington DC.
41. Gälli, R., and Leisinger, T. (1985) Specialized bacterial strains for the removal of dichloromethane from industrial waste, Conservation and Recycling 8, 91-100.
42. Gälli, R. (1987) Biodegradation of dichloromethane in wastewater using a fluidized bed bioreactor, Appl. Microbiol. Biotechnol. 27, 206-213.
43. Gälli, R., and Leisinger, T. (1988) Plasmid analysis and cloning of the dichloromethane-utilization genes of *Methylobacterium sp.* DM4, J. Gen. Microbiol. 134, 943-952.
44. Gillespie, J.H. (1991) The causes of molecular evolution, Oxford University Press, Oxford.
45. Gisi, D., Willi, L., Traber, H., Leisinger, T., and Vuilleumier, S. (1998) Effects of bacterial host and dichloromethane dehalogenase on the competitiveness of methylotrophic bacteria growing with dichloromethane, Appl. Environ. Microbiol. 64, 1194-1202.
46. Gisi, D., Leisinger, T., and Vuilleumier, S. (1999) Enzyme-mediated dichloromethane toxicity and mutagenicity of bacterial and mammalian dichloromethane-active glutathione *S*-transferases, Arch. Toxicol. 73, 71-79.
47. Gisi, D., Maillard, J., Flanagan, J.U., Rossjohn, J., Chelvanayagam, G., Board, P.G., Parker, M.W., Leisinger, T., and Vuilleumier, S. (2001) Dichloromethane mediated *in vivo* selection and functional characterization of rat glutathione *S*-transferase theta 1-1 variants, Eur. J. Biochem. 268 (14), in press.
48. Graves, R.J., and Green, T. (1996) Mouse liver glutathione *S*-transferase mediated metabolism of methylene chloride to a mutagen in the CHO/HPRT assay, Mutat. Res. 367, 143-150.
49. Green, T. (1983) The metabolic activation of dichloromethane and chlorofluoromethane in a bacterial mutation assay using *Salmonella typhimurium*, Mutat. Res. 118, 277-288.
50. Green, T. (1997) Methylene chloride induced mouse liver and lung tumours: an overview of the role of mechanistic studies in human safety assessment, Hum. Exp. Toxicol. 16, 3-13.
51. Gutman, P.D., P., F., and Minton, K.W. (1994) Restoration of the DNA damage resistance of *Deinococcus radiodurans* DNA polymerase mutants by *Escherichia coli* DNA polymerase I and Klenow fragment, Mutat. Res. 314, 87-97.
52. Halden, K., and Chase, H.A. (1991) Methanotrophs for clean-up of polluted aquifers, Water Sci. Technol. 24(11), 9-17.
53. Hanson, D.J. (1996) Toxics release inventory report shows chemical emissions continuing to fall, Chem. Eng. News 74(29), 29-46.
54. Hansson, L.O., Widersten, M., and Mannervik, B. (1999) An approach to optimizing the active site in a glutathione transferase by evolution *in vitro*, Biochem. J. 344, 93-100.
55. Hansson, L.O., Bolton Grob, R., Massoud, T., and Mannervik, B. (1999) Evolution of differential substrate specificities in Mu class glutathione transferases probed by DNA shuffling, J. Mol. Biol. 287, 265-276.
56. Harrington, C.A., Rosenow, C., and Retief, J. (2000) Monitoring gene expression using DNA microarrays, Curr. Op. Microbiol. 3, 285-291.
57. Hartmans, S., and Tramper, J. (1991) Dichloromethane removal from waste gases with a trickle-bed bioreactor, Bioprocess Eng. 6, 83-92.
58. Heraty, L.J., Fuller, M.E., Huang, L., Abrajano, T., and Sturchio, N.C. (1999) Isotopic fractionation of carbon and chlorine by microbial degradation of dichloromethane, Org. Geochem. 30, 793-799.

59. Inman, M.A., Butler, M.A., Connor, T.H., and Matney, T.S. (1983) The effects of excision repair and the plasmid pKM101 on the induction of His$^+$ revertants by chemical agents in *Salmonella typhimurium*, Teratogen. Carcinogen. Mutagen. 3, 491-501.

60. Isidorov, V.A. (1990) Organic chemistry of the earth's atmosphere, Springer Verlag, Berlin.

61. Janssen, D.B., van den Wijngaard, A.J., van der Waarde, J.J., and Oldenhuis, R. (1991) Biochemistry and kinetics of aerobic degradation of chlorinated aliphatic hydrocarbons, in R.E. Hinchee and R.F. Olfenbuttel (ed.), On-site bioreclamation, Butterworth-Heinemann, Boston, pp. 92-112.

62. Jemth, P., and Mannervik, B. (1997) Kinetic characterization of recombinant human glutathione transferase T1-1, a polymorphic detoxication enzyme, Arch. Biochem. Biophys. 348, 247-254.

63. Kästner, M. (1989) Biodegradation of volatile hydrocarbons., Dechema Biotechnology Conferences, vol. 3B, VCH Verlagsgesellschaft, Frankfurt am Main, pp. 909-912.

64. Kayser, M.F., Stumpp, M.T., and Vuilleumier, S. (2000) DNA polymerase I is essential for growth of *Methylobacterium dichloromethanicum* DM4 with dichloromethane, J. Bacteriol. 182, 5433-5439.

65. Kayser, M.F., and Vuilleumier, S. (2001) Dehalogenation of dichloromethane by dichloromethane dehalogenase/glutathione *S*-transferase leads to the formation of DNA adducts, J. Bacteriol. 183 (17), in press.

66. Keene, W.C., Khalil, M.A.K., Erickson, D.J., McCulloch, A., Graedel, T.E., Lobert, J.M., Aucott, M.L., Gong, S.L., Harper, D.B., Kleiman, G., Midgley, P., Moore, R.M., Seuzaret, C., Sturges, W.T., Benkovitz, C.M., Koropalov, V., Barrie, L.A., and Li, Y.F. (1999) Composite global emissions of reactive chlorine from anthropogenic and natural sources: Reactive Chlorine Emissions Inventory, J. Geophys. Res. 104, 8429-8440.

67. Keith, L.H., and Telliard, W.A. (1979) Priority pollutants I - a perspective view, Env. Sci. Technol. 13, 416-423.

68. Klecka, G.M. (1982) Fates and effects of methylene chloride in activated sludge, Appl. Environ. Microbiol. 44, 701-707.

69. Ko, Y., Koch, B., Harth, V., Sachinidis, A., Thier, R., Vetter, H., Bolt, H.M., and Bruning, T. (2000) Rapid analysis of GSTM1, GSTT1 and GSTP1 polymorphisms using real-time polymerase chain reaction, Pharmacogenetics 10, 271-274.

70. Kohler-Staub, D., and Leisinger, T. (1985) Dichloromethane dehalogenase of *Hyphomicrobium sp.* strain DM2, J. Bacteriol. 162, 676-681.

71. Kohler-Staub, D., Hartmans, S., Gälli, R., Suter, F., and Leisinger, T. (1986) Evidence for identical dichloromethane dehalogenases in different methylotrophic bacteria, J. Gen. Microbiol. 132, 2837-2844.

72. Kohler-Staub, D., Frank, S., and Leisinger, T. (1995) Dichloromethane as the sole carbon source for *Hyphomicrobium* sp. strain DM2 under denitrification conditions, Biodegradation 6, 229-235.

73. Kraulis, J.P. (1991) MOLSCRIPT: A program to produce both detailed & schematic plots of protein structures, J. Appl. Crystallogr. 24, 946-950.

74. La Roche, S.D., and Leisinger, T. (1990) Sequence analysis and expression of the bacterial dichloromethane dehalogenase structural gene, a member of the glutathione *S*-transferase supergene family, J. Bacteriol. 172, 164-171.

75. La Roche, S.D., and Leisinger, T. (1991) Identification of *dcmR*, the regulatory gene governing expression of dichloromethane dehalogenase in *Methylobacterium* sp. DM4, J. Bacteriol. 173, 6714-6721.

76. Landi, S. (2000) Mammalian class theta GST and differential susceptibility to carcinogens: a review, Mutat. Res. 463, 247-283.

77. LaPat-Polaska, L.T., McCarty, P.L., and Zehnder, A.J.B. (1984) Secondary substrate utilization of methylene chloride by an isolated strain of *Pseudomonas* sp., Appl. Environ. Microbiol. 47, 825-830.

78. Leisinger, T., Bader, R., Hermann, R., Schmid-Appert, M., and Vuilleumier, S. (1994) Microbes, enzymes and genes involved in dichloromethane utilization, Biodegradation 5, 237-248.

79. Lewontin, R.C. (1989) Inferring the number of evolutionary events from DNA coding sequence differences, Mol. Biol. Evol. 6, 15-32.

80. Liteplo, R.G., Long, G.W., and Meek, M.E. (1998) Relevance of carcinogenicity bioassays in mice in assessing potential health risks associated with exposure to methylene chloride, Hum. Exp. Toxicol. 17, 84-87.

81. Maduke, M., Pheasant, D.J., and Miller, C. (1999) High-level expression, functional reconstitution, and quaternary structure of a prokaryotic ClC-type chloride channel, J. Gen. Physiol. 114, 713-722.

82. Mägli, A., Rainey, F.A., and Leisinger, T. (1995) Acetogenesis from dichloromethane by a two-component mixed culture comprising a novel bacterium, Appl. Environ. Microbiol. 61, 2943-2949.

83. Mägli, A., Wendt, M., and Leisinger, T. (1996) Isolation and characterization of *Dehalobacterium formicoaceticum* gen. nov. sp. nov., a strictly anaerobic bacterium utilizing dichloromethane as source of carbon and energy, Arch. Microbiol. 166, 101-108.

84. Mägli, A., Messmer, M., and Leisinger, T. (1998) Metabolism of dichloromethane by the strict anaerobe *Dehalobacterium formicoaceticum*, Appl. Environ. Microbiol. 64, 646-650.

85. Mckay, D., Shiu, W.Y., and Ma, K.C. (1993) Volatile organic chemicals, vol. 3, Lewis Publishers, Boca Raton.

86. McLaggan, D., Naprstek, J., Buurman, E.T., and Epstein, W. (1994) Interdependence of K^+ and glutamate accumulation during osmotic adaptation of *Escherichia coli*, J. Biol. Chem. 269, 1911-1917.

87. Messmer, M., Reinhardt, S., Wohlfarth, G., and Diekert, G. (1996) Studies on methyl chloride dehalogenase and *O*-demethylase in cell extracts of the homoacetogen strain MC based on a newly de veloped coupled enzyme assay, Arch. Microbiol. 165, 18-25.

88. Meyer, D.J., Coles, B., Pemble, S.E., Gilmore, K.S., Fraser, G.M., and Ketterer, B. (1991) Theta, a new class of glutathione transferases purified from rat and man, Biochem. J. 274, 409-414.

89. Mills, J., Wyborn, N.R., Greenwood, J.A., Williams, S.G., and Jones, C.W. (1997) An outer-membrane porin inducible by short-chain amides and urea in the methylotrophic bacterium *Methylophilus methylotrophus*, Microbiology 143, 2373-2379.

90. Mills, J., Wyborn, N.R., Greenwood, J.A., Williams, S.G., and Jones, C.W. (1998) Characterisation of a binding-protein-dependent, active transport system for short-chain amides and urea in the methylotrophic bacterium *Methylophilus methylotrophus*, Eur. J. Biochem. 251, 45-53.

91. Nicolaidis, A.A., and Sargent, A.W. (1987) Isolation of methane monooxygenase-deficient mutants from *Methylosinus trichosporium* Ob3b using dichloromethane, FEMS Microbiol. Lett. 41, 47-52.

92. Oda, Y., Yamazaki, H., Thier, R., Ketterer, B., Guengerich, F.P., and Shimada, T. (1996) A new *Salmonella typhimurium* NM5004 strain expressing rat glutathione *S*-transferase 5-5: use in detection of genotoxicity of dihaloalkanes using an SOS/*umu* test system, Carcinogenesis 17, 297-302.

93. O'Donovan, M.R., and Mee, M.D. (1993) Formaldehyde is a bacterial mutagen in a range of *Salmonella* and *Escherichia* indicator strains, Mutagenesis 8, 577-581.

94. Oldenhuis, R., Vink, R.L.J.M., Janssen, D.B., and Witholt, B. (1989) Degradation of chlorinated aliphatic hydrocarbons by *Methylosinus trichosporium* Ob3b expressing soluble methane monooxygenase, Appl. Environ. Microbiol. 55, 2819-2826.

95. Osterman-Golkar, S., Hussain, S., Walles, S., Anderstam, B., and Sigvardsson, K. (1983) Chemical reactivity and mutagenicity of some dihalomethanes, Chem.-Biol. Interactions 46, 121-130.

96. Pegram, R.A., Andersen, M.E., Warren, S.H., Ross, T.M., and Claxton, L.D. (1997) Glutathione *S*-transferase-mediated mutagenicity of trihalomethanes in *Salmonella typhimurium*: contrasting results with bromodichloromethane and chloroform, Toxicol. Appl. Pharmacol. 144, 183-188.

97. Penninckx, M.J., and Elskens, M.T. (1993) Metabolism and functions of glutathione in microorganisms, Adv. Microbial Physiol. 34, 239-301.

98. Riccillo, P.M., Muglia, C.I., de Bruijn, F.J., Roe, A.J., Booth, I.R., and Aguilar, O.M. (2000) Glutathione is involved in environmental stress responses in *Rhizobium tropici*, including acid tolerance, J. Bacteriol. 182, 1748-1753.

99. Rittmann, B.E., and McCarty, P.L. (1980) Utilization of dichloromethane by suspended and fixed-film bacteria, Appl. Environ. Microbiol. 39, 1225-1226.

100. Roberts, A.L., Sanborn, P.N., and Gschwend, P.M. (1992) Nucleophilic substitution reactions of dihalomethanes with hydrogen sulfide species, Environ. Sci. Technol. 26, 2263-2274.

101. Roe, A.J., McLaggan, D., Davidson, I., O'Byrne, C., and Booth, I.R. (1998) Perturbation of anion balance during inhibition of growth of *Escherichia coli* by weak acids, J. Bacteriol. 180, 767-772.

102. Rossjohn, J., Polekhina, G., Feil, S.C., Allocati, N., Masulli, M., Di Ilio, C., and Parker, M.W. (1998) A mixed disulfide bond in bacterial glutathione transferase: functional and evolutionary implications, Structure 6, 721-734.

103. Rossjohn, J., McKinstry, W.J., Oakley, A.J., Verger, D., Flanagan, J., Chelvanayagam, G., Tan, K.-L., Board, P.G., and Parker, M.W. (1998) Human Theta class glutathione transferase: the crystal structure reveals a sulfate-binding pocket within the buried active site, Structure 6, 309-322.

104. Sancar, A. (1998) DNA excision repair, Annu. Rev. Biochem. 65, 43-81.

105. Schmid-Appert, M., Zoller, K., Traber, H., Vuilleumier, S., and Leisinger, T. (1997) Association of newly discovered IS elements with the dichloromethane utilization genes of methylotrophic bacteria, Microbiology 143, 2557-2567.

106. Scholtz, R., Wackett, L.P., Egli, C., Cook, A.M., and Leisinger, T. (1988) Dichloromethane dehalogenase with improved catalytic activity isolated from a fast-growing dichloromethane-utilizing bacterium, J. Bacteriol. 170, 5698-5704.
107. Sherratt, P.J., Manson, M.M., Thomson, A.M., Hissink, E.A.M., Neal, G.E., vanBladeren, P.J., Green, T., and Hayes, J.D. (1998) Increased bioactivation of dihaloalkanes in rat liver due to induction of class Theta glutathione S-transferase T1-1, Biochem. J. 365, 619-630.
108. Sies, H. (1999) Glutathione and its role in cellular functions, Free Rad. Biol. Medicine 27, 916-921.
109. Stromeyer, S.A., Winkelbauer, W., Kohler, H., Cook, A.M., and Leisinger, T. (1991) Dichloromethane utilized by an anaerobic mixed culture: acetogenesis and methanogenesis, Biodegradation 2, 129-137.
110. Stucki, G., Gälli, R., Ebersold, H.R., and Leisinger, T. (1981) Dehalogenation of dichloromethane by cell extracts of Hyphomicrobium DM2, Arch. Microbiol. 130, 366-371.
111. Stucki, G. (1990) Biological decomposition of dichloromethane from a chemical process effluent, Biodegradation 1, 221-228.
112. Studer, A., Stupperich, E., Vuilleumier, S., and Leisinger, T. (2001) Chloromethane:tetrahydrofolate methyl transfer by two proteins from Methylobacterium chloromethanicum strain CM4, Eur. J. Biochem. 268, 2931-2938.
113. Thier, R., Taylor, J.B., Pemble, S.E., Humphreys, W.G., Persmark, M., Ketterer, B., and Guengerich, F.P. (1993) Expression of mammalian glutathione S-transferase 5-5 in Salmonella typhimurium TA1535 leads to base-pair mutations upon exposure to dehalomethanes, Proc. Natl. Acad. Sci. USA 90, 8576-8580.
114. Thier, R., Wiebel, F.A., Hinkel, A., Burger, A., Brüning, T., Morgenroth, K., Senge, T., Wilhelm, M., and Schulz, T.G. (1998) Species differences in the glutathione transferase GSTT1-1 activity towards the model substrates methyl chloride and dichloromethane in liver and kidney, Arch. Toxicol. 72, 622-629.
115. Thompson, J.D., G., H.D., and Gibson, T.J. (1994) CLUSTAL W: improving the sensitivity of progressive multiple sequence alignment through sequence weighting, position-specific gap penalties and weight matrix choice, Nucl. Acids Res. 22, 4673-4680.
116. Uotila, L., and Koivusalo, M. (1974) Formaldehyde dehydrogenase from human liver, J. Biol. Chem. 249, 7653-7663.
117. Urhahn, T., and Ballschmiter, K. (1998) Chemistry of the biosynthesis of halogenated methanes: C1-organohalogens as pre-industrial chemical stressors in the environment?, Chemosphere 37, 1017-1032.
118. van Agteren, M.H., Keuning, S., and Janssen, D.B. (1998) Handbook on biodegradation and biological treatment of hazardous organic compounds, Kluwer, Dordrecht.
119. van Hylckama Vlieg, J.E.T., de Koning, W., and Janssen, D.B. (1996) Transformation kinetics of chlorinated ethenes by Methylosinus trichosporium OB3b and detection of unstable epoxides by on-line gas chromatography, Appl. Environ. Microbiol. 62, 3304-3312.
120. van Spanning, R.J.M., de Vries, S., and Harms, N. (2000) Coping with formaldehyde during C1 metabolism of Paracoccus denitrificans, J. Molec. Catal. B 8, 37-50.
121. Vannelli, T., Messmer, M., Studer, A., Vuilleumier, S., and Leisinger, T. (1999) A corrinoid-dependent catabolic pathway for growth of a Methylobacterium strain with chloromethane, Proc. Natl. Acad. Sci. USA 96, 4615-4620.
122. Viña, J.E. (1990) Glutathione: metabolism and physiological functions, CRC Press, Boca Raton.
123. Vuilleumier, S., and Leisinger, T. (1996) Protein engineering studies of dichloromethane dehalogenase/glutathione S-transferase from Methylophilus sp. strain DM11. Ser12 but not Tyr6 is required for enzyme activity, Eur. J. Biochem. 239, 410-417.
124. Vuilleumier, S. (1997) Bacterial glutathione S-transferases: what are they good for?, J. Bacteriol. 179, 1431-1441.
125. Vuilleumier, S., Sorribas, H., and Leisinger, T. (1997) Identification of a novel determinant of glutathione affinity in dichloromethane dehalogenase/glutathione S-transferases, Biochem. Biophys. Res. Commun. 238, 452-456.
126. Vuilleumier, S. (2001) Bacterial dichloromethane dehalogenases and the detoxification of xenobiotics: dehalogenation through glutathione conjugation and beyond, in J.C. Hall, R.E. Hoagland, and R.E. Zablotowicz (ed.), Biotransformations in plants and microorganisms, ACS Symposium Series, vol. 777, Oxford University Press, Oxford, pp. 240-252.
127. Vuilleumier, S., Ivoš, N., Dean, M., and Leisinger, T. (2001) Sequence variation in dichloromethane dehalogenases/glutathione S-transferases, Microbiology 147, 611-616.
128. Whittington, A.T., Vichai, V., Webb, G.C., Baker, R.T., Pearson, W.R., and Board, P.G. (1999) Gene structure, expression and chromosomal localization of murine Theta class glutathione transferase mGSTT1-1, Biochem. J. 337, 141-151.

129. Wilce, M.C.J., Board, P.G., Feil, S.C., and Parker, M.W. (1995) Crystal structure of a theta-class glutathione transferase, EMBO J. 14, 2133-2143.
130. Winkelbauer, W., and Kohler, H. (1991) Biologischer Abbau von Dichlormethan unter anaeroben Bedingungen in einer Aktivkohle-Anlage, Das Gas- und Wasserfach Wasser/Abwasser 132, 425-432.
131. Wohlfarth, G., and Diekert, G. (1997) Anaerobic dehalogenases, Curr. Op. Biotechnol. 8, 290-295.
132. Wood, J.M. (1999) Osomosensing by bacteria: signals and membrane-bases sensors, Microbiol. Molec. Biol. Rev. 63, 230-262.
133. Zijlstra, J.A. (1989) Liquid holding increases mutation induction by formaldehyde and some other cross-linking agents in *Escherichia coli* K12, Mutat. Res. 210, 255-261.
134. Zuber, L. (1995) Trickling filter and three-phase airlift bioreactor for the removal of dichloromethane from air. Ph. D. thesis No. 11202, ETH Zürich, Switzerland.
135. Zuber, L., Dunn, I.J., and Deshusses, M.A. (1997) Comparative scale-up and cost estimation of a biological trickling filter and an airlift reactor for the removal of methylene chloride from polluted air, J. Air Waste Manag. Assoc. 47, 969-975.

MICROBIAL DEGRADATION OF POLLUTANTS AT LOW CONCENTRATIONS AND IN THE PRESENCE OF ALTERNATIVE CARBON SUBSTRATES: EMERGING PATTERNS

THOMAS EGLI
Swiss Federal Institute for Environmental Science and Technology (EAWAG), CH-8600 Dübendorf, Switzerland email: egli@eawag.ch

1. Introduction

Two contrasting approaches are typically followed to study the (bio)degradation of organic chemicals: On one hand, a "real world approach" is used where the disappearance of a compound is followed in the environment (or a soil or water sample brought into the laboratory). On the other hand the degradation is studied "in the ivory tower" by investigating the growth of pure cultures of pollutant-degrading microbes. The first approach gives certainly realistic information about the disappearance of a compound in a particular environment (but usually only limited information about its final fate). However, being a "black box" approach it can give little information about principles and mechanisms that affect and govern the degradation of a compound. The "pure culture" approach again can provide detailed information about types of organisms that can do the job, quantitative aspects of growth and degradation of the chemical, degradation pathways, enzymes involved, intermediates formed, genetic information on the genes involved and their regulation. Definitely both approaches are needed but it is most often difficult, if not impossible, to link information collected in the laboratory with that obtained in the environment.

In this contribution principles of microbial pollutant degradation behaviour now emerging from defined laboratory studies will be reviewed which may help to link the two opposite approaches to study the biodegradation of chemicals. Two aspects will be considered in particular, namely the fact that microorganisms grow in the environment in the presence of a) low concentrations, and b) mixtures of carbon substrates, including pollutants.

S.N. Agathos and W. Reineke (eds.),
Biotechnology for the Environment: Strategy and Fundamentals, 131–139.
© 2002 *Kluwer Academic Publishers. Printed in the Netherlands.*

2. Carbon-limited growth as the key factor simultaneous utilisation of mixtures of carbon substrates

In the laboratory, studies on the degradation of pollutants by pure cultures of heterotrophic microorganisms are usually carried out in batch cultures where the organism is supplied with high concentrations of a particular chemical that serves as the only source of carbon and energy. This contrasts strongly with the conditions microbes experience in the environment. In most ecosystems, heterotrophic microbial growth occurs in a dilute, usually carbon/energy-limited environment (Morita, 1988), where - in addition to the chemical - a multiplicity of other alternative, easily degradable carbon substrates of natural origin are present (systems heavily polluted with carbonaceous compounds, where other factors limit biodegradation, e.g., availability of terminal electron acceptors, nitrogen or phosphorus, will not be considered here). There is now much evidence that under such conditions microorganisms do not specialise on growth with a particular carbon substrate but that they simultaneously take up as many of the different available carbon compounds as possible (Egli, 1995). This behaviour is usually referred to as "mixed substrate growth" (Harder and Dijkhuizen, 1976). However, very little is known with respect to the consequences of the interaction of natural substrates with pollutants, although it has always been assumed that their presence affects pollutant biodegradation in the environment (Alexander, 1994).

Recently, a number of studies done with carbon-limited chemostat cultures have provided information from which general kinetic and physiological patterns can be deduced with respect to the interaction of natural carbon sources with pollutants. Mixed substrate growth with two or more carbon substrates under carbon-limited growth conditions has been demonstrated for numerous combinations of substrate mixtures and microbes (compiled in Egli, 1995). Many of these combinations of substrates are known to provoke diauxic growth patterns when supplied at high concentrations in batch culture, such as the classical mixture or glucose and lactose in *E. coli*. Similarly, the simultaneous utilisation of pollutants together with alternative carbon substrates under carbon-limited growth conditions has been demonstrated for a number of cases. For example, in our laboratory *Chelatobacter heintzii* was shown to simultaneously utilise nitrilotriacetate (NTA) and glucose (Bally and Egli, 1996), *Comamonas testosteroni* used mixtures of acetate plus *p*-toluenesulfonate simultaneously (Tien, 1997), *Methylobacterium* DM4 acetate in combination with dichloromethane (Tien, 1997), or *Escherichia coli* show mixed substrate growth with glucose and 3-phenypropionate (Kovárová et al., 1997). It should be pointed out that these examples include both organisms and pollutants that are phylogenetically and structurally only distantly related. A number of other examples can be found in the literature (see Egli, 1995).

From these mixed substrate studies two aspects are particularly interesting because they are probably generally valid and applicable, namely those concerning the expression of pollutant-degrading enzyme systems under steady-state and dynamic conditions, and those with respect to the kinetics of growth.

3. Expression patterns of pollutant-degrading enzyme systems

Most of the enzyme systems catalysing the breakdown of organic pollutants are inducible. Hence, the question arises as to how these enzyme systems are regulated in an environment where a cell grows with mixtures of pollutants plus alternative carbon substrates and where the availability of both natural carbon substrates and pollutants varies. Experimental data obtained in our laboratory with carbon-limited continuous cultures fed with mixtures of pollutants plus easily degradable carbon substrates suggest that essentially two different regulation patterns for pollutant-degrading enzyme systems can be recognised:

- Enzyme systems that are essentially inducible, but which, even in the absence of the pollutant are, are nevertheless expressed at a (usually low) background level, i.e., the show some degree of derepression under such conditions.

- Enzyme systems that are not expressed (at any detectable level) during growth of the cells in the absence of the pollutant but are dependent on a certain threshold concentration of this compound for expression.

Furthermore, experiments in which the dynamics of the expression pollutant-degrading enzyme systems was studied clearly demonstrated the positive influence of mixed substrate growth on the adaptive response of an organism to substrates that suddenly become available in the environment and to synthesise quickly the proteins necessary for its utilisation.

3.1. DEREPRESSION OF ENZYME SYSTEMS

The derepression of the pollutant-degrading enzyme system during growth in C-limited chemostat cultures with easily degradable substrates (note, in the absence of the pollutant) was observed for NTA-, p-toluenesulfonate-, dichloromethane-, and methanol-utilising microorganisms (Egli et al., 1980; Bally et al., 1994; Tien, 1997). For all these micro-organisms, a low background expression level of the chemical-degrading enzymes was detected when they were cultivated at low dilution rates in a carbon-limited chemostat with an easily degradable carbon source such as glucose or acetate. In the p-toluenesulfonate-degrading *Comamonas testosteroni*, the dichloromethane-utilising *Methylobacterium* DM4 (Tien, 1997) and the two methanol-yeasts *Hansenula polymorpha* and *Kloeckera* sp. 2201 (Egli et al., 1980), the derepression level was in the range of 1-5% of the maximum specific enzyme activities found in fully induced cells. In the NTA-degrader *Chelatobacter heintzii* the background level was close to the detection limit (Bally et al., 1994). Hence, although grown in the absence of the pollutant these cells exhibited a low capacity to utilise the pollutant.

The consequence of this background expression level of pollutant-degrading enzymes was that the cells were able to consume the pollutant immediately, and even when it was added to the growth medium in low amounts. Hence, no apparent threshold concentration for the utilisation of the pollutant was observed. An example is shown in

Table 1 for growth *Chelatobacter heintzii* with different mixtures of glucose and NTA in a continuous culture. Even when the cells were cultivated with mixtures containing only a small fraction of the total carbon in the feed and the key enzyme of NTA-metabolism was not significantly induced, they were able to degrade it together with glucose. As long a NTA contributed to less than 1% of the total carbon consumed no significant induction of NTA-monooxygenase over the derepression level was observed (Tab. 1). When NTA contributed to 1-20% of the simultaneously utilised carbon, NTA-monooxygenase expression increased with increasing proportions of NTA in the feed. It was in this range where the expression of NTA-monooxygenase was most strongly stimulated. Similar enzyme expression and degradation patterns were observed for *C. testosteroni* and *Methylobacterium* DM4 during growth in carbon-limited chemostat culture with mixtures of acetate plus p-toluenesulfonate, or acetate plus dichloromethane, respectively (Tien, 1997). In the case of *Methylobacterium*, dichloromethane was even consumed in the presence of synthetic sewage.

Table 1. Degradation of nitrilotriacetete (NTA) by Chelatobacter heintzii *during growth with glucose/NTA mixtures and regulation of NTA-monooxygenase. Adapted from Bally et al. (1994).*

NTA/glucose conc. in feed		NTA/glucose conc. in the culture	Expression of NTA-monooxygenase
(mg C/L)	% NTA	(mg C/L)	
0 / 727	0	0.017 / <0.01	< 1%
0.262 / 727	0.036	0.012 / <0.01	< 1%
2.62 / 724	0.36	0.009 / <0.01	< 1%
26.2 / 693	3.6	0.012 / <0.01	~ 3%
131 / 569	18	0.027 / <0.01	~ 10%
727 / 0	100	0.050 / <0.01	100%

Cells were cultivated in carbon-limited chemostat culture at a dilution rate of 0.06 h^{-1}. Expression of the NTA-monooxygenase protein was quantified immunologically.

3.2. THRESHOLD SYSTEMS

For the utilisation of the aromatic compound 3-phenylpropionic acid (3ppa) by *E. coli* growing with glucose a second regulation pattern was documented recently in our laboratory (Kovárová et al., 1997a,b). In contrast to the examples discussed above, a culture of *E. coli* cultivated in a glucose-limited chemostat exhibited no detectable 3ppa-degrading activity (Kovárová, 1996). When low concentrations of 3ppa (0.3 and 3.0 mg L^{-1}) were added to the glucose feed medium (S_0 of glucose was 100 mg L^{-1}) no degradation of 3ppa was observed and a perfect wash-in curve for 3ppa in the culture was measured upon addition of the 3ppa to the medium reservoir. Obviously the cells were unable to induce the enzymes necessary to catabolise this compound. However, when the concentration of 3ppa was raised in the feed to 5 mg L^{-1} or more, degradation was induced. The data obtained indicate a threshold concentration in the range of 3 mg L^{-1} for induction of the 3ppa-degrading pathway. Most interestingly, once induced, the culture was able to degrade 3ppa below the threshold concentration required for induction and the degradation of 3ppa continued although the actual concentration of 3ppa in the reactor was always below 1 mg L^{-1}.

3.3. DYNAMICS OF INDUCTION

It is of course of much interest how fast enzymes required for the degradation of a particular pollutant can be induced once this compound becomes available (or, also, how quickly these enzymes will be lost in the absence of the pollutant). Some of these dynamic aspects have been studied recently in our laboratory for the NTA-degrading bacterium *C. heintzii* (Bally and Egli, 1996). When a carbon-limited culture growing at a constant dilution rate with glucose as the only substrate was subjected to a medium shift where all the carbon from glucose was suddenly replaced by NTA the cells required more than 20 hours before they were able to metabolise NTA (Fig. 1). As a consequence NTA accumulated in the culture and almost complete wash-out of the cells was observed during the time before induction started. However, when the culture was shifted to a medium containing 1% glucose plus 99% NTA the time required for induction was reduced from approximately 24 hours to some 10 hours. This surprising positive influence of glucose on the ability to induce NTA-catabolic enzymes was investigated also for mixtures containing higher proportions of glucose. For example, an increase of the proportion of glucose carbon in the feed medium to 90% glucose / 10% NTA reduced the time needed for induction to some 30 min and no accumulation of NTA was observed in the culture. These observations indicate that under carbon-limited conditions the availability of alternative substrates does actually not inhibit but support the induction of other catabolic enzymes, probably by supplying energy and building blocks for the synthesis of the proteins to be newly synthesised.

With respect to the metabolic flexibility it should be pointed out that a general and quantitatively significant derepression of catabolic enzymes under carbon-limited conditions confers to a cell a considerable degree of metabolic freedom and flexibility. It allows exchanging substrates quickly according to their availability in the environment. A good example for such a metabolic flexibility is the case of *E. coli* with respect to sugar utilisation investigated recently in our laboratory (Lendenmann and Egli, 1995). Cells grown in glucose-limited chemostat culture exhibited high levels of mannose-, maltose-, ribose-, or galactose-transporting and catabolising capacity. As a consequence, glucose-growing cells were able to immediately utilise these sugars and replace all glucose carbon by e.g. maltose and to continue growth without apparent lag.

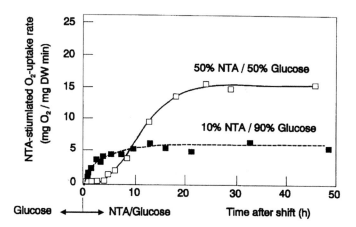

Figure 1. *Dynamics of induction of NTA metabolism in Chelatobacter heintzii ATCC 29600 growing in continuous culture after switching the feed from a medium containing glucose as the only source of carbon and energy to a medium containing either NTA, or to mixtures of glucose plus NTA containing proportions of either 10%, or 50% of NTA. Induction of enzymes involved in the metabolism of NTA was measured via the specific oxygen consumption rate of washed cells exposed to excess NTA in a Clark oxygen electrode [given in μmol O_2 (g dry cell weight x min)$^{-1}$]. Throughout the experiment the dilution rate was kept constant at 0.05 h^{-1}. The total carbon concentration in the feed was 60mM. Adapted from Bally and Egli, 1996).*

4. Kinetics of mixed substrate growth

Evidence for the fact that simultaneous utilisation of carbon substrates has kinetic consequences was obtained in the mid-seventies (reviewed in Kovárová and Egli, 1998). This information has recently been extended and first principles of steady-state mixed substrate kinetics in continuous culture are now firmly established (Lendenmann et al., 1996; Kovárová et al., 1997b).

Lendenmann et al. (1996) investigated steady-state sugar concentrations in carbon-limited chemostat cultures of *E. coli* with defined mixtures of up to six sugars. It was found that at a particular growth (dilution) rate the steady-state concentrations of sugars were consistently lower during the simultaneous utilisation of mixtures of sugars than during growth with single sugars. The steady-state concentrations of individual sugars depended approximately linearly on their contribution to the total sugar consumption rate. An example is shown in Fig. 2a for the growth of *E. coli* with a mixture of three sugars.

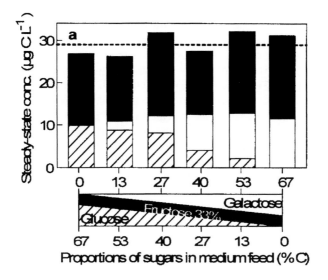

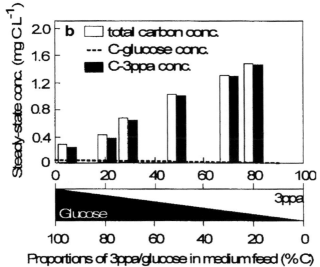

Figure 2. Mixed substrate kinetics during growth of Escherichia coli *in carbon-limited chemostat culture. Adapted from Kovárová and Egli, 1998.*
a) Growth with mixtures of glucose, fructose and galactose at a dilution rate of 0.3 h⁻¹.
b) Growth with mixtures of glucose and 3-phenylpropionic acid (3ppa) at a dilution rate 0.6

This reduction of steady-state substrate concentrations during mixed substrate utilisation has recently also demonstrated for a combination of structurally unrelated substrates feeding into different catabolic pathways (Kovárová et al., 1997b). In this investigation *E. coli* was cultivated in carbon-limited culture with mixtures of glucose plus 3ppa and the steady-state concentrations of the two substrates was measured as a function of the mixture composition (Fig. 2b). During growth with all mixtures the steady-state concentrations of glucose and 3ppa were lower than during growth with either glucose or 3ppa alone. Interestingly, the data indicate that the growth rate of the culture (held constant at D = 0.6 h^{-1}) was controlled by the concentrations of the individual substrates and not by the concentration of the steady-state concentration of total organic carbon. This contrasts from the example given above for the mixed sugar utilisation where the steady-state concentration of total carbon from all sugars remained approximately constant.

Although the number of examples of mixed substrate kinetics are still limited the data presently available indicate that the simultaneous utilisation of mixtures of carbon sources generally results in reduced steady-state concentrations of individual substrates. Considering the biological treatment of pollutants the latter example (Fig. 2) is particularly interesting. The information obtained may be applied for the reduction of outflow concentrations of particular pollutants in industrial wastewater treatment plants by addition of cheap and easily degradable carbon sources.

5. Outlook

The information summarised here highlights the important differences in substrate utilisation patterns, enzyme regulation and kinetics during growth at low environmental concentrations of carbon substrates. Certainly, under such conditions mixed substrate growth is the rule rather than the exception and this behaviour has an enormous influence on the degree of expression of pollutant-catabolising enzyme systems. Although we are only beginning to unravel the patterns that govern the growth behaviour of microorganisms during carbon-limited mixed substrate growth some of the basic principles are now emerging and they indicate a great potential for manipulation of the cells for environmental (and industrial) biotechnology.

6. References

Alexander, M. (1994) Biodegradation and Bioremediation, Academic Press, San Diego.

Bally, M., Wilberg, E., Kühni, M., and Egli, T. (1994) Growth and regulation of enzyme synthesis in the nitrilotriacetic acid (NTA) degrading bacterium *Chelatobacter heintzii* sp. ATCC 29600, Microbiology 140, 1927-1936.

Bally, M., and Egli, T. (1996) Dynamics of substrate consumption and enzyme synthesis in *Chelatobacter heintzii* during growth in carbon-limited continuous culture with different mixtures of glucose and nitrilotriacetate, Applied and Environmental Microbiology 62, 133-140.

Egli, T. (1995) The ecological and physiological significance of microbial growth with mixtures of substrates, Advances in Microbial Ecology 14, 305-386.

Egli, T., van Dijken, J. P., Veenhuis, M., Harder, W., and Fiechter, A. (1980) Methanol metabolism in yeasts: regulation of the synthesis of catabolic enzymes, Archives of Microbiology 124, 115-121.

Harder, W., and Dijkhuizen, L. (1976) Mixed substrate utilization, in A. C. R. Dean, D. C. Ellwood, C. G. T. Evans, and I. Melling (eds.), Continuous Culture 6. Applications and New Fields Ellis Horwood, Chichester, pp. 297-314.

Kovárová, K. (1996) Growth kinetics of Escherichia coli: effect of temperature, mixed substrate utilization, and adaptation to carbon-limited growth, PhD thesis No 11727, Swiss Federal Institute of Technology, Zürich, Switzerland.

Kovárová, K., Käch, A., and Egli, T. (1997a) Cultivation of Escherichia coli with mixtures of 3-phenylpropionic acid and glucose: dynamics of growth and substrate consumption, Biodegradation 7, 445-453.

Kovárová, K., Käch, A., Chaloupka, V., Zehnder, A.J.B., and Egli, T. (1997b) Cultivation of *Escherichia coli* with mixtures of 3-phenylpropionic acid and glucose: steady state growth kinetics, Applied and Environmental Microbiology 63, 2619-2624.

Lendenmann, U., and Egli, T. (1995) Is *Escherichia coli* growing in glucose-limited chemostat culture able to utilize other sugars without lag? Microbiology 141, 71-78.

Lendenmann, U., Snozzi, M., and Egli, T. (1996) Kinetics of simultaneous utilization of sugar mixtures by Escherichia coli in continuous culture, Applied and Environmental Microbiology 62, 1493-1499.

Morita, R. Y. (1988) Bioavailability of energy and its relationship to growth and starvation survival in nature, Journal of Canadian Microbiology 43, 436-441.

Tien, A. (1997) The physiology of a defined four membered mixed bacterial culture during continuous cultivation with mixtures of three pollutants in synthetic sewage, PhD thesis No 11905, Swiss Federal Institute of Technology, Zürich, Switzerland

TOWARDS A BETTER UNDERSTANDING OF ENHANCED PESTICIDE BIODEGRADATION

SÉBASTIEN J. GOUX[1,2], SPIROS N. AGATHOS[2] AND LUC D. PUSSEMIER[1]

[1] *Veterinary and Agrochemical Research Centre, Leuvensesteenweg 17, B-3080 Tervuren, Belgium Phone: + 32 2 769 22 47; Fax: + 32 2 769 23 05; e-mail: L. Pussemier@terv. var. fgov. be.* [2] *Unit of Bioengineering, Catholic University of Louvain, Place Croix du Sud 2/19, B-1348 Louvain-la-Neuve, Belgium*

Abstract

The biodegradation of three pesticides (2,4-D, atrazine, and carbofuran) by soil microorganisms and its genetic and enzymatic basis are reviewed. For each pesticide, two types of biodegradation are reported to occur in soils. Nonspecific degradation is mainly due to fungi. Some metabolic pathways only lead to partial biodegradation. Complete mineralisation pathways are elucidated for 2,4-D and atrazine. Genetic bases of 2,4-D mineralisation and carbofuran metabolism are diverse whereas atrazine catabolic genes are, until now, reported to be remarkably conserved. Catabolic genes encoding a given biodegradation pathway are most generally carried on plasmids and often located close to each other. New hypotheses grounded on the current knowledge of pesticide biodegradation are proposed regarding the mechanisms and the maintenance of enhanced pesticide biodegradation in the soil environment. New approaches to improve soil bioremediation via inoculation of xenobiotic degraders are also suggested.

1. Introduction

Since their introduction in the late forties, pesticides have been extensively used and remain key-components in modern agriculture. Their success led some of them to their loss. Besides the well-known phenomenon of resistance, another natural adaptation may occur to significantly reduce the efficacy of soil-applied pesticides: enhanced biodegradation. Two chlorinated herbicides will be especially presented and compared. 2,4-D (2,4-dichlorophenoxyacetic acid), is a relatively easily metabolised xenobiotic. Even microflora from non-previously treated soils were shown to rapidly metabolise 2,4-D. On the contrary atrazine, and especially its *s*-triazine ring, was considered to be

141

S.N. Agathos and W. Reineke (eds.),
Biotechnology for the Environment: Strategy and Fundamentals, 141–156.
© 2002 *Kluwer Academic Publishers. Printed in the Netherlands.*

recalcitrant to biodegradation. However, during the last decade, numerous reports indicated a rapid mineralisation of atrazine and this phenomenon appears nowadays to be widespread. Carbofuran, a carbamate soil-applied pesticide, is another compound for which soil microflora has been shown to rapidly develop rapid degradation abilities, but this time mainly through hydrolytic cleavage of the N-methyl carbamate moiety.

In this paper, genetic studies that provides a better understanding of the mechanisms leading to enhanced biodegradation are presented. These results are then discussed from an evolutionary point of view and an approach of the bioremediation process focused on the fate of plasmids is suggested.

2. A phenoxyacetic acid: 2,4-D

2. 1. INTRODUCTION

In 1950, Audus isolated the first soil bacteria able to degrade 2,4-D. This was the ultimate proof that pesticides could be subject to biological degradation in the soil environment. Since this time, 2,4-D has been considered as a model xenobiotic. As a consequence, its degradation pathway(s), the microorganisms involved as well as the genetic factors supporting its catabolism were extensively studied. Although major advances in the understanding of 2,4-D were achieved in the previous decade, numerous additional breakthroughs were realised recently and underline the variety of mechanisms that may appear in the xenobiotic biodegradation by soil microorganisms.

2. 2. FUNGAL DEGRADATION

Although fungi are generally considered to play an important role in xenobiotic degradation, their capacity to degrade 2,4-D is poorly documented. Donnelly et al. [23] investigated the 2,4-D biodegradation properties of 10 fungal strains. Effective 2,4-D mineralisation (30% in 8 weeks) was only recorded with Phanerochaete chrysosporium 1767 whereas less than 1% mineralisation was recorded within the same time for the other tested strains.

2. 3. METABOLIC PATHWAY ENCODED BY THE TFD GENES

Ralstonia eutrophia (frequently reported as Alcaligenes eutrophus) JMP134 (pJP4) is able to use 2,4-D as sole carbon and energy source. Its metabolic pathway encoded by the tfd genes carried by the plasmid pJP4 is the most extensively studied and is reported in figure 1. The three first steps lead to the opening of the aromatic ring. 2,4-D is firstly transformed to 2,4-dichlorophenol by a α-ketoglutarate dependent dioxygenase, TfdA [28]. The aromatic ring is then hydroxylated in position 2 by the phenol hydroxylase TfdB to form 3,5-dichlorocatéchol. A single operon with the tfdCDEF genes encodes for the formation of 2,4-dichloromuconate, the first aliphatic metabolite, and further metabolism to β-ketoadipate which is then metabolised by chromosome-encoded enzymes [73]. The uptake of 2,4-D has been shown to be an active process, the uptake system being inducible and encoded by another gene located on plasmid pJP4, *tfdK* [42].

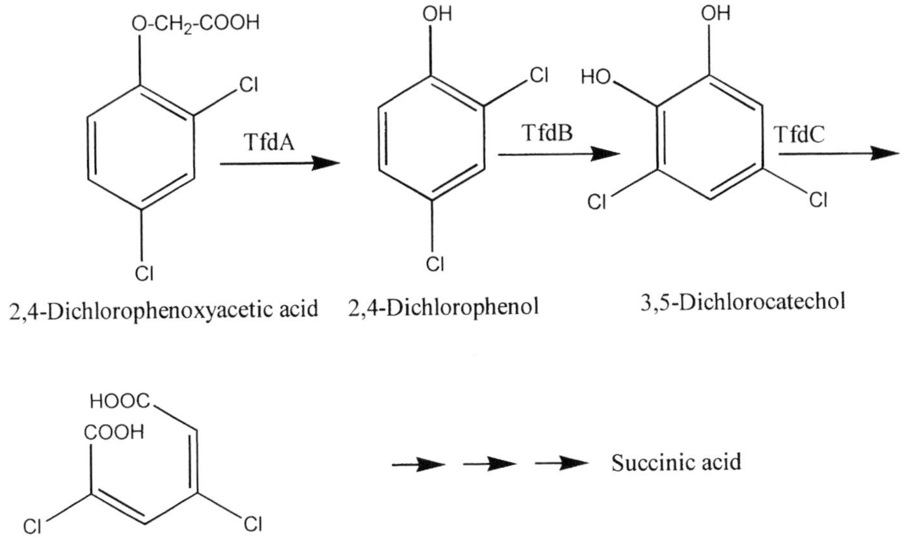

| 2,4-Dichlorophenoxyacetic acid | 2,4-Dichlorophenol | 3,5-Dichlorocatechol |

3,5-Dichloromuconate

Figure 1 Methabolic pathway of 2,4,D by Ralstonia eutrophia JMP (pJP4)

Different genera of microorganisms able to degrade 2,4-D have been isolated. The similarity of the degrading genes was found to be linked to the phylogeny of the degrading microorganisms [29, 36]. While looking for numerically dominant 2,4-D degraders, Ka *et al.* [36] observed that 57% of their strains were belonging to one of the three following genera: *Sphingomonas*, *Pseudomonas*, or *Alcaligenes*. *Sphingomonas* was the most numerous isolated genus with 38% of the total isolates. This group is reported to carry the most dissimilar genes with only in some cases some similarity to the *tfdB* gene but not to *tfdA* and *tfdC* whereas *Pseudomonas* and *Alcaligenes* species generally carry genes similar to *tfd-A*, *-B*, and *-C*. *Burkholderia* strains generally exhibited genes highly similar to *tfdB* and *tfdC* but only weakly similar to *tfdA*. *Rhodoferax fermentans* strains exhibited as a whole very good similarity with *tfdA* but weak similarities to *tfd-B* and *-C* [29]. A wealth of 2,4-D degraders that showed similarity neither to the *tfd* genes nor to *Spa* (a gene that has been designed for the detection of other 2,4-D degrading microorganisms [37]) was also obtained by Ka *et al.* [36]. The high diversity of 2,4-D degrading genes implies a variety of processes in gene recruitment and pathway assembly [29]. Working directly on two soils that were only different regarding their 2,4-D exposure (no prior exposure or repeated exposure for 42 years), Holben *et al.* [32] observed the same 2,4-D population in both soils. The numbering of the *tfd-A* and *-B* genes were in good agreement with MPN (Most Probable

Number) estimation of the degraders whereas none of the *tfd-C*, *-D*, *-E*, and *-F* genes could be detected. Several non-*tfd* sequences of pJP4 were also detected, suggesting common plasmid "backbone" features. 2,4-D degrading bacteria were also isolated from pristine environments. One isolate was identified as a *Variovorax* while the others were close to the *Bradyrhizobium* group [39]. Only the *Variovorax* isolate carried the *tfdA* gene and all the isolates were characterised by very slow growth rates and developed only on poor (10%) media. The genes encoding 2,4-D degradation appear thus to be linked to the phylogeny of the degrading organism.

The chromosomal *tftA1* and *tftA2* genes encode for both degradation of 2,4-D and 2,4,5-T (hydroquinone pathway) in *Burkholderia* (previously identified as *Pseudomonas) cepacia* AC1100 [31]. Although exhibiting similar functional and mechanistic properties than TfdA [17], no extensive similarity was found with this enzyme, but well with dioxygenase systems encoding for benzoate and toluate catabolism [18].

In the last decade, many reports have pointed out that 2,4-D metabolism could be coded by other genes than the *tfd* genes and that different plasmids could also be involved.

2. 4. INVOLVEMENT OF MOBILE ELEMENTS IN 2,4-D BIODEGRADATION

In the 80-kb plasmid pJP4 of *Ralstonia eutrophus* JMP134, the *tfd* genes encoding the 2,4-D catabolic pathway are contiguous and in the order *ACDEFB*. Two regulatory genes (*tfdR* and *tfdS*), a gene coding for an uptake system of the herbicide (*tfdK*), as well as copies of *tfdA* and *tfdC* are also carried by this plasmid. Inoculation of *Ralstonia eutrophus* JMP134 in a nonsterile soil amended with 2,4-D resulted in the spreading of pJP4 throughout the indigenous soil microflora [22].

Chaudhry and Huang [14] reported a 45-kb degradative plasmid (pRC10) hosted by a *Flavobacterium* sp. (strain 5001). Besides 2,4-D metabolism, this strain shared numerous properties with *R. eutrophus*: metabolism of 3-chlorobenzoate (3-CBA) and 2-4-methylchlorophenoxyacetic acid (MCPA) as well as $HgCl_2$ resistance. It was also resistant to several antibiotics as tetracycline, kanamycin, and ampicillin but these properties were not encoded on the pRC10 plasmid. Interestingly, similarities of pRC10 and pJP4 were limited to the regions encoding *tfd* genes but not with pJP4 regions that harbour the conjugative, incompatibility, replication, and maintenance functions.

The capacity of utilisation of 2,4-D as sole carbon and energy source in *Pseudomonas cepacia* CSV90 was due to a 90-kb plasmid, pMAB1 [7]. The *tfdCD* genes were located on a 20-kb fragment of this plasmid that could be naturally deleted from pMAB1 while cultivating CSV90 on a nonselective medium. Although it is highly probable that pMAB1 carries a *tfdCDEF* operon similar to that of pJP4, its physical map did not exhibit other similar regions to pJP4 [7]. Ka *et al.* [35] reported four different degradative plasmids. pBS5 and pKO51 both carried the four *tfd-A*, *-B*, *-C*, *-D* genes. pKA4 was harboured from *Pseudomonas pikettii* 712 and only carried *tfdA*. Finally, pBS3 is a large plasmid from *Sphingomonas paucimobilis* 1443 on which none of the *tfd* genes were detected. Interestingly, these plasmids and the interactions with their hosts were found to confer different competitiveness (determined by the lag time after inoculation and the growth rate) in axenic cultures and in soils [35]. In axenic culture, strains carrying pJP4 had the highest growth rates, whereas pKA4 confered the least competitiveness.

A highly similar plasmid to 40. 9-kb pKA4, 42. 9-kb pKA2 was isolated from an *Alcaligenes paradoxus* 2811P isolated from the same agricultural soil [38]. These two plasmids had a very high genetic similarity, they only hybridised with *tfdA* and shared similar functional traits as high self-transmissibility. They are thought to result from a natural horizontal gene transfer. Integration of pKA2 to the host chromosome was observed without loss of the 2,4-D degradation activity.

The genotypic evolution of the large (200-kb) plasmid pTFD41 and of its host *Ralstonia* sp. strain TFD41 was monitored by Nakatsu *et al.* [52]. Duplication of a region carrying the *tfdA* gene was observed in all the independently propagated populations. This region was found to be flanked by IS elements. Other rearrangements within the plasmid are also reported but they were much less frequent and did not involve DNA fragments carrying *tfd* genes.

Seven different plasmids were isolated from soil by complementation [66]. As a general rule, the isolated plasmids demonstrated high homologies to *tfdA*, moderate degrees of homology with *tfdB* and low degree of homology with *tfdC* of pJP4. Only one plasmid exhibited high homologies to all the catabolic genes of pJP4, but tfdA and tfdR were located much closer to the *CDEFB* operon than in pJP4. In another plasmid, a highly homologous *tfdA* was mapped before a moderately homologous *tfdB* and a weakly homologous *tfdC* while weakly homologous *tfd-D* and *-F* were not mapped and that homology to *tfdE* was not detectable. These genes were thus not in the same order than in pJP4.

3. A *s*-triazine: atrazine

3. 1. INTRODUCTION

For a long time, atrazine has been considered as a moderately persistent compound. The removal of the chloro-substituent was exclusively attributed to abiotic processes whereas the biotic degradation was mainly attributed to fungi and limited to N-dealkylation. The biodegradation of its *s*-triazine ring was considered to be very slow and bound-residue formation an important way of dissipation [77].

It is only forty years after its introduction on the market that rapid atrazine degradation by soil microorganisms was observed. This rapid biodegradation appeared to be coupled with a new catabolic pathway leading to extensive mineralisation of the *s*-triazine ring. It can be assumed that the outbreak of this accelerated biodegradation in agricultural fields occurred nearly simultaneously in different countries from different continents [2, 33, 55, 57, 68, 75].

3. 2. THE N-DEALKYLATION DEAD-END CYTOCHROME P450 PATHWAY

Fungal degradation of atrazine appears to be limited to the side-chains of the *s*-triazine ring. In liquid culture, *Phanerochaete chrysosporium* only partially biotransformed atrazine, producing mainly CIAT but CEAT, OEIT (figure 2) as well as unidentified very polar compound(s) were also detected by Mougin *et al.* [47]. This activity was attributed to a P_{450} monooxygenase [46]. Masaphy *et al.* [44, 45] reported different results while working with *Pleurotus pulmonaris*, another white rot fungus. Besides N-dealkylation (leading to CIAT, CEAT and CAAT), these authors did not detect OEIT but well another hydroxylated metabolite: Hydroxyisopropylatrazine [2-chloro-4-ethylamino-6-(1-hydroxyisopropyl)amino-1,3,5-triazine]. The formation of this latter compound is attributed to unspecific oxidation of the alkyl chains, the incorporation of a single oxygen atom into the C2-position leading to unstable compounds and dealkylation whereas oxidation at the methyl group level gives rise to a stable hydroxylated-isopropyl chain. Here also, it was demonstrated that the atrazine biotransformation was due to the cytochrome P_{450} activity. It must be pointed out that fungal biodegradation studies of atrazine were always carried out with species that were not isolated or enriched for this purpose. So that their action is to be attributed to their wide-spectrum enzymes that they produce "naturally". Different actinomycetes were also reported to be able to catabolise the substituents of the atrazine ring. A *Streptomyces* strain was reported to transform atrazine as well as 11 other herbicides. Atrazine metabolism by *Rhodoccocus* strains has been extensively studied by Behki and co-workers [3-6, 64] and Nagy and co-workers [50, 51]. Interestingly, the *Rhodococcus* strain TE1 (firstly reported as an *Arthrobacter* sp.) was initially isolated thanks to enrichment grounded on EPTC metabolism as sole carbon source [64]. Furthermore, atrazine N-dealkylation (but also the other *s*-triazine propazine, simazine, and cyanazine) was linked to the 77-kb plasmid that was required for EPTC degradation [5]. In this study, 33% of the initial atrazine was transformed to unidentified compound(s) that did not belong to the generally reported metabolites (CAAT, OIET, OIAT, OEAT, and OAAT). Another *Rhodococcus* sp. (strain N186/21) exhibited the same metabolism regarding EPTC and atrazine [50, 51]. It was established that the same cytochrome P_{450} system was involved in both strains [60] and that hydroxyisopropylatrazine was one of the metabolite formed [50]. This cytochrome P_{450} system was shown to cause de degradation of N-methylcarbamate insecticides (carbofuran, propoxur, and carbaryl) [6] as well as organophosphorous insecticides.

Although not directly effective on atrazine, another member of the *Rhodococcus* group (*Rhodococcus corralinus* NRRL B-15444R) was shown to be able to carry out dechlorination and deamination on both deisopropylatrazine and deethylatrazine (CEAT and CIAT) [15]. This time, the enzymatic activity appeared to be much more specific and due to an *s*-triazine hydrolase [48]. This inducible TrzA enzyme was inhibited by atrazine and simazine.

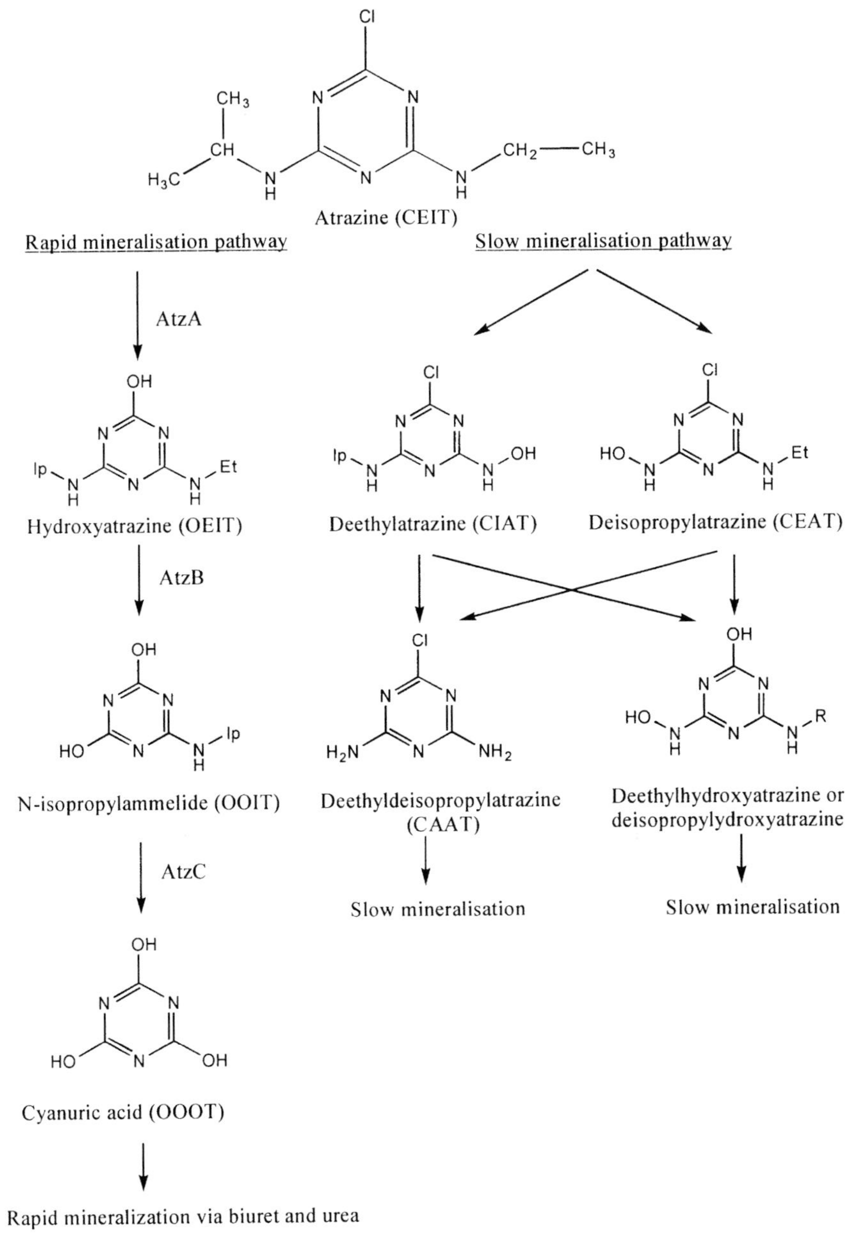

Figure 2. Different metabolic pathways of atrazine (Ip = isopropyl, Et = ethyl, R = isopropyl or ethyl).

3. 3. THE MINERALISATION PATHWAY

It is only recently that bacterial isolates able to mineralise atrazine were reported [43, 56, 62, 78]. All the successful isolations were obtained while using atrazine as sole N source in the enrichment process. The three first steps of atrazine mineralisation (Figure 2) by *Pseudomonas* ADP were identified by the team of Wackett [10, 19, 59]. AtzA, the first enzyme of the pathway, converts atrazine to hydroxyatrazine via hydrolytic dechlorination [19]. Hydroxyatrazine transformation is then carried out by two amidases (AtzB and AtzC) to successively produce N-isopropylammelide [10] and cyanuric acid [59], the central metabolite to all *s*-triazine mineralisation pathways reported. Surprisingly, this pathway differed from the catabolic reactions previously reported. Firstly, N-dealkylation was previously shown to be a prerequisite to dechlorination [3, 15]. Secondly, the two N-alkyl substituents undergo deamidation and not N-dealkylation. Thirdly, the three first reactions are here of hydrolytic nature whereas oxidative process were generally reported. It is also remarkable to quote that this is the most direct way to cyanuric acid, the metabolite allowing the mineralisation of the *s*-triazine ring [16, 34]. This new pathway is followed by diverse atrazine mineralising microorganisms belonging to different genera (*Pseudomonas*, *Ralstonia*, *Alcaligenes*, *Agrobacterium*, and *Clavibacter*) and that shared highly conserved degradative genes [20]. A protein similar to AtzA was also reported from a *Rhizobium* sp. isolate capable of atrazine dechlorination [11]. A comparison of a 24 amino-acid sequence, demonstrated 92% sequence identity with AtzA. This observation was confirmed by the detection of the *atzA* gene within a group of *Rhizobium* sp. (loti) that mineralised atrazine [70]. However, this gene was not detected in *Nocardioides* sp. strains [70] despite the fact that these strains mineralised also atrazine via hydroxyatrazine.

The degradation pathway of cyanuric acid by a *Pseudomonas* sp. has been described by Cook *et al.* [16]. The same pathway involving a similar cyanuric acid amidohydrolase (TrzD) was shown in another *Pseudomonas* strain [25] and in a *Klebsellia* strain [41]. TrzD has been recently sequenced by Karns [40]. Cyanuric acid is transformed in biuret, an easily metabolisable compound that is degraded in urea and ultimately to carbon dioxide and ammonia. The *s*-triazine ring is thus essentially a N-source since the C atoms of the cycle are already at their maximum oxidation stage. The same pathway is thought to be followed by the atrazine mineralisation bacteria since biuret and urea production by *Ralstonia* M91 was reported by Radosevich *et al.* [56].

3. 4. INVOLVEMENT OF MOBILE ELEMENTS IN ATRAZINE BIODEGRADATION

The three genes *atzABC* coding for atrazine degradation have been shown to be carry by a sole 96-kb self-transmissible plasmid (pADP-1) in *Pseudomonas* ADP [20]. However, the genes encoding for the end of the pathway (cyanuric acid catabolism) were not located on pADP-1. Another plasmid detected in the same isolate (pADP-2) did not contain any of the three *atz* genes. A group of isolated bacteria catabolising atrazine via hydroxyatrazine was also reported to carry a plasmid of approximately 97-kb but the role of this plasmid and its link to pADP-1 remain to be established [69]. Similarly to the *atz* genes, the *trzCDE* genes coding for the three amidohydrolases that catabolise ammelide to urea are located on a large Inclα plasmid (> 114-kb) in *Klebsiella pneumoniae* [41]. Since the Inclα plasmids can not replicate in *Pseudomonas* and that the *trzC* and *trzD* genes were reported in two strains of this group [25], the dissemination of these genes is likely to be due to their location on other plasmids.

The likelihood of IS sequences around the genes coding for *s*-triazine degradation was first reported by Eaton and Karns [25]. These authors mapped a structure having the properties of a transposon since a cluster of the three *trz* genes were bordered by a repeated 2. 2-kb sequence in inverted orientation. The mapping of the surroundings of the *trzCD* cluster from two other strains did not exhibit such IS elements. The sequencing of upstream sequences of *atz-A* and *-B* did not lead to the detection of IS elements [10]. However, the presence of a conserved sequence upstream of these two genes for more than 600 nucleotides coupled to the detection of a truncated *pdhB* gene suggests genetic rearrangement [10]. Topp *et al.* [70] detected in several strains carrying *atzB* the presence of IS1071.

4. A carbamate pesticide: carbofuran

Carbofuran is a soil-applied pesticide that was used extensively to control insects and nematodes in potato, sugar beet, corn, rice, and other crops. Although it has been used since the seventies, first indications of enhanced carbofuran dissipation were only observed in the mid-eighties and were attributed to repeated soil treatments [63].

The different catabolic pathways of carbofuran are presented in figure 3. Two different hydrolytic enzymes degrading carbofuran to carbofuran-phenol and methylamine have been reported. The first hydrolase was purified from an *Achromobacter* sp. strain WM111 [21] and the *mcd* gene encoding this enzyme is localised on the >100-kb plasmid pDL11 [65]. The second carbamate hydrolase was produced by the *Pseudomonas* strain CRL-OK [49]. Both enzymes also transformed carbaryl and aldicarb, but the *Pseudomonas* enzyme exhibited a somewhat narrower substrate range. The metabolite methylamine appear to be the primary substrate that can be used as nitrogen or carbon source by a number of degrading *Pseudomonas* or *Flavobacterium* isolates [12]. A methylotrophic bacterium was found to contain a 120-kb plasmid that was very similar to pDL11 and that carried the *mcd* gene [67]. The *mcd* gene was not detected on the second plasmid (130-kb) of this strain. The striking similarity of two plasmids encoding carbofuran metabolism, although those were contained by different strains originating from different geographical area, suggests that this genetic arrangement may be a widespread conserved unit in soils exhibiting rapid carbofuran hydrolysis. This was further confirmed by the detection of very similar plasmids in 23 geographically and phenotypically different soil bacteria that contained very similar plasmids to pDL11 and sequences homologous to *mcd* [54]. However, Parekh *et al.* [53] found that 60% of their carbofuran-degrading isolates did not have genes homologous to *mcd*. It seems thus that the *mcd*-like genes are structurally associated with pDL11-like plasmids but that at least one different gene may encode carbofuran hydrolysis.

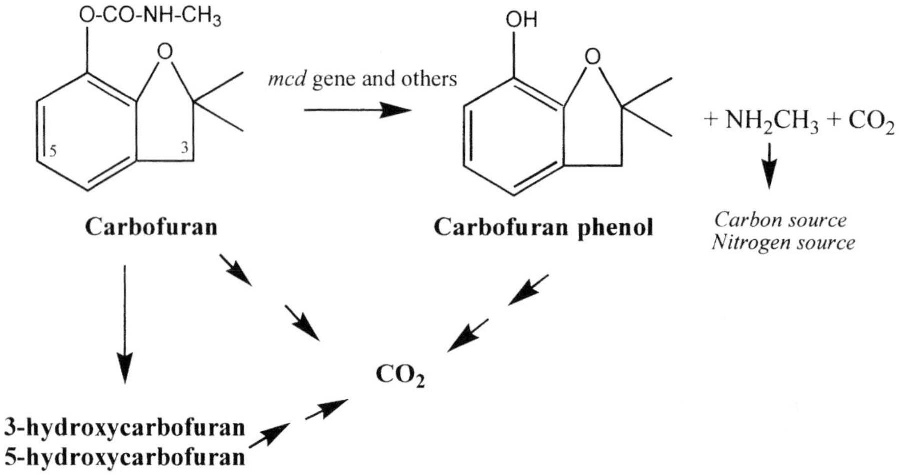

Figure 3. Microbial degradation of carbofuran

Carbofuran-phenol is more recalcitrant to biodegradation and its catabolism is less often reported. However, mineralisation of the carbofuran ring by a *Pseudomonas* sp. strain was reported by Chaudhry and Ali [12]. These authors suggest that mineralisation to CO_2 is carried out via an oxidative pathway. Strains able to mineralise atrazine were only obtained from a waste disposal site whereas other isolates were obtained from previously treated agricultural soils. The higher selective pressure in the heavily contaminated soil could be the origin of the recruitment of genes allowing more rapid and more extensive biodegradation. On the contrary, the carbofuran degrading character appeared to be more "diluted" and less efficient in the microflora of the agricultural soils [12]. In contrast to the *Pseudomonas* sp. strain, a mineralising *Arthrobacter* sp. strain catabolises carbofuran via carbofuran-phenol [58]. This metabolite was transiently detected during carbofuran degradation and the strain was also able to metabolise carbofuran-phenol as main substrate. More recently, carbofuran mineralisation by *Sphingomonas* sp. strain CF06 was reported [27]. This strain contains five plasmids on which the degrading phenotype is encoded. Out of the five plasmids, two pairs have large regions of similarity but none of them hybridized with the *mcd* gene. Furthermore, the authors were not able to obtain cured strains with less plasmids in order to localise more accurately the plasmid(s) carrying the gene of interest. The presence of extensive duplication regions between plasmids as well as of several active IS elements suggests that this system may still evolve towards a more efficient and better organised system [27].

Rhodococcus TE1 was found to cometabolise carbofuran as well as two other N-methylcarbamates: propoxur and carbaryl [6]. The end-product of carbofuran transformation was 5-hydroxycarbofuran. Thus, the insecticide could not serve as a carbon or a nitrogen source. The gene encoding this transformation was located on the same 77-kb plasmid that is involved in the biodegradation of EPTC (a thiocarbamate) and atrazine (an *s*-triazine). To our knowledge, no attempts to form consortia cultures with the simultaneous presence of both bacteria producing 7-phenol carbofuran and 5-hydroxycarbofuran were done. Since substitution of aromatic rings by hydroxy groups generally ease mineralisation, the production of 5-hydroxy-7-phenol carbofuran might allow to go further in the metabolism of the carbofuran ring.

5. Discussion

Soil biodegradation of xenobiotics may be nonspecific or, on the contrary, highly substrate-dedicated. Nonspecific soil metabolism is mainly carried out by fungi, a typical example being *Phanerochaete chrysosporium*. This microorganism is famous for its production of broad-range enzymes able to transform many recalcitrant compounds. But nonspecific systems enabling the metabolism of a wide range of pesticides from different families are also reported in the bacteria domain. For example, the single *Rhodococcus* strain TE1 could transform compounds from four different pesticide families: *s*-triazine and thiocarbamate herbicides, as well as N-methylcarbamate and organophosphorous insecticides [6]. In these two microorganisms, peculiar cytochrome-P_{450} systems are responsible to the catabolic activity. These transformations are most generally limited to the side-chains of aromatic compounds and extended mineralisation of the whole compound is thus not achieved. However, the end-products often present features (as hydroxy groups) that result in a large drop in biological activity and that are more prone to form bound-residues, a major alternative to mineralisation for soil pesticide detoxification [8, 9]. In Sweden, this approach was largely supported throughout installations of biobeds in numerous farms [71]. These set-ups aim to prevent spillage of pesticides during filling and rinsing spraying equipment thanks to a mixture of soil, peat mould and straw designed to optimise the balance between adsorption and degradation [72].

Dedicated enzymatic systems for the degradation of xenobiotics are most often reported from gram-negative bacteria. This is the case for the metabolism of the three pesticides that have been more extensively reviewed in this paper. In the three cases, *Pseudomonas* strains able to extensively mineralise the parent-compound have been isolated. In each case, they were found to possess extremely efficient and dedicated plasmid-encoded enzymatic systems. Genes coding for the degradation of atrazine did not exhibit large similarities with any other genes. However, limited conserved sequences suggest that they arose from a purine or pyrimidine catabolic genes [59, 76]. Similar sequences upstream of *atz-A* and *-B* also suggest recent duplication and rearrangements. While studying the still limited number of strains [5] able to mineralise atrazine, it was found that the three genes were highly conserved. This is in contrast with 2,4-D for which many variants to the *tfd* genes have been reported. Furthermore, the *tfd* genes are less unique than the *atz genes.* Similar operonic structures than *tfdCDEF* (*clcABD* and *tcbCDEF*) were found to be involved in the degradation of other chlorinated compounds (3-chlorobenzoic acid and 1,2,4-trichlorobenzene, respectively) for which the metabolic pathway passes via a chlorocatechol [74]. Furthermore, hybridisation studies suggest that the plasmids carrying these genes arose from a common ancestor [13].

For each of the three pesticides reviewed, metabolic pathways are reported. In the case of atrazine, it seems that the capacity of hydrolytic dechlorination was a key-step to open a new and rapid pathway for mineralisation. For 2,4-D, no such key-step seemed to be needed since the opening of the benzene cycle is possible even with two chloro substituents.

The last aromatic intermediate in the biodegradation pathway of 2,4-D is 3,5 dichlorocatechol. The substitution of the benzenoic ring by two hydroxyls is a common treats of many biodegradation pathways of chlorinated aromatic compounds [13]. This is not the case for the *s*-triazine herbicides. For these compounds, cyanuric acid is the central metabolite that precedes ring cleavage, hydroxyls being in positions 1, 3, and 5. Indeed, the three nitrogen atoms present within the ring structure do not allow ring substitution in position 2,4,6 without loss of aromaticity.

For the isolation of degrading soil microorganisms, a prerequisite is their culturability. The number of soil microorganisms able to grow in laboratory conditions is generally estimated to be between 0. 1 and 1% of the total diversity. The use of 16S-rDNA based techniques allowing to detect microorganisms without pre-cultivation led to the detection of a wealth of new species. Some of them appear to be widely spread as, for example, microorganisms belonging to the *Acidobacterium* kingdom [1]. In a synthetic table on the biodegradation of chlorinated aromatic compounds, 50% of the reported strains were belonging to the *Pseudomonas* genus [13]. *Sphingomonas* strains were isolated while studying the degradation of 2,4-D and Carbofuran. In both cases, catabolic enzymes of these strains strongly diverged from the well characterised and more spread enzymatic systems. In order to be isolated as a xenobiotic degrader by the conventional techniques, a given microorganism should possess several properties. First, it should easily acquire the catabolic phenotype while submitted to high selective pressure. The origin of catabolic genes is still vastly unknown. However, either the recruitment of existing and dormant gene or mutations leading to a new gene from a remote ancestor require intensive genetic rearrangements and instability. Plasmids are the most indicated location where such rearrangements may occur without causing too much cell perturbation. These mobile elements have been shown to be ubiquitous in environmental bacteria [30]. Thus, bacteria carrying plasmids or being good plasmid recipients are prone to either directly develop catabolic phenotypes or to readily acquire them from other strains. To be isolated, the degrading strains must also support cultivation conditions and be able to grow rapidly under these conditions. Direct plating allowed to observe a much wider diversity of 2,4-D degrading genes than a more conventional liquid enrichment process [24]. In a different way, El Fantroussi *et al.* [26] presented some evidence that strains able to degrade linuron in liquid culture did not survive while plated on convential media. Since plasmids may affect the growth rate of their hosts, a kind of symbiosis between the degrading microorganism and the catabolic plasmid(s) must exist. Many authors did already insist on the several drawbacks linked to the cultivation of microorganisms in vitro. Those were still underlined by the arising of molecular techniques. It is thus hard to pretend and to evaluate the relevance of the isolated microorganisms in the environment. However, the selection of strains that are only competitive *in* vitro is a necessary step for numerous biodegradation applications.

The fact that pesticide catabolic genes are plasmid-borne may lead to some speculations regarding soil ecology. Plasmid-borne characters are known to be less stable than chromosome-encoded characters. In rich cultivation conditions, plasmid losses are very frequently observed. However, enhanced pesticide biodegradation has been shown in some cases to be maintained for several years. The appearance of the phenomenon itself is only due to annual xenobiotic application. Hypothesising that microorganisms with short generation times are more prone to lose their mobile elements than slow growing microorganisms, catabolic plasmids conservation may well be ensured by these latest microorganisms. A new pesticide application, would first promote these microorganisms but also the transfer of the catabolic plasmids to more competitive strains that would rapidly develop and ensure the most of the biodegradation process.

The localisation of the catabolic genes on plasmids may also be an incentive to modify the approach of environmental bioremediation. It is no more the fate of a degrading strain that has to be optimised but well the fate of the catabolic genes and thus of the catabolic plasmid. In this case, genetic engineering should focus on the cloning of the catabolic genes on one or several plasmids that are susceptible to spread into the well-adapted endogenous microflora. Self-transmissible, broad-host range plasmids that do not impede the development of their hosts appear thus to be ideal shuttles to ensure efficient bioremediation.

References

1. Barns, S. M. , S. L. Takala, and C. R. Kuske. 1999. Wide distribution and diversity of members of the bacterial kingdom *Acidobacterium* in the environment. Appl. Environ. Microbiol. 65:1731-1737.
2. Barriuso, E. , and S. Houot. 1996. Rapid mineralization of the *s*-triazine ring of atrazine in soils in relation to soil management. Soil Biol. Biochem. 28:1341-1348.
3. Behki, R. M. , and S. U. Khan. 1986. Degradation of atrazine by *Pseudomonas*: N-dealkylation and dehalogenation of atrazine and its metabolites. J. Agric. Food Chem. 34:746-749.
4. Behki, R. M. , and S. U. Khan. 1994. Degradation of atrazine, propazine, and simazine by *Rhodococcus* strain B-30. J. Agric. Food Chem. 42:1237-1241.
5. Behki, R. M. , E. Topp, W. Dick, and P. Germon. 1993. Metabolism of the herbicide atrazine by *Rhodococcus* strains. Appl. Environ. Microbiol. 59:1955-1959.
6. Behki, R. M. , E. E. Topp, and B. A. Blackwell. 1994. Ring hydroxylation of N-methylcarbamate insecticides by *Rhodococcus* TE1. J. Agric. Food Chem. 42:1375-1378.
7. Bhat, M. A. , M. Tsuda, K. Horiike, M. Nozaki, C. S. Vaidyanathan, and T. Nakazawa. 1994. Identification and characterization of a new plasmid carrying genes for degradation of 2,4-dichlorophenoxyacetate from *Pseudomonas cepacia* CSV90. Appl. Environ. Microbiol. 60:307-312.
8. Bollag, J. -M. 1992. Decontaminating soil with enzymes. Environ. Sci. Technol. 26:1876-1881.
9. Bollag, J. -M. 1999. Immobilization of pesticides in soil through enzymatic reactions. 9th European Congress on Biotechnology Proceedings (on CD-ROM) ISBN80521 (C 1999-2000) Ed. M. Hofman Branche Belge de la Société de Chimie Industrielle Brussels -Belgium
10. Boundy-Mills, K. L. , M. L. de Souza, R. T. Mandelbaum, L. P. Wackett, and M. J. Sadowsky. 1997. The *atzB* gene of *Pseudomonas* sp strain ADP encodes the second enzyme of a novel atrazine degradation pathway. Appl. Environ. Microbiol. 63:916-923.
11. Bouquard, C. , J. Ouazzani, J. C. Prome, Y. Michel-Briand, and P. Plesiat. 1997. Dechlorination of atrazine by a *Rhizobium* sp. isolate. Appl. Environ. Microbiol. 63:862-866.
12. Chaudhry, G. R. , and A. N. Ali. 1988. Bacterial metabolism of carbofuran. Appl. Environ. Microbiol. 54:1414-1419.
13. Chaudhry, G. R. , and S. Chapalamadugu. 1991. Biodegradation of halogenated organic compounds. Microbiol. Rev. 55:59-79.
14. Chaudhry, G. R. , and G. H. Huang. 1988. Isolation and characterization of a new plasmid from a *Flavobacterium* sp. which carries the genes for degradation of 2,4- dichlorophenoxyacetate. J. Bacteriol. 170:3897-3902.
15. Cook, A. H. , and R. Hütter. 1984. Deethylsimazine: bacterial dechlorination, deamination, and complete degradation. J. Agric. Food Chem. 32:581-585.
16. Cook, A. M. , P. Beilstein, H. Grossenbacher, and R. Hütter. 1985. Ring cleavage and degradative pathway of cyanuric acid in bacteria. Biochem. J. 231:25-30.

17. Danganan, C. E. , S. Shankar, R. W. Ye, and A. M. Chakrabarty. 1995. Substrate diversity and expression of the 2,4,5-trichlorophenoxyacetic acid oxygenase from *Burkholderia cepacia* AC1100. Appl. Environ. Microbiol. 61:4500-4504.

18. Danganan, C. E. , R. W. Ye, D. L. Daubaras, L. Xun, and A. M. Chakrabarty. 1994. Nucleotide sequence and functional analysis of the genes encoding 2,4,5- trichlorophenoxyacetic acid oxygenase in *Pseudomonas cepacia* AC1100. Appl. Environ. Microbiol. 60:4100-4106.

19. de Souza, M. L. , M. J. Sadowsky, and L. P. Wackett. 1996. Atrazine chlorohydrolase from *Pseudomonas* sp. strain ADP: Gene sequence, enzyme purification, and protein characterization. J. Bacteriol. 178:4894-4900.

20. de Souza, M. L. , J. Seffernick, B. Martinez, M. J. Sadowsky, and L. P. Wackett. 1998. The atrazine catabolism genes *atzABC* are widespread and highly conserved. J. Bacteriol. 180:1951-1954.

21. Derbyshire, M. K. , J. S. Karns, P. C. Kearney, and J. O. Nelson. 1987. Purification and characterization of an N-methylcarbamate pesticide hydrolizing enzyme. J. Agric. Food Chem. 35:871-877.

22. DiGiovanni, G. D. , J. W. Neilson, I. L. Pepper, and N. A. Sinclair. 1996. Gene transfer of *Alcaligenes eutrophus* JMP134 plasmid pJP4 to indigenous soil recipients. Appl. Environ. Microbiol. 62:2521-2526.

23. Donnelly, P. K. , J. A. Entry, and D. L. Crawford. 1993. Degradation of atrazine and 2,4-dichlorophenoxyacetic acid by mycorrhizal fungi at three nitrogen concentrations in vitro. Appl. Environ. Microbiol. 59:2642-2647.

24. Dunbar, J. , S. White, and L. Forney. 1997. Genetic diversity through the looking glass: effect of enrichment bias. Appl. Environ. Microbiol. 63:1326-1331.

25. Eaton, R. W. , and J. S. Karns. 1991. Cloning and comparison of the DNA encoding ammelide aminohydrolase and cyanuric acid amidohydrolase from three *s*-triazine degrading bacterial strains. J. Bacteriol. 173:1363-1366.

26. El Fantroussi, S. , L. Verschuere, W. Verstraete, and E. M. Top. 1999. Effect of phenylurea herbicides on soil microbial communities. 9th European Congress on Biotechnology Proceedings (on CD-ROM) ISBN80521 (C 1999-2000) Ed. M. Hofman Branche Belge de la Société de Chimie Industrielle Brussels -Belgium

27. Feng, X. , L. T. Ou, and A. Ogram. 1997. Plasmid-mediated mineralization of carbofuran by *Sphingomonas* sp. strain CF06. Appl. Environ. Microbiol. 63:1332-1337.

28. Fukumori, F. , and R. P. Hausinger. 1993. *Alcaligenes eutrophus* JMP134 "2,4-dichlorophenoxyacetate monooxygenase" is an α-ketoglutarate-dependent dioxygenase. J. Bacteriol. 175:2083-2086.

29. Fulthorpe, R. R. , C. McGowan, O. V. Maltseva, W. E. Holben, and J. M. Tiedje. 1995. 2,4-Dichlorophenoxyacetic acid-degrading bacteria contain mosaics of catabolic genes. Appl. Environ. Microbiol. 61:3274-3281.

30. Götz, A. , R. Pukall, E. Smit, E. Tietze, R. Prager, H. Tschäpe, J. D. van Elsas, and K. Smalla. 1996. Detection and characterization of broad-host-range plasmids in environmental bacteria by PCR. Appl. Environ. Microbiol. 62:2621-2628.

31. Haugland, R. A. , D. J. Schlemm, R. P. d. Lyons, P. R. Sferra, and A. M. Chakrabarty. 1990. Degradation of the chlorinated phenoxyacetate herbicides 2,4- dichlorophenoxyacetic acid and 2,4,5-trichlorophenoxyacetic acid by pure and mixed bacterial cultures. Appl. Environ. Microbiol. 56:1357-1362.

32. Holben, W. E. , B. M. Schroeter, V. G. Calabrese, R. H. Olsen, J. K. Kukor, V. O. Biederbeck, A. E. Smith, and J. M. Tiedje. 1992. Gene probe analysis of soil microbial populations selected by amendment with 2,4-dichlorophenoxyacetic acid. Appl. Environ. Microbiol. 58:3941-3948.

33. Issa, S. , M. Wood, L. Pussemier, V. Vanderheyden, C. Douka, S. Vizantinopoulos, Z. Gyori, M. Borbely, and J. Katai. 1997. Potential dissipation of atrazine in the soil unsaturated zone: A comparative study in four European countries. Pestic. Sci. 50:99-103.

34. Jutzi, K. , A. M. Cook, and R. Hütter. 1982. The degradative pathway of the *s*-triazine melamine. Biochem. J. 208:679-684.

35. Ka, J. O. , W. E. Holben, and J. M. Tiedje. 1994. Analysis of competition in soil among 2,4-dichlorophenoxyacetic acid- degrading bacteria. Appl. Environ. Microbiol. 60:1121-1128.

36. Ka, J. O. , W. E. Holben, and J. M. Tiedje. 1994. Genetic and phenotypic diversity of 2,4-dichlorophenoxyacetic acid (2,4- D)-degrading bacteria isolated from 2,4-D-treated field soils. Appl. Environ. Microbiol. 60:1106-1115.

37. Ka, J. O. , W. E. Holben, and J. M. Tiedje. 1994. Use of gene probes to aid in recovery and identification of functionally dominant 2,4-dichlorophenoxyacetic acid-degrading populations in soil. Appl. Environ. Microbiol. 60:1116-1120.

38. Ka, J. O. , and J. M. Tiedje. 1994. Integration and excision of a 2,4-dichlorophenoxyacetic acid-degradative plasmid in *Alcaligenes paradoxus* and evidence of its natural intergeneric transfer. J. Bacteriol. 176:5284-5289.
39. Kamagata, Y. , R. R. Fulthorpe, K. Tamura, H. Takami, L. J. Forney, and J. M. Tiedje. 1997. Pristine environments harbor a new group of oligotrophic 2,4- dichlorophenoxyacetic acid-degrading bacteria. Appl. Environ. Microbiol. 63:2266-2272.
40. Karns, J. S. 1999. Gene Sequence and Properties of an *s*-Triazine Ring-Cleavage Enzyme from *Pseudomonas* sp. Strain NRRLB-12227. Appl. Environ. Microbiol. 65:3512-3517.
41. Karns, J. S. , and R. W. Eaton. 1997. Genes encoding *s*-triazine degradation are plasmid-borne in *Klebsellia pneumoniae* strain 99. J. Agric. Food Chem. 45:1017-1022.
42. Leveau, J. H. , A. J. Zehnder, and J. R. van der Meer. 1998. The *tfdK* gene product facilitates uptake of 2,4-dichlorophenoxyacetate by *Ralstonia eutropha* JMP134(pJP4). J. Bacteriol. 180:2237-2243.
43. Mandelbaum, R. T. , D. L. Allan, and L. P. Wackett. 1995. Isolation and characterization of a *Pseudomonas* sp. that mineralizes the *s*-triazine herbicide atrazine. Appl. Environ. Microbiol. 61:1451-1457.
44. Masaphy, S. , Y. Henis, and D. Levanon. 1993. Isolation and characterization of a novel atrazine metabolite produced by the fungus *Pleurotus pulmonaris*, 2-chloro-4-ethylamino-6-(1-hydroxyisopropyl)amino-1,3,5-triazine. Appl. Environ. Microbiol. 59:4342-4346.
45. Masaphy, S. , Y. Henis, and D. Levanon. 1996. Manganese-enhanced biotransformation of atrazine by the white rot fungus *Pleurotus pulmonarius* and its correlation with oxidation activity. Appl. Environ. Microbiol. 62:3587-3593.
46. Mougin, C. , C. Laugero, M. Asther, and V. Chaplain. 1997. Biotransformation of *s*-triazine herbicides and related degradation products in liquid cultures by the white rot fungus *Phanerochaete chrysosporium*. Pestic. Sci. 49:169-177.
47. Mougin, C. , C. Laugero, M. Asther, J. Dubroca, P. Frasse, and M. Asther. 1994. Biotransformation of the herbicide atrazine by the white rot fungus *Phanerochaete chrysosporium*. Appl. Environ. Microbiol. 60:705-708.
48. Mulbry, W. W. 1994. Purification and characterization of an inducible *s*-triazine hydrolase from *Rhodococcus corallinus* NRRL B-15444R. Appl. Environ. Microbiol. 60:613-618.
49. Mulbry, W. W. , and R. W. Eaton. 1991. Purification and characterization of the N-methylcarbamate hydrolase from *Pseudomonas* strain CRL-OK. Appl. Environ. Microbiol. 57:3679-3682.
50. Nagy, I. , F. Compernolle, K. Ghys, J. Vanderleyden, and R. De Mot. 1995. A single cytochrome P-450 system is involved in degradation of the herbicides EPTC (S-ethyl dipropylthiocarbamate) and atrazine by *Rhodococcus* sp. strain NI86/21. Appl. Environ. Microbiol. 61:2056-2060.
51. Nagy, I. , G. Schoofs, F. Compernolle, P. Proost, J. Vanderleyden, and R. De Mot. 1995. Degradation of the thiocarbamate herbicide EPTC (*s*-ethyl-dipropylcarbamothioate) and biosafening by *Rhodococcus* sp. strain N186/21 involve an inducible cytochrome P-450 system and aldehyde dehydrogenase. J. Bacteriol. 177:676-687.
52. Nakatsu, C. H. , R. Korona, R. E. Lenski, F. J. de Bruijn, T. L. Marsh, and L. J. Forney. 1998. Parallel and divergent genotypic evolution in experimental populations of *Ralstonia* sp. J. Bacteriol. 180:4325-4331.
53. Parekh, N. R. , A. Hartmann, M. -P. Charnay, and J. -C. Fournier. 1995. Diversity of carbofuran-degrading soil bacteria and detection of plasmid encoded sequences homologous to the *mcd* gene. FEMS Microbiol. Ecol. 17:149-160.
54. Parekh, N. R. , A. Hartmann, and J. -C. Fournier. 1996. PCR detection of the *mcd* gene and evidence of sequence homology between the degradative genes and plasmids from diverse carbofuran-degrading bacteria. Soil Biol. Biochem. 28:1797-1804.
55. Pussemier, L. , S. Goux, V. Vanderheyden, P. Debongnie, I. Tresinie, and G. Foucart. 1997. Rapid dissipation of atrazine in soils taken from various maize fields. Weed Res. 37:171-179.
56. Radosevich, M. , S. J. Traina, Y. L. Hao, and O. H. Tuovinen. 1995. Degradation and mineralization of atrazine by a soil bacterial isolate. Appl. Environ. Microbiol. 61:297-302.
57. Radosevich, M. , S. J. Traina, and O. H. Tuovinen. 1996. Biodegradation of atrazine in surface soils and subsurface sediments collected from an agricultural research farm. Biodegradation. 7:137-149.
58. Ramanand, K. , M. Sharmila, and N. Sethunathan. 1988. Mineralization of carbofuran by a soil bacterium. Appl. Environ. Microbiol. 54:2129-2133.
59. Sadowsky, M. J. , Z. Tong, M. de Souza, and L. P. Wackett. 1998. AtzC is a new member of the amidohydrolase protein superfamily and is homologous to other atrazine-metabolizing enzymes. J. Bacteriol. 180:152-158.

60. Shao, Z. Q. , and R. Behki. 1996. Characterization of the expression of the *thcB* gene, coding for a pesticide-degrading cytochrome P-450 in *Rhodococcus* strains. Appl. Environ. Microbiol. 62:403-407.

61. Shapir, N. , S. Goux, R. T. Mandelbaum, and L. Pussemier. The potential of soil microorganisms to mineralize atrazine by MCH-PCR followed by nested-PCR. Can. J. Microbiol. 46: 425-432.

62. Struthers, J. K. , K. Jayachandran, and T. B. Moorman. 1998. Biodegradation of atrazine by *Agrobacterium radiobacter* J14a and use of this strain in bioremediation of contaminated soil. Appl. Environ. Microbiol. 64(9):3368-3375.

63. Suett, D. L. , J. -C. Fournier, E. Papadopoulou-Mourkidou, L. Pussemier, and J. Smelt. 1996. Accelerated degradation: the European dimension. Soil Biol. Biochem. 28:1741-1748.

64. Tam, A. C. , R. M. Behki, and S. U. Khan. 1987. Isolation and characterization of an *s*-ethyl-N,N-dipropylthiocarbamate- degrading *Arthrobacter* strain and evidence for plasmid-associated *s*-ethyl-N,N-dipropylthiocarbamate degradation. Appl. Environ. Microbiol. 53:1088-1093.

65. Tomasek, P. H. , and J. S. Karns. 1989. Cloning of a carbofuran hydrolase gene from *Achromobacter* sp. strain WM111 and its expression in gram-negative bacteria. J. Bacteriol. 171:4038-4044.

66. Top, E. M. , W. E. Holben, and L. J. Forney. 1995. Characterization of diverse 2,4-dichlorophenoxyacetic acid-degradative plasmids isolated from soil by complementation. Appl. Environ. Microbiol. 61:1691-1698.

67. Topp, E. , R. S. Hanson, D. B. Ringelberg, D. C. White, and R. Wheatcroft. 1993. Isolation and characterization of an N-methylcarbamate insecticide- degrading methylotrophic bacterium. Appl. Environ. Microbiol. 59:3339-3349.

68. Topp, E. , L. Tessier, and E. G. Gregorich. 1996. Dairy manure incorporation stimulates rapid atrazine mineralization in an agricultural soil. Can. J. Soil Sci. 76:403-409.

69. Topp, E. , L. Tessier, and M. Lewis. 1997. Characterization of atrazine-degrading bacteria isolated from agricultural soil. Presented at the 97th General Meeting of the American Society for Microbiol. , Washington, D. C.

70. Topp, E. , H. Zhu, M. Lewis, and D. Cuppels. 1999. Characterization of soil bacteria which rapidly degrade the herbicide atrazine. 9th European Congress on Biotechnology Proceedings (on CD-ROM) ISBN80521 (C 1999-2000) Ed. M. Hofman Branche Belge de la Société de Chimie Industrielle Brussels -Belgium .

71. Torstensson, L. 1999. Biobeds can protect pesticide pollution of waters. 9th European Congress on Biotechnology Proceedings (on CD-ROM) ISBN80521 (C 1999-2000) Ed. M. Hofman Branche Belge de la Société de Chimie Industrielle Brussels -Belgium

72. Torstensson, L. , and M. d. P. Castillo. 1997. Use of biobeds in Sweden to minimize environmental spillages from agricultural spraying equipment. Pestic. Outlook. 8:24-27.

73. van der Meer, J. R. , W. M. de Vos, S. Harayama, and A. J. B. Zehnder. 1992. Molecular mechanisms of genetic adaptation to xenobiotic compounds. Microbiol. Rev. 56:677-694.

74. van der Meer, J. R. , R. I. Eggen, A. J. Zehnder, and W. M. de Vos. 1991. Sequence analysis of the *Pseudomonas* sp. strain P51 *tcb* gene cluster, which encodes metabolism of chlorinated catechols: evidence for specialization of catechol 1,2-dioxygenases for chlorinated substrates. J. Bacteriol. 173:2425-2434.

75. Vanderheyden, V. , P. Debongnie, and L. Pussemier. 1997. Accelerated degradation and mineralization of atrazine in surface and subsurface soil materials. Pestic. Sci. 49:237-242.

76. Wackett, L. P. 1999. Microbial enzymes in biodegradation. Presented at the 42nd Oholo Conference, Eilat, Israel, May 3-7, 1998.

77. Winkelmann, D. A. , and S. J. Klaine. 1991. Degradation and bound residue formation of four atrazine metabolites, deethylatrazine, deisopropylatrazine, dealkylatrazine and hydroxyatrazine, in a western Tennessee soil. Environ. Toxicol. Chem. 10:347-354.

78. Yanze Kontchou, C. , and N. Gschwind. 1994. Mineralization of the herbicide atrazine as a carbon source by a *Pseudomonas* strain. Appl. Environ. Microbiol. 60:4297-4302.

MICROBIAL DEGRADATION OF CHLORINATED AROMATIC COMPOUNDS

The meta-cleavage pathway

Walter Reineke[1], Astrid E. Mars[2], Stefan R. Kaschabek[1] And Dick B. Janssen[2]

[1] *Bergische Universität - Gesamthochschule Wuppertal, Chemische Mikrobiologie, Gaußstraße 20, D-42097 Wuppertal, Germany. reineke@uni-wuppertal.de* [2] *Department of Biochemistry, University of Groningen, Nijenborgh 4, NL-9747 AG Groningen, The Netherlands*

The aerobic microbial degradation of various chloroaromatics usually occurs via chlorocatechols as central intermediates. These are further degraded through the modified *ortho*-cleavage pathway. Dechlorination takes place during cycloisomerization of chloromuconates and reduction of chloromaleylacetates. In contrast, the degradation of haloaromatics via *meta*-cleavage was thought to be impossible due to toxicity of cleavage intermediates, whereas the *meta* route is more effective for methylaromatics. Recently, *Pseudomonas putida* strain GJ31 was shown to be able to degrade toluene and chlorobenzene simultaneously. Strain GJ31 rapidly degrades chlorobenzene via 3-chlorocatechol using the *meta*-cleavage pathway without any apparent toxic effects. An unusual chlorocatechol 2,3-dioxygenase oxidizes 3-chlorocatechol. Stoichiometric displacement of chloride then leads to the production of 2-hydroxymuconate, which is a common metabolite of the *meta*-cleavage pathway. In contrast to other catechol 2,3-dioxgenases, which are subject to inactivation when exposed to 3-chlorocatechol, the chlorocatechol 2,3-dioxygenase is resistant. The gene encoding the chlorocatechol 2,3-dioxygenase (*cbzE*) of strain GJ31 was cloned and sequenced. CbzE was most similar to catechol 2,3-dioxygenases of the 2.C subfamily of type 1 extradiol dioxygenases. Hybrid enzymes, which were made of CbzE and the 3-methylcatechol 2,3-dioxygenase of strain *P. putida* UCC2, showed that the resistance of CbzE to suicide inactivation and the substrate specificity were mainly determined by the C-terminal region of the protein. Establishing whether the *meta*-cleavage pathway of strain GJ31 can function as pathway segment in the construction of novel chloroaromatics-degraders is a future task. Organisms such as *P. putida* strain GJ31 and its derivatives may be useful for effective treatment of waste streams containing various methyl- and chloroaromatics.

S.N. Agathos and W. Reineke (eds.),
Biotechnology for the Environment: Strategy and Fundamentals, 157–168.
© 2002 *Kluwer Academic Publishers. Printed in the Netherlands.*

1. Introduction

Chlorinated aromatic compounds represent an important class of environmental pollutants. Most of these compounds can be biologically degraded aerobically and/or anaerobically. Besides being subject to cometabolic turnover, a process which is unproductive for microorganisms, because it is not coupled to energy conservation and biomass production, chloroaromatics can serve for growth. While a variety of microbial populations that mediate dechlorination of chloroaromatics under anaerobic conditions have been reported, only few studies on dechlorination with pure cultures of anaerobic bacteria are known. These organisms use chlorinated compounds such as chlorobenzoates, chlorophenols, or chlorohydroxyphenylacetate as an electron acceptor in a so-called dehalorespiration process leading to dechlorination. In contrast, a larger number of aerobic bacteria are known which use chloroaromatics as sole carbon and energy source. Dechlorination reactions are essential for mineralization which finally results in the formation of chloride, carbon dioxide and biomass.

Aerobic organisms growing with chloroaromatics can be differentiated on the basis of the catabolic pathways used for removal of the chlorine substituents. Chlorine substituents can be cleaved off prior to ring-cleavage by oxygenolytic, reductive or hydrolytic reactions. Further conversion of the chlorine-free metabolites can then occur via classical pathways for the metabolism of aromatic compounds such as the 3-oxoadipate, the protocatechuate, or the *meta*-cleavage pathway. However, the majority of the organisms that are able to mineralize chlorinated aromatics do not possess enzymes capable of initial dechlorination prior to ring-cleavage. They use oxygenases to convert chlorinated aromatics to chlorocatechols, and dechlorination takes place after ring-cleavage. The enzymatic reactions leading to formation of chlorocatechols are similar to those used for the degradation of non-chlorinated aromatic compounds (Figure 1).

Here we discuss the different routes for the degradation of chlorocatechols. A description of the well-known modified *ortho*-cleavage pathway will be followed by a discussion of the new discovery that also the *meta*-cleavage pathway may be used for the degradation of chlorocatechols.

2. Degradation of chlorocatechols via the modified *ortho*-cleavage pathway

It is well established that in most organisms that degrade chloroaromatics, the chlorocatechol degradative pathway starts with *ortho*-cleavage by chlorocatechol 1,2-dioxygenases which introduce molecular oxygen and produce the corresponding chloro-*cis,cis*-muconate (Figure 2). The pathways used for the degradation of 3-chloro-, 4-chloro-, and 3,5-dichlorocatechol have mostly been studied with the 3-chlorobenzoate-degrading *Pseudomonas* sp. strain B13 and the 2,4-dichlorophenoxyacetate-degrading *Ralstonia eutropha* strain JMP134. Elimination of a chlorine substituent occurs spontaneously when 3-chloro- and 2,4-dichloro-*cis,cis*-muconate are subsequently converted by chloromuconate cycloisomerases to 4-chloro- and 2,4-dichloromuconolactone, respectively. The dienelactones are formed by *anti*-elimination of hydrogen chloride and formation of an exocyclic double bond (Schmidt and Knackmuss 1980). For example, *cis*-dienelactone results from 3-chloro-*cis,cis*-

muconate. 2,4-Dichloro-*cis,cis*-muconate, the intermediate in the 3,5-dichlorocatechol degradation, is converted to 2-chlorodienelactone probably in the *cis*-configuration. Only recently it was realized that the dechlorination of 2-chloro-*cis,cis*-muconate, the intermediate in the degradation of 3-chlorocatechol, is catalyzed by the chloromuconate cycloisomerase instead of being a spontaneous step. Both cycloisomerization, leading to (+)5-chloromuconolactone, and dechlorination are carried out by the same enzyme to give *trans*-dienelactone (Vollmer and Schlömann 1995; Vollmer *et al.* 1994). The dienelactones are further converted to the respective maleylacetates by dienelactone hydrolases.

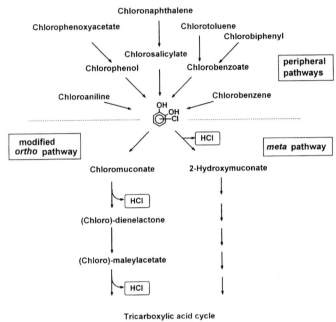

Figure 1. Schematic presentation of the mineralization of chloroaromatics with chlorocatechols as key metabolites

An additional dechlorination takes place with 2-chloromaleylacetate, the intermediate in the degradation of 3,5-dichlorocatechol, by a reductive reaction (Chapman 1979; Kaschabek and Reineke 1992, 1995; Müller *et al.* 1996; Sander *et al.* 1991; Vollmer *et al.* 1993). The elimination is thought to occur spontaneously during the maleylacetate reductase reaction (Kaschabek and Reineke 1995).

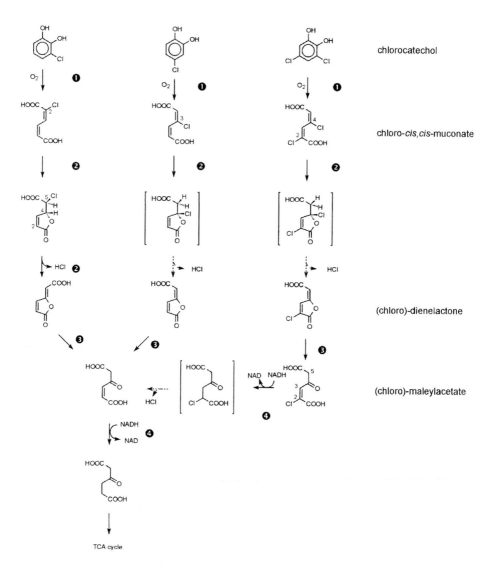

Figure 2. Modified ortho-cleavage pathway for the degradation of 3-chloro-, 4-chloro-, and 3,5-dichlorocatechol. ❶ chlorocatechol 1,2-dioxygenase, ❷ chloromuconate cycloisomerase, ❸ dienelactone hydrolase, ❹ maleylacetate reductase. The broken arrows indicate a spontaneous elimination of chlorine substituents from intermediates in parentheses.

The modified *ortho*-cleavage pathway of *Pseudomonas chlororaphis* strain RW71 tolerates substitution at the aromatic ring up to four chlorine atoms in the degradation of 1,2,3,4-tetrachlorobenzene via tetrachlorocatechol (Potrawfke *et al.* 1998). Two dechlorination steps have been described so far. The steps leading to removal of the third and fourth chlorine substituent are unknown.

3. Degradation of chlorocatechols via the *meta*-cleavage pathway

3.1. PROBLEMS WITH THE *META*-CLEAVAGE PATHWAY IN THE CASE OF CHLOROAROMATICS

It was assumed for a long time that organisms cannot use the *meta*-cleavage pathway for the degradation of chlorinated aromatic compounds. The reasons for this statement were the following:

- Studies on the degradation of chlorocatechols with strains harboring both the modified *ortho-* and *meta*-cleavage pathway indicated that channelling of chlorocatechols into the *meta*-cleavage pathway was unproductive or even counterproductive. The cells exclusively used the modified *ortho*-cleavage pathway for the degradation of 3-chloro-, 4-chloro-, and 3,5-dichlorocatechol (Reineke *et al.* 1982).

- 3-Chlorocatechol has often been observed to act as a strong inhibitor of extradiol dioxygenases. Bartels *et al.* (1984) showed that the inactivation of catechol 2,3-dioxygenase of *Pseudomonas putida* PaW1 (XylE) by 3-chlorocatechol is an irreversible process that requires oxygen. It was proposed that the enzyme cleaves 3-chlorocatechol to an acyl chloride that acts as a suicide product by reacting with nucleophilic groups in the enzyme (Figure 3). The proposed acyl chloride, which has not yet been identified, can also react with water, yielding 2-hydroxymuconate. Small amounts of this product were identified during conversion of 3-chlorocatechol by the catechol 2,3-dioxygenase of strain PaW1. In contrast, Klečka and Gibson (1981) observed that the catechol 2,3-dioxygenase of *P. putida* strain F1 was inactivated by 3-chlorocatechol in a reversible way. This was explained by chelation of the ferrous iron cofactor of the enzyme by 3-chlorocatechol. The absence of oxygen uptake by whole cells when confronted with 3-chlorocatechol might suggest that the catechol 2,3-dioxygenase from strain F1 does not convert any 3-chlorocatechol. The absence of a suicide inactivation mechanism was also indicated by experiments with ^{14}C-labelled 3-chlorocatechol. Labeling was found at various sites when the catechol 2,3-dioxygenase from strain F1 was proteolytically separated into peptides after the incubation with ^{14}C-labelled 3-chlorocatechol (D. T. Gibson, pers. comm.). With a suicide inactivation mechanism, labeling at a single site would be expected. At present, the precise mechanism of inactivation is still unknown. Preliminary experiments with the catechol 2,3-dioxygenase from strain PaW1 showed that the homodimeric protein band of native enzyme migrated slightly faster on SDS-polyacrylamide gels than that of enzyme that was inactivated with 3-chlorocatechol, provided that the samples were not boiled prior to application on the gel (Mars *et al.* 1999). Modification of the catechol 2,3-dioxygenase by 3-chlorocatechol also made the enzyme less susceptible to digestion with proteinase

LysC. Complete digestion of dialyzed, inactivated XylE was only possible in the presence of 1 M urea, while the native enzyme could be digested in the absence of urea. These results indicate that inactivation by 3-chlorocatechol involves a labile covalent modification which influences the multimeric form of XylE.

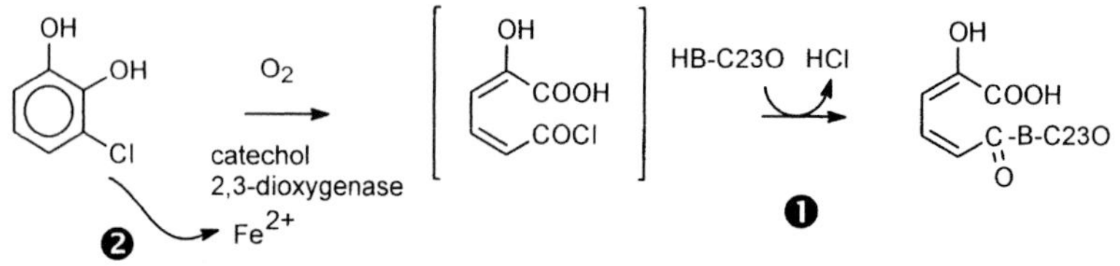

Figure 3. *Inactivation of catechol 2,3-dioxygenases by 3-chlorocatechol.* ❶ *Suicide inactivation due to the formation of an acyl chloride,* ❷ *inactivation due to chelation of the ferrous iron.*

• In contrast to the situation of catechol 2,3-dioxygenases, when exposed to 3-chlorocatechol, various extradiol dioxygenases can convert catechols that are chlorinated at the 4-position (Bartels *et al.* 1984; Hirose *et al.* 1994; Moon *et al.* 1996; Nozaki *et al.* 1970; Rast *et al.* 1980; Sala-Trepat and Evans 1971). However, growth on substrates via 4-chlorinated catechols and a *meta*-cleavage pathway was usually absent.

The initial observations on toxicity of the cleavage products led to the old dogma that chloroaromatics cannot be degraded by the *meta*-cleavage pathway (Knackmuss 1981). In addition, it was generally accepted that *ortho*-cleavage pathways are not suitable for the degradation of methylsubstituted aromatics. Therefore, it has long been thought that simultaneous conversion of halo- and methylcatechols is not possible and can only be obtained in engineered strains (Rojo *et al.* 1987). However, more recent results showed that microorganisms exist which grow with chlorinated aromatics via a *meta*-cleavage pathway.

3.2. PRODUCTIVE USE OF THE *META*-CLEAVAGE PATHWAY FOR CHLORINATED CATECHOLS SUBSTITUTED IN *PARA*-POSITION

In recent years some bacterial strains were described that utilize chlorinated compounds like 4-chlorobenzoate, 4-chlorobiphenyl, 3-chloro-2-methylbenzoate, and 4-chlorophenol through a *meta*-cleavage pathway (Arensdorf and Focht 1994, 1995; Higson and Focht 1992; Hollender *et al.* 1994, 1997; Kim *et al.* 1997, 1998; McCullar *et al.* 1994; Seo *et al.* 1997). The chlorinated substrates are converted to 4-chlorinated catechols, which are cleaved by catechol 2,3-dioxygenases. The chlorine substituents are removed after cleavage of the aromatic ring, although it is unknown where the dechlorination occurs.

3.3. PRODUCTIVE DEGRADATION OF 3-CHLOROCATECHOL VIA THE *META*-CLEAVAGE PATHWAY

An exception to the general rule that the *meta*-cleavage pathway is insufficient for the degradation of 3-chlorocatechol and that productive simultaneous degradation of methyl- and chloroaromatics is not possible by natural isolates was reported about ten years ago. *Pseudomonas putida* strain GJ31, which was obtained by conventional batch enrichment, i. e. without prolonged adaptation time, from sediment samples taken from the river Rhine using chlorobenzene as sole carbon source, was described to utilize toluene and chlorobenzene simultaneously (Oldenhuis *et al.* 1989). More recently, it was found that *P. putida* GJ31 rapidly degrades chlorobenzene via 3-chlorocatechol. However, instead of using the modified *ortho*-cleavage pathway as do other chlorobenzene-degraders described in the literature, strain GJ31 uses a *meta*-cleavage pathway (Mars *et al.* 1997) (Figure 4). The pathway is initiated by proximal cleavage between C-2 and C-3.

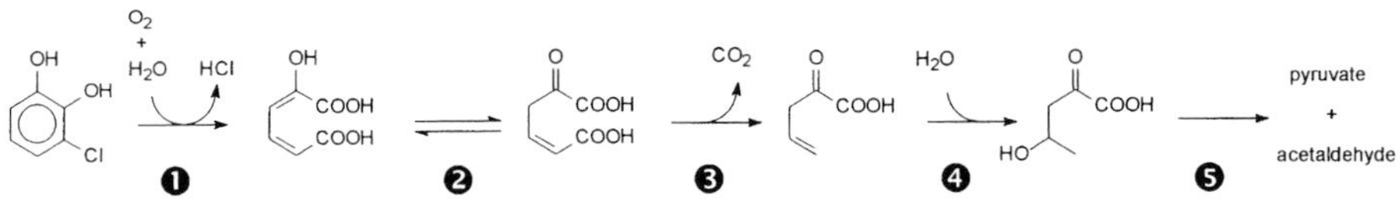

Figure 4. Degradative pathway for 3-chlorocatechol used by strain GJ31 during degradation with chlorobenzene. ❶ chlorocatechol 2,3-dioxygenase, ❷ 4-oxalocrotonate isomerase, ❸ 4-oxalocrotonate decarboxylase, ❹ 2-oxopent-4-enoate hydratase, ❺·4-hydroxy-2-oxovalerate aldolase.

In contrast to other catechol 2,3-dioxygenases, which are subject to inactivation, the 2,3-dioxygenase of strain GJ31 productively converts 3-chlorocatechol. Stoichiometric displacement of chlorine subsequently leads to the production of 2-hydroxymuconate (Kaschabek *et al.* 1998). Because of that property the enzyme is termed chlorocatechol 2,3-dioxygenase. Since 2-hydroxymuconate is a common metabolite in the degradation of catechol through the *meta*-cleavage pathway, further degradation runs accordingly via 4-oxalocrotonate, 2-oxopent-4-enoate, 4-hydroxy-2-oxovalerate and finally pyruvate and acetaldehyde (Figure 4).

3.4. THE CHLOROCATECHOL 2,3-DIOXYGENASE OF *PSEUDOMONAS PUTIDA* STRAIN GJ31

To establish whether the capacity of productive conversion of 3-chlorocatechol and dechlorination to 2-hydroxymuconate was restricted to chlorocatechol 2,3-dioxygenase or whether it was a feature which was shared with other catechol 2,3-dioxygenases, cloning and sequence analysis of the respective gene, purification of the enzyme, and determination of the substrate range were performed.

The purified chlorocatechol 2,3-dioxygenase of strain GJ31 is similar to other catechol 2,3-dioxygenases such as that of *P. putida* PaW1. It is a typical Fe^{2+}-dependent extradiol dioxygenase. The enzyme has a native molecular mass of 135 ± 10 kDa and exists as a tetramer of identical subunits (Kaschabek *et al.* 1998).

A 3.1 kb region of genomic DNA of strain GJ31 containing the gene for the chlorocatechol 2,3-dioxygenase (*cbzE*) was cloned and sequenced (Mars *et al.* 1999). The *cbzE* gene was found in a region that harbors genes encoding a transposase, a ferredoxin, an open reading frame with similarity to a protein of the *meta*-cleavage pathway with unknown function, and a 2-hydroxymuconic semialdehyde dehydrogenase.

By comparing the amino acid sequence of the chlorocatechol 2,3-dioxygenase with other 2,3-dioxygenases it was found that it was most similar to the two-domain, iron containing catechol 2,3-dioxygenases, which form the 2.C subfamily of type 1 extradiol dioxygenases (Eltis and Bolin 1996). The amino acid sequence identity of chlorocatechol 2,3-dioxygenase ranged from 72% for the 3-methylcatechol 2,3-dioxygenase of *P. putida* UCC2 (TdnC) and the catechol 2,3-dioxygenase of *Burkholderia cepacia* AA1 to 51% for the enzyme of *Ralstonia (Pseudomonas) pickettii* strain PKO1 (TbuE) (Kukor and Olsen 1996). The TOL plasmid encoded catechol 2,3-dioxygenase of *P. putida* strain PaW1 (XylE) (Nakai *et al.* 1983) and the 2,3-dihydroxybiphenyl dioxygenase of *B. cepacia* strain LB400 (BphC) (Hofer *et al.* 1993) are 38 and 15% identical to the chlorocatechol 2,3-dioxygenase, respectively.

The chlorocatechol 2,3-dioxygenase of strain GJ31 was the only *meta*-cleaving enzyme among a set of related enzymes that could productively convert 3-chlorocatechol. Even the 3-methylcatechol 2,3-dioxygenases of *P. putida* strain UCC2 (TdnC) (McClure and Venables 1986) and strain MT15 (Keil *et al.* 1985), which show the highest degree of sequence identity, failed to cleave 3-chlorocatechol without inactivation. Thus, the GJ31-enzyme is not only a 2,3-dioxygenase with preference for catechols substituted in position 3.

While most catechol 2,3-dioxygenases cleave 3-chlorocatechol in a proximal manner, an alternative cleavage at the distal position between C-1 and C-6 has also been reported (Heiss *et al.* 1995; Riegert *et al.* 1998; Sala-Trepat and Evans 1971). This leads to the formation of 3-chloro-2-hydroxymuconic semialdehyde without production of an acyl chloride. However, the conversions by the catechol 2,3-dioxygenase of the *Azotobacter vinelandii* and the 2,3-dihydroxybiphenyl dioxygenase of *Sphingomonas* sp. BN6 are only cooxidative reactions. This means that the chlorocatechol 2,3-dioxygenase of strain GJ31 is the only catechol 2,3-dioxygenase so far that productively converts 3-chlorocatechol as part of a *meta*-cleavage pathway.

Further attempts to elucidate the reasons for the resistance of the chlorocatechol 2,3-dioxygenase were made by creating hybrid enzymes between the dioxygenase of strain GJ31 and the 3-methylcatechol 2,3-dioxygenase of strain UCC2 (TdnC, Figure 5). One monomer of catechol 2,3-dioxygenase consists of two separate domains that are thought to arise from a genetic duplication. The active site is located in the C-terminal domain (Eltis and Bolin 1996). The exchanges in hybrids H1 and H2 were made just in front of the start of domain 2, which is formed by the C-terminal half of the protein. Hybrids H1 and H2 yielded active enzymes. The substrate ranges and turnover capacities with 3-chlorocatechol of H1 and H2 were comparable to the corresponding values of chlorocatechol 2,3-dioxygenase and 3-methylcatechol 2,3-dioxygenase, respectively, indicating that both the substrate specificity and the susceptibility to inactivation were indeed determined by the C-terminal domains of both enzymes. In the hybrids H3 and H4, made by exchanges in the beginning of the C-terminal domains, the replacement of amino acids 148 to 189 influenced the substrate specificity of the enzymes in such a way that both hybrid enzymes lost their preference for substituted catechols. When the middle parts of the C-terminal domains (amino acids 190 to 240) were exchanged in H5 and H6, no activity could be detected. This was also the case when the middle and last part of the domain (amino acids 190 to 314) were exchanged together in H9 and H10. When the last regions (amino acids 241 to 314) were exchanged, both hybrids H7 and H8 had very low activities. Hybrid H7 was very rapidly inactivated when it converted methylated catechols. In contrast, hybrid H8, which was equal to the 3-methylcatechol 2,3-dioxygenase except for the last region, maintained a high relative activity with 3-methylcatechol, while it appeared to have an improved relative rate for 3-chlorocatechol. The turnover capacity measurement also suggested that the ability to convert 3-chlorocatechol was improved for H8 as compared to the 3-methylcatechol 2,3-dioxygenase.

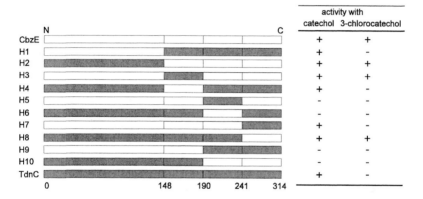

Figure 5. Schematic representation of the hybrid catechol 2,3-dioxygenases. Open bars: polypeptide fragments derived from CbzE; solid bars: fragments from TdnC. Activities with catechol and 3-chlorocatechol are indicated.

The data with the hybrid enzymes indicated that the resistance of chlorocatechol 2,3-dioxygenase to suicide inactivation and the substrate specificity were mainly determined by the C-terminal region of the protein.

4. Conclusion, future prospects

Although it was assumed for many years that it was impossible to productively metabolize chlorinated catechols via the *meta*-cleavage pathway and that simultaneous degradation of chloro- and methylaromatics was not possible, the results recently obtained clearly showed that a biochemical generalization based on the study of individual pathways and organisms is problematic. Whether the *meta*-cleavage pathway is an alternative to the degradation of chloroaromatics via the modified *ortho*-cleavage pathway or whether strain GJ31 is only a rarity has to be answered in the future. At present no other organism is known that exhibits the unusual property.

Although our results clearly showed that the substrate specificities and degree of resistance to inactivation are determined by the C-terminal domains of catechol 2,3-dioxygenases, it remains to be established what mechanism underlies the inactivation with 3-chlorocatechol and which biochemical properties are responsible for the resistance of the chlorocatechol 2,3-dioxygenase.

We are presently testing if the GJ31-*meta*-cleavage pathway can function as pathway segment in the construction of novel chloroaromatics-degraders. Organisms such as *P. putida* strain GJ31 and its derivatives may then be used for effective treatment of waste streams containing various methyl- and chloroaromatics.

Acknowledgment

The studies in Groningen and Wuppertal were financed by the European Union, under contract EV5V-CT92-0192, by a grant from the Dutch IOP Environmental Biotechnology program and from the Deutsche Forschungsgemeinschaft (Re 659/7-1).

References

Arensdorf JJ and Focht DD (1994) Formation of chlorocatechol *meta* cleavage products by a pseudomonad during metabolism of monochlorobiphenyls. Applied and Environmental Microbiology 60:2884-2889.

Arensdorf JJ and Focht DD (1995) A *meta* cleavage pathway for 4-chlorobenzoate, an intermediate in the metabolism of 4-chlorobiphenyl by *Pseudomonas cepacia* P166. Applied and Environmental Microbiology 61:443-447.

Bartels I, Knackmuss H-J and Reineke W (1984) Suicide inactivation of catechol 2,3-dioxygenase from *Pseudomonas putida* mt-2 by 3-halocatechols. Applied and Environmental Microbiology 47:500-505.

Chapman PJ (1979) Degradation mechanisms. In: Bourquin AW and Pritchard PH (Eds) Microbial degradation of pollutants in marine environments. EPA-600/9-79-012 (pp. 28-66) Environmental Protection Agency, Gulf Breeze, Fla. USA.

Eltis LD and Bolin JT (1996) Evolutionary relationship among extradiol dioxygenases. Journal of Bacteriology 178:5930-5937.

Heiss G, Stolz A, Kuhm AE, Müller C, Klein J, Altenbuchner J and Knackmuss H-J (1995) Characterization of a 2,3-dihydroxybiphenyl dioxygenase from the naphthalenesulfonate-degrading bacterium strain BN6. Journal of Bacteriology 177:5865-5871.

Higson FK and Focht DD (1992) Utilization of 3-chloro-2-methylbenzoic acid by *Pseudomonas cepacia* MB2 through the *meta* fission pathway. Environmental Microbiology 58:2501-2504.

Hirose J, Kimura N, Suyama A, Kobayashi A, Hayashida S and Furukawa K (1994) Functional and structural relationship of various extradiol aromatic ring-cleavage dioxygenases of *Pseudomonas* origin. FEMS Microbiology Letters 118:273-278.

Hofer B, Eltis LD, Dowling DN and Timmis KN (1993) Genetic analysis of a *Pseudomonas* locus encoding a pathway for biphenyl/polychlorinated biphenyl degradation. Gene 130:47-55.

Hollender J, Dott W and Hopp J (1994) Regulation of chloro- and methylphenol degradation in *Comamonas testosteroni* JH5. Applied and Environmental Microbiology 60:2330-2338.

Hollender J, Hopp J and Dott W (1997) Degradation of 4-chlorophenol via the *meta* cleavage pathway by *Comamonas testosteroni* JH5. Applied and Environmental Microbiology 63:4567-4572.

Kaschabek SR, Kasberg T, Müller D, Mars AE, Janssen DB and Reineke W (1998) Degradation of chloroaromatics: Purification and characterization of a novel type of chlorocatechol 2,3-dioxygenase of *Pseudomonas putida* GJ31. Journal of Bacteriology 180:296-302.

Kaschabek SR and Reineke W (1992) Maleylacetate reductase of *Pseudomonas* sp. strain B13: dechlorination of chloromaleylacetates, metabolites in the degradation of chloroaromatic compounds. Archives of Microbiology 158:412-417.

Kaschabek SR and Reineke W (1995) Maleylacetate reductase of *Pseudomonas* sp. strain B13: Specificity of substrate conversion and halide elimination. Journal of Bacteriology 177:320-325.

Keil H, Lebens MR and Williams PA (1985) TOL plasmid pWW15 contains two nonhomologous, independently regulated catechol 2,3-dioxygenase genes. Journal of Bacteriology 163:248-255.

Kim K-P, Seo D-I, Lee D-H, Kim Y and Kim C-K (1998) Cloning and expression in *E. coli* of the genes responsible for degradation of 4-chlorobenzoate and 4-chlorocatechol from *Pseudomonas* sp. strain S-47. Journal of Microbiology (Korea) 36:99-105.

Kim KP, Seo D-I, Min KH, Ka JO, Park YK and Kim C-K (1997) Characterization of catechol 2,3-dioxygenase produced by 4-chlorobenzoate-degrading *Pseudomonas* sp. S-47. Journal of Microbiology (Korea) 35:295-299.

Klecka GM and Gibson DT (1981) Inhibition of catechol 2,3-dioxygenase from *Pseudomonas putida* by 3-chlorocatechol. Applied and Environmental Microbiology 41:1159-1165.

Knackmuss H-J (1981) Degradation of halogenated and sulfonated hydrocarbons. In: Leisinger T, Cook AM, Hütter R and Nüesch J (Eds) Microbial degradation of xenobiotics and recalcitrant compounds. pp. 189-212. Academic Press, London, UK.

Kukor JJ and Olsen RH (1996) Catechol 2,3-dioxygenases functional in oxygen-limited (hypoxic) environments. Applied and Environmental Microbiology 62:1728-1740.

Mars AE, Kasberg T, Kaschabek SR, van Agteren MH, Janssen DB and Reineke W (1997) Microbial degradation of chloroaromatics: Use of the *meta*-cleavage pathway for mineralization of chlorobenzene. Journal of Bacteriology 179:4530-4537.

Mars AE, Kingma J, Kaschabek SR, Reineke W and Janssen DB (1999) Conversion of 3-chlorocatechol by various catechol 2,3-dioxygenases and sequence analysis of the chlorocatechol dioxygenase region of *Pseudomonas putida* GJ31. Journal of Bacteriology 181:1309-1318.

McClure NC and Venables WA (1986) Adaptation of *Pseudomonas putida* mt-2 to growth on aromatic amines. Journal of General Microbiology 132:2209-2218.

McCullar MV, Brenner V, Adams RH and Focht DD (1994) Construction of a novel polychlorinated biphenyl-degrading bacterium: utilization of 3,4′-dichlorobiphenyl by *Pseudomonas acidovorans* M3GY. Applied and Environmental Microbiology 60:3833-3839.

Moon J, Min KR, Kim C-K, Min K-H and Kim Y (1996) Characterization of the gene encoding catechol 2,3-dioxygenase of *Alcaligenes* sp. KF711: overexpression, enzyme purification, and nucleotide sequencing. Archives of Biochemistry and Biophysics 332:248-254.

Müller D, Schlömann M and Reineke W (1996) Maleylacetate reductases in chloroaromatic-degrading bacteria using the modified *ortho* pathway: Comparison of catalytic properties. Journal of Bacteriology 178:298-300.

Nakai C, Kagamiyama H, Nozaki M, Nakazawa T, Inouye S, Ebina Y and Nakazawa A (1993) Complete nucleotide sequence of the metapyrocatechase gene on the TOL plasmid of *Pseudomonas putida* mt-2. Journal of Biological Chemistry 258:2923-2928.

Nozaki M, Kotani S, Ono K and Senoh S (1970) Metapyrocatechase. III. Substrate specificity and mode of ring fission. Biochimica et Biophysica Acta 220:213-223.

Oldenhuis R, Kuijk K, Lammers A, Janssen DB and Witholt B (1989) Degradation of chlorinated and non-chlorinated aromatic solvents in soil suspensions by pure bacterial cultures. Applied Microbiology and Biotechnology 30:211-217.

Potrawfke T, Timmis KN and Wittich RM (1998) Degradation of 1,2,3,4-tetrachlorobenzene by *Pseudomonas chlororaphis* RW71. Applied and Environmental Microbiology 64:3798-3806.

Rast HG, Engelhardt G and Wallnöfer P (1980) 2,3-Cleavage of substituted catechols in *Nocardia* sp. DSM 43251 (*Rhodococcus rubrus*). Zentralblatt für Bakteriologie, Mikrobiologie und Hygiene, 1. Abteilung, Original C 1:224-236.

Reineke W, Jeenes DJ, Williams PA and Knackmuss H-J (1982) TOL plasmid pWW0 in constructed halobenzoate-degrading *Pseudomonas* strains: Prevention of *meta* pathway. Journal of Bacteriology 150:195-201.

Riegert U, Heiss G, Fischer P and Stolz A (1998) Distal cleavage of 3-chlorocatechol by an extradiol dioxygenase to 3-chloro-2-hydroxymuconic semialdehyde. Journal of Bacteriology 180:2849-2853.

Rojo F, Pieper DH, Engesser K-H, Knackmuss H-J and Timmis KN (1987) Assemblage of *ortho* cleavage route for simultaneous degradation of chloro- and methylaromatics. Science 238:1395-1398.

Walter Reineke, Astrid E. Mars, Stefan R. Kaschabek and Dick B. Janssen

Sala-Trepat JM and Evans WC (1971) The *meta* cleavage of catechol by *Azotobacter* species: 4-oxalocrotonate pathway. European Journal of Biochemistry 20:400-413.
Sander P, Wittich R-M, Fortnagel P, Wilkes H and Francke W (1991) Degradation of 1,2,4-trichloro- and 1,2,4,5-tetrachlorobenzene by *Pseudomonas* strains. Applied and Environmental Microbiology 57:1430-1440.
Schmidt E and Knackmuss H-J (1980) Chemical structure and biodegradability of halogenated aromatic compounds. Conversion of chlorinated muconic acids into maleoylacetic acid. Biochemical Journal 192:339-347.
Seo D-I, Lim JY, Kim YC, Min KH and Kim C-K (1997) Isolation of *Pseudomonas* sp. S-47 and its degradation of 4-chlorobenzoic acid. Journal of Microbiology (Korea) 35:188-192.
Vollmer MD and Schlömann M (1995) Conversion of 2-chloro-*cis,cis*-muconate and its metabolites 2-chloro- and 5-chloromuconolactone by chloromuconate cycloisomerases of pJP4 and pAC27. Journal of Bacteriology 177:2938-2941.
Vollmer MD, Stadler-Fritzsche K and Schlömann M (1993) Conversion of 2-chloromaleylacetate in *Alcaligenes eutrophus* JMP134. Archives of Microbiology 159:182-188.
Vollmer MD, Fischer P, Knackmuss H-J and Schlömann M (1994) Inability of muconate cycloisomerases to cause dehalogenation during conversion of 2-chloro-*cis,cis*-muconate. Journal of Bacteriology 176:4366-4375.

THE ROLE OF ENVIRONMENTAL CONDITIONS AND BIOTIC INTERACTIONS BETWEEN MICROBIAL SPECIES IN DEGRADATION OF CHLORINATED POLLUTANTS

OLIVER DRZYZGA[1], JANNEKE KROONEMAN[2], JAN GERRITSE[3], AND JAN C. GOTTSCHAL[4]

[1]Department of Marine Microbiology, University of Bremen, Leobener Strasse, D-28359 Bremen, Germany; [2]Bioclear bv., P.O. Box 2262, 9704 CG, Groningen, The Netherlands; [3]TNO Institute of Environmental Sciences, P.O. Box 342, 7300 AH Apeldoorn, The Netherlands; [4]Centre for Ecological and Evolutionary Studies, Laboratory of Microbial Ecology, University of Groningen, P.O. Box 14, 9750 AA Haren, The Netherlands, Email: J.C.Gottschal@Biol.rug.nl, Fax: +31-50.363.2154.

Summary

Degradation of chlorinated aromatic and aliphatic organic compounds in natural environments depends on the interplay of a multitude of both abiotic and biotic factors. Hence the extent and rate of these processes can only be understood if, in addition to monitoring the individual microbial processes under the prevailing environmental conditions, detailed knowledge of the interactions between the various physiologically different microbial populations is obtained as well.

1. Chlorinated contaminants

Organochlorine compounds represent a major group of environmental pollutants which in many cases reach such high concentrations in soils, sediments, groundwater, surface waters or the atmosphere that they can adversely affect health of humans, wildlife and plants.

The range of different types of chlorinated organic compounds is extraordinarily large, with great differences in abundance in the environment, chemical structure, toxicity, biodegradability and, hence, persistence in nature (Anderson and Lovley, 1997; Fetzner, 1998; Mohn and Tiedje, 1992). These compounds range from simple chlorinated ethenes, widely used as degreasing agents and solvents, via (poly)chlorinated benzenes, phenols and phenoxyacetates to very complex organochlorine pesticides. Although many of these pollutants have entered the biosphere as a result of human industrial activities, it has also become clear that

169

S.N. Agathos and W. Reineke (eds.),
Biotechnology for the Environment: Strategy and Fundamentals, 169–175.

naturally produced halogenated compounds occur abundantly in marine environments and soils alike. Over 2000 different compounds have so far been identified as being produced by plants, insects, marine algae, mammals and fungi (Gribble, 1994; Gribble, 1996). Some of these naturally produced compounds even exceed the levels produced by human industrial activities. Not surprisingly, this enormous diversity of different naturally produced halogenated compounds has resulted in the evolution of pathways in microorganisms which in principle can now also be exploited, adapted and enhanced for the degradation of pollutants in environments which have been highly contaminated by human activities.

The large differences in the nature of chemical pollutants in our environment confronts microbial ecologists and bioremediation engineers with the difficult task to predict and create environmental conditions which will accommodate their degradation. Quite often the various microbes involved in these processes may have opposing requirements. This makes studying microbial communities and individual populations involved in bioremediations a real challenge and requires thorough understanding of the metabolic processes involved. Perhaps the most striking and explicit difference in requirements for microbes is their need for the presence or absence of oxygen or alternative external electron acceptors. A general pattern seen in most aquatic sediments and soils is that organic matter is metabolized through a succession of microbial processes. As long as oxygen is available organic matter will be oxidized by aerobic microbes. After its depletion this organic matter is metabolized by a succession of different anaerobic electron accepting processes. This will usually start with nitrate reduction, followed by manganese and iron reduction, sulphate reduction and finally methanogenesis. Mono- and dihalogenated compounds can be degraded rather easily under oxic conditions and many bacterial species capable of doing so have been isolated and studied extensively. In contrast, under anoxic conditions the opposite is true: anaerobic transformation processes seem to be more effective with higher halogenated aromatics and aliphatics (Gerritse and Gottschal, 1992). Yet, in spite of the current awareness of the enormous potential of microorganisms capable of degrading halogenated compounds under both oxic and anoxic conditions there is still a wealth of unanswered questions concerning the ways this potential can be put to work in polluted soils and sediments. Some of the following questions have formed the basis for an important part of our laboratory studies over the past few years:

- Which are the environmental conditions leading to significant degradation of simple chlorinated compounds, especially with respect to the availability of oxygen?

- Which organisms are most likely involved in degradation of chlorinated compounds *in situ*, as opposed to those also present but possibly playing only a minor role?

- What is the influence of the presence of alternative terminal electron acceptors on the degradation of chlorinated compounds, especially under anoxic conditions?

- What is the influence of the presence of microbial species not directly involved in the process of dechlorination itself, e.g. competition for common electron donors or hydrogen transfer between members of anaerobic microbial communities?

2. Chlorobenzoate-degradation and the influence of different oxygen concentrations

In nature a significant part of the microbial activity is concentrated at or near oxic/anoxic interfaces, where oxygen concentrations are often low. Bacteria possessing different kinetic characteristics for oxygen and employing different metabolic pathways for the degradation of (halo)aromatic substrates may have to compete with each other in such environments. As an example we have studied the competitiveness of *Pseudomonas* sp. strain A3 relative to *Alcaligenes* sp. strain L6 in batch and in continuous culture (Krooneman *et al.*, 1998). While both of these strains are able to metabolise 3-chlorobenzoate (3CBa), the former was isolated under air saturating conditions and employs the "catechol"-pathway, whereas the latter was isolated under reduced partial pressures of oxygen and was capable of metabolising 3CBa via the "gentisate"-pathway. Competition experiments in batch culture resulted in pure cultures of *Pseudomonas* sp. strain A3 under air saturating conditions. However, if reduced partial pressures of oxygen (2%) were used, *Alcaligenes* sp. strain L6 remained present in substantial numbers after three transfers. As seen in Figure 1, continuous culture experiments demonstrated that *Alcaligenes* sp. strain L6 was able to outcompete *Pseudomonas* sp. strain A3 under oxygen- and under carbon-limiting conditions as long as the dilution rate remained below 0.136 h^{-1} (low oxygen) and below 0.178 h^{-1} (high oxygen).

These results lend preliminary support to the hypothesis that organisms metabolizing chlorobenzoate via the "gentisate"-pathway in particular may play a significant role in natural ecosystems where xenobiotic compounds occur at very low concentrations and in combination with limiting oxygen tensions. The outcome of further chemostat enrichments and subsequent characterization of the isolates supported this notion (Krooneman *et al.*, 2000). This is a most significant result because our knowledge on the metabolism of dechlorinating aerobes is predominantly based on the study of organisms using the "catechol"-type pathways.

A particularly interesting situation occurs in the uppermost few millimetres of aquatic sediments where light may be available to support growth of photoheterotrophic bacteria. From such environments we have isolated 3CBa degrading strains of *Rhodopseudomonas palustris* (van der Woude *et al.*, 1994, Oda *et al.*, 2000). Although such facultatively anaerobic organisms may seem eminently suitable to play a role in degrading chlorinated benzoates in such habitats, laboratory studies have now indicated that their importance may be restricted to zones of the sediment which are anoxic. In cocultures in chemostats with *Alcaligenes* sp. strain L6, grown under oxygen limitation in the light, all 3CBa was eventually metabolized by the aerobic heterotrophs, probably due to inhibition of the anaerobic (chloro)benzoate degrading pathway in *Rps. palustris* (Krooneman *et al.*, 1999). Further studies to enumerate photoheterotrophic bacteria in polluted surface sediments support this view, because direct enumeration with 3Cba as the sole growth substrate in the light did not yield any isolates. Nevertheless, up to approximately $5*10^2$ viable chlorobenzoate degrading photoheterotrophs per gram of sediment were encountered in such experiments if benzoate was supplied as a second substrate. And, most interestingly, when these isolates were incubated with 3Cba alone (in the light) still no growth or degradation was observed for up to 2 months, but after further incubation for up to 4 months the ability to grow on 3Cba alone was acquired by most of the isolates tested. This new ability appeared to be a stable property, expressed instantaneously upon subsequent transfers (Oda et al., 2000).

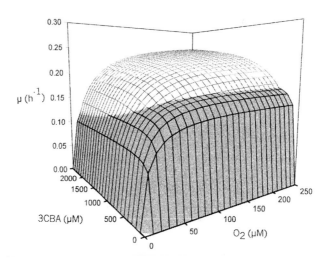

Figure 1: *The 3-chlorobenzoate and oxygen concentrations as a function of the growth rate according to the double substrate Monod-equation in which both oxygen and 3CBa limit the specific growth rate of the culture. The grey area represents Alcaligenes sp. strain L6 and the white area corresponds to Pseudomonas sp. strain A3.*

3. Dehalorespiration of chloroethenes by strictly anaerobic bacteria

Several strictly anaerobic bacteria are capable of dehalorespiration, a metabolic process in which halogenated compounds act as terminal electron acceptors. Among the best-studied organisms known to date are strains of *Desulfitobacterium* spp. For such organisms to play a significant role in removal of halogenated compounds it is important to know whether alternative electron acceptors, which are always present in anoxic environments, interfere with dehalorespiration. For *Desulfitobacterium frappieri* strain TCE1 (Gerritse et al., 1999) we found that, in tetrachloroethene (PCE)-limited continuous cultures with excess of lactate, nitrate and fumarate were both reduced along with PCE and did not affect the rate of PCE dehalogenation. In contrast, PCE dechlorination was strongly inhibited by sulphite. Addition of 2 mM sulphite to the PCE-limited culture resulted in an immediate accumulation of PCE. Sulphite was reduced to sulphide and only after all sulphite had been reduced dechlorination resumed and the culture recovered to its initial steady-state situation. Sulphate cannot be used as an alternative electron acceptor by *D. frappieri* strain TCE1. However, the presence of sulphate does affect dechlorination by this organism in a more indirect way. Competition for these substrates occurs and particularly under sulphate limiting conditions hydrogen equivalents from substrates used by sulphate reducing bacteria can be channelled to the dechlorinating anaerobe. We examined this type of interaction in chemostat cultures of *D. frappieri* strain TCE1 and *Desulfovibrio* strain SULF1, both isolated from soil of a chloroethene-contaminated site of a former laundry in Breda (The Netherlands). *Desulfitobacterium frappieri* strain TCE1 was shown to reductively dechlorinate tetrachloroethene (PCE) and trichloroethene (TCE) to mainly cis-dichloroethene (cis-DCE) with different electron donors such as H_2, lactate, pyruvate, formate, ethanol, and butyrate, whereas *Desulfovibrio* strain SULF1 reduces sulphate to sulphide by using H_2, lactate, pyruvate, formate, and malate as electron donors.

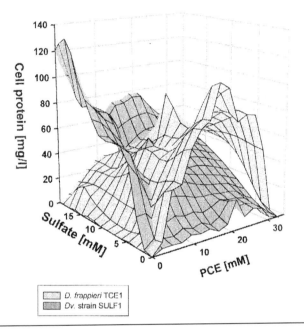

Figure 2: Chemostat coculture composition of Desulfitobacterium *strain TCE1 and* Desulfovibrio *strain SULF1 at different steady states with various PCE and sulphate concentrations in the reservoir medium. Growth conditions: pH 7.2, 35°C, dilution rate = 0.02 h⁻¹, electron donor = 40 mM lactate.*

A coculture of these two strains, grown in a chemostat in which lactate was given in excess (> 40 mM) and PCE and sulphate were added in different concentrations to investigate their influence on the coculture composition (Figure 2) and on the dehalogenating activity of the *D. frappieri* strain TCE1 (Drzyzga *et al.*, 1999). In the presence of excess of sulphate, the *Desulfovibrio* species outcompeted the dehalogenating bacterium, whereas under limitation of both electron acceptors (< 10 mM in the reservoir medium) relatively stable populations of both strains were observed. The coculture experiments also showed that the hydrogen concentration in the absence of sulphate is about 20 fold higher than with sulphate in excess. It is likely that strain TCE1 consumed the bulk of the hydrogen produced by strain SULF1 in the coculture, thereby enabling strain SULF1 to grow without sulphate. In other words, strain TCE1 served as a biological electron acceptor for strain SULF1. This interaction is shown schematically in Figure 3 for a chemostat coculture, supplied with lactate, PCE, and no sulphate. In this steady state, *D. frappieri* strain TCE1 remained the numerically dominant organism.

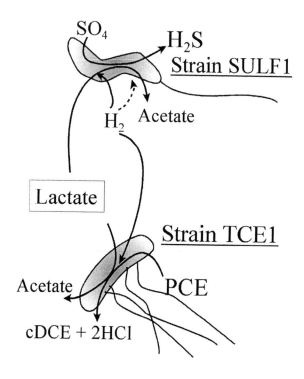

Figure 3: Schematic presentation of some interactions between Desulfitobacterium frappieri *strain TCE1 and* Desulfovibrio *strain SULF1 in a chemostat coculture with 40 mM of lactate, 33 mM of PCE and no sulphate in the reservoir medium.*

Our experiments indicate that hydrogen can be a key electron donor used in dehalorespiration of chloroethenes with various organic electron donors mainly serving as primary precursors for the supply of H_2 via other microbial activities (e.g., fermentative bacteria, sulphate -reducing bacteria). Environmentally such consortia may be important with respect to the clean-up of organochlorine-contaminated ecosystems. In particular because many of the dehalorespiring bacteria isolated so far can use only a restricted set of substrates as electron donors - whereas most of them can use hydrogen - interspecies hydrogen transfer between sulphate-reducing bacteria and dehalorespiring bacteria may add enormously to the substrate availability for the dehalogenating process. Especially in freshwater and soil environments in which sulphate may often be depleted, this situation may be advantageous to both the sulphate reducing and dehalogenating populations. In combination with the relatively high affinity for H_2, determined for strain TCE1 (data not shown) and for other dehalogenating strains, this leads to the conclusion that interspecies hydrogen transfer adds significantly to the competitive success of dehalogenating anaerobes in occupying niches in anoxic environments similar to that occupied by H_2-utilizing methanogenic, homo-acetogenic, and sulphidogenic bacteria.

References

Anderson, R. T. and Lovley, D. R. Ecology and biogeochemistry of in situ groundwater bioremediation. (1997) Adv. Microb. Ecol. 15: 289-350.

Drzyzga, O., J. Gerritse, H. Elissen, J. Dijk, and J.C. Gottschal (2000) Coexistence of a sulphate reducing *Desulfovibrio* species strain SULF1 and the dehalorespiring *Desulfitobacterium frappieri* TCE1 in defined chemostat cultures grown with various combinations of sulphate and tetrachloroethene. Environ. Microbiol. 3:92-99.

Fetzner, S. (1998) Bacterial dehalogenation. Appl. Microbiol. Biotechnol. 50:633-657.

Gerritse, J., O. Drzyzga, G. Kloetstra, M. Keijmel, L.P. Wiersum, R. Hutson, M.D. Collins, and J.C. Gottschal (1999) Influence of different electron donors and acceptors on the dehalorespiration of tetrachloroethene and trichloroethene by *Desulfitobacterium frappieri* strain TCE1. Appl. Environ. Microbiol. 65:5212-5221.

Gerritse, J. and J. C. Gottschal (1992) Mineralisation of the herbicide 2,3,6-trichlorobenzoic acid by a co-culture of anaerobic and aerobic bacteria. FEMS Microbiol. Ecol. 101:89-98.

Gribble, G. W. (1994) The natural production of chloringated compounds. Environ. Sci. Technol. 28:310-319.

Gribble, G. W. (1996) Naturally occurring organohalogen compounds - a comprehensive survey. Prog. Org. Nat. Prod. 68:1-498.

Krooneman, J. , E. R. B. Moore, J. C. L. van Velzen, R. A. Prins, L. J. Forney, and J. C. Gottschal (1998) Competition for oxygen and 3-chlorobenzoate between two aerobic bacteria using different degradation pathways. FEMS Micobiol. Ecol. 26:171-179.

Krooneman, J, S. van den Akker, T. M. Dias Gomes, L. J. Forney, and J. C. Gottschal (1999) Degradation of 3-chlorobenzoate under low oxygen conditions in pure and mixed cultures of the anoxygenic photoheterotroph Rhodopseudomonas palustris DCP3 and an aerobic *Alcaligenes* species. Appl. Environ. Microbiol. 65:131-137.

Krooneman, J, A. O. Sliekers, T. M. Pedro-Gomes, L. J. Forney, and J. C. Gottschal (2000) Characterisation of 3-chlorobenzoate degrading aerobic bacteria isolated under various environmental conditions. FEMS Micobiol. Ecol. 32:53-59.

Mohn, W. W. and J. M. Tiedje (1992) Microbial reductive dehalogenation. Microbiol. Rev. 56:482-507.

Oda, Y, de Vries, Y. P, Forney, L. J, and Gottschal, J. C. (2001) Acquisition of the ability to degrade chlorinated benzoic acids by the purple non-sulfur bacterium *Rhodopseudomonas palustris*. (In preparation.)

van der Woude, B. J. , M. de Boer, N. M. J. van der Put, F. M. van der Geld, R. A. Prins, and J. C. Gottschal (1994) Anaerobic degradation of halogenated benzoic acids by photoheterotrophic bacteria. FEMS Microbiol. Lett. 119:199-207.

CHARACTERISED REACTIONS IN AEROBIC AND ANAEROBIC UTILISATION OF LINEAR ALKYLBENZENESULPHONATE (LAS)

WENBO DONG, SASKIA SCHULZ, DAVID SCHLEHECK AND ALASDAIR M. COOK
Department of Biology, The University, D-78457 Konstanz, Germany
email: Alasdair.Cook@uni-konstanz.de.

Abstract

Commercial linear alkylbenzenesulphonate surfactant (LAS) is known to be fully degradable aerobically, as a carbon and energy source for bacteria, but up till now, no detail on any reaction has been available. Data on the microbiology of degradative consortia and on degradative reactions are now becoming available. LAS is also known to be desulphonated by aerobic bacteria under sulphate-limiting conditions. LAS was thought to be non-biodegradable under anoxic conditions; here too, a biotransformation reaction (desulphonation) has been discovered.

1. Introduction

A biochemist aiming to purify a membrane-protein uses a quality detergent, i.e. a pure compound, to solubilise the protein. Her husband, at home minding house and child, uses detergents to wash the clothes, the dishes and the floor; in this case, however, the word 'detergent' does not mean a pure compound, but a complex formulation including one or more surfactants to separate 'soil' from what is soiled. Different parts of the chemical industry thus use the same word for totally different products. The basic phenomenon is the same in both cases, however. The 'surface-acting' properties of individual chemicals, or groups of chemicals, are required, and these compounds are called 'surfactants'.

Surfactants come in three large families, anionics, cationics and non-ionics, and many other smaller categories [27], of which the anionics are the major class on a world-wide tonnage basis [36]. The major contributor to this tonnage, at about 2,500,000 tonnes/annum and rising, is LAS, the linear alkylbenzenesulphonates [36].

LAS is not a single compound, but, ideally, a mixture of 20 compounds, all sub-terminally-substituted, linear, alkyl chains (C10-C13) carrying a 4-sulphophenyl moiety (cf. Fig. 1). Of these 20 compounds, 18 are optically active, so there are 38 structures in the ideal mixture. The reasons for this mixture involve solubility, environmental factors

177

S.N. Agathos and W. Reineke (eds.),
Biotechnology for the Environment: Strategy and Fundamentals, 177–184.
© 2002 *Kluwer Academic Publishers. Printed in the Netherlands.*

and production techniques, which, together with information on impurities, can be found elsewhere [e.g. 27].

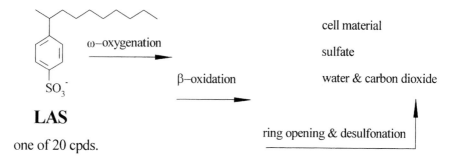

Figure 1. A representative linear alkylbenzenesulphonate (LAS) and its putative major degradative pathway.

One of many reasons for the predominance of LAS is the fact that this group of compounds itself is fully biodegradable [cf. 24-26], as has been known for many years [e.g. 13, 31, 33, 38, 41]. Nonetheless, exhaustive work has gone into confirming this repeatedly [e.g. 34]. If these studies reflect the limit of (government) regulatory imagination and cost effectiveness, the latter are limited, because we still know little more about the degradation of LAS than we did when Swisher first published 30 years ago [39; cf. 4]. Essentially, LAS is removed to about 99.9 % by a functional sewage treatment plant, but the compounds can accumulate in (anaerobic) sediment, if they are allowed into a river [34, 40]. From a less negative standpoint, the data allow a thorough ecological risk assessment of surfactants [1].

The interest of the authors is, thus, not whether LAS is degraded, but under which conditions, and how. And because said authors are still working with many unpublished data, relatively little information will be presented in detail here.

2. What LAS offers microorganisms

Organosulphonates can be regarded as carbon and energy or as sulphur sources for aerobic microorganisms [5], whereby the former represents the conditions in an activated sludge wastewater treatment plant. Anaerobic microorganisms also conserve energy and derive carbon from the degradation of organosulphonates or they can assimilate sulphonate sulphur [5, 28].

We want to explore each of these four possibilities:

- LAS as carbon and energy source for aerobes (Section 3)

- LAS as sole source of sulphur for aerobes (Section 4)

- LAS as carbon and energy source for anaerobic bacteria (Section 5)

- LAS as sole source of sulphur for anaerobic bacteria (Section 6)

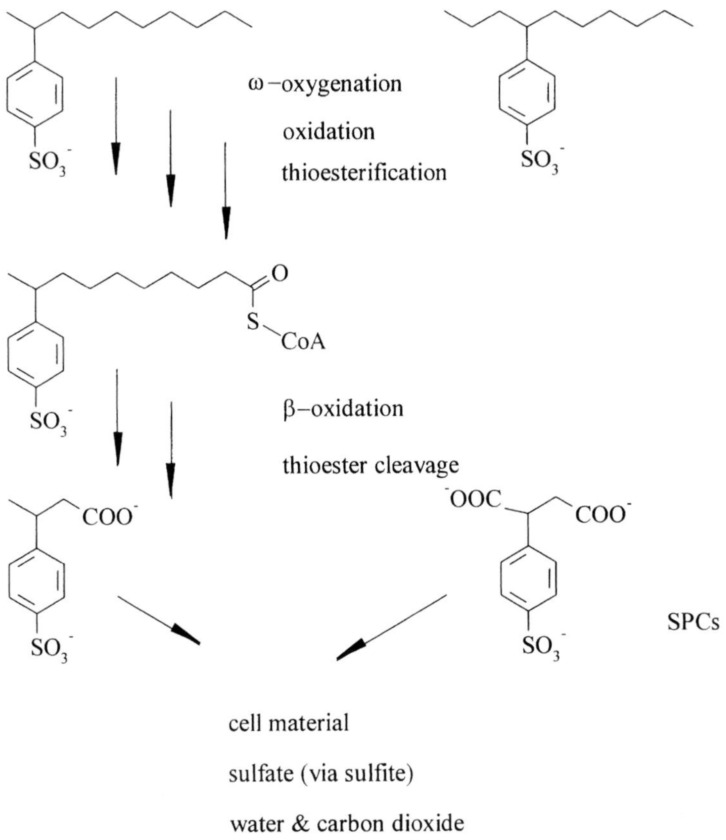

ω–oxygenation

oxidation

thioesterification

SO₃⁻

SO₃⁻

β–oxidation

thioester cleavage

SO₃⁻

SO₃⁻

SPCs

cell material

sulfate (via sulfite)

water & carbon dioxide

Figure 2. Some implications of the general pathway shown in Fig. 1. The ω-oxygenation of a methyl group is presumably followed by oxidation steps to yield the corresponding sulphophenylcarboxylate [14], which would then have to be converted to the CoA ester for β-oxidation.
Free, short-chain sulphophenylcarboxylates (SPC) are routinely observed [e.g. 9, 12], and individual species are fully degraded [e.g. 25].

3. LAS as carbon and energy source for aerobes

The sewage works is the targeted destiny of LAS, where it is essentially quantitatively eliminated [34], some of this being achieved by sorption to biomass [13]. There are many data to support a range of attacks on the LAS molecule [26, 33, 38], but a generally accepted pathway is found in all reviews, and it is shown in Fig. 1 [4].

3.1. THE NUMBER OF BACTERIA INVOLVED

The simple picture in Fig. 1 masks a wealth of phenomena. One major problem is the number of organisms involved. Whereas one recent review anticipates all reactions in a single microbe [41], others anticipate a consortium [cf. 4], as is observed in at least two environments [19, 37]. Our own data [35] indicate that a bacterium able to degrade 2-(4-sulphophenyl)butyrate can degrade the 'model LAS', 1-octyl(4-phenylsulphonate), but not commercial LAS (Fig. 1), so we suspect that 'model LAS' is not representative of commercial LAS. Correspondingly, the one review [41] cites work with model LAS, whereas the consortia involved commercial LAS [19, 37].

3.2. ω-OXYGENATION AND ß-OXIDATION

The nature of the ω-oxygenation is still unclear, but a key suggestion has been made, namely that it is methane monooxygenase [16-18]. Thus a methanotroph, which is presumably unable to further degrade the product of oxygenation, would oxygenate LAS and leave the further metabolism to other members of the consortium. Environmental data showing sulphophenylcarboxylates with no shortening of the alkyl sidechain would support this idea [14].

Our unpublished data with trickling filters [10] indicate the absence of methanotrophs (PCR methods) but the presence of ammonia oxidisers. Ammonia monooxygenase resembles particulate methane monooxygenase [15, 29]. We thus postulated that ammonia oxidisers attack LAS, at both available terminal methyl groups, if appropriate (Fig. 2), with heterotrophic bacteria catalysing other reactions.

The trickling filter was too complex an ecosystem for the follow-up experiments, so we tried to establish our LAS-degrading bacteria in suspended culture. This functioned only if we supplied a support in the liquid medium, and we often used polyester fleece [cf. 25]. Support for the hypothesis involving ammonia oxidisers came when we replaced the ammonia in the growth medium with nitrate and observed no degradation. Complementary experiments involving an inhibitor of ammonia monooxygenase showed merely a retardation of degradation, so we currently believe that both ammonia oxidisers and heterotrophic bacteria could initiate the attack on LAS.

We have isolated a pure culture of a heterotroph able to convert LAS to sulphophenylcarboxylates (cf. Fig. 2) [32]. The same range of products was formed from LAS by an enrichment culture grown in media to enrich ammonia oxidisers.

Our working hypothesis is, thus, that the general scheme in Fig. 2 is correct. ß-oxidation, proven for the degradation of model LAS [2], has recently been confirmed for degradation of commercial LAS [10]. In the case of the heterotroph, we presume that a second organism is required for desulphonation and ring cleavage, but the system may be more complex.

3.3. RING CLEAVAGE AND DESULPHONATION

Recent observations indicate the importance of chirality in the degradation of environmental chemicals [21]. In our hands, the degradation of a representative sulphophenylcarboxylate, 2-(4-sulphophenyl)butyrate, by a pure culture involves enantiomer-specific degradation of the racemic substrate, whose degradation involves

4-sulphocatechol [35]. The degradative pathway elucidated by Feigel and Knackmuss [11] is used [35] and seems to be widespread [10].

3.4. OUTLOOK

We hope to generate a multi-component defined mixed culture to degrade commercial LAS, but given the possibility that we have also ammonia oxidisers involved in attacking LAS, we suspect that many alternative enzymes could be involved [10, 17]. When these alternatives are characterised, and some of the enzymology understood, a start can be made to explore the other reactions recognised by earlier reviewers [34, 38].

4. LAS as sole source of sulphur for aerobes

We anticipate that a sewage treatment plant contains relatively high concentrations of sulphate, a preferred sulphur source for microbes [23], in part from the general water supply and in part from the degradation of sulphate esters and sulphonates. Other environments, predominantly soils, are known to contain little free sulphate [cf. 23]. Under conditions of starvation for sulphate and cysteine, a scavenging system for sulphur is expressed which has been termed the sulphate-starvation induced (SSI) response [23], which Kertesz has reviewed in the light of new data [22].

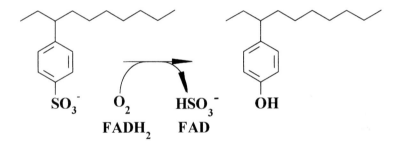

Figure 3. Presumed reaction in the monooxygenolytic desulphonation of LAS.

As far as LAS is concerned, the enzymes of the SSI response monooxygenate the compounds with concomitant loss of the sulphite moiety (Fig. 3).

Fig. 3 is deduced from the known reactions with 4-toluenesulphonate [20] and the product detected from LAS [23]. It represents the best understood biotransformation of LAS known to date.

5. LAS as carbon and energy source for anaerobic bacteria

This literature has been reviewed elsewhere, with the conclusion that no published report provided support for any dissimilation of LAS under anoxic conditions [30].

6. LAS as sole source of sulphur for anaerobic bacteria

The conclusion that no data support the dissimilation of LAS under anoxic conditions need not mean that the desulphonation of LAS by anaerobic bacteria is impossible. In a series of papers, Denger et al. have shown that inert alkylsulphonates and arylsulphonates, including LAS, are desulphonated [6-8; see also 3]. The nature of the reaction(s) catalysed is still unknown.

7. Membrane transport of organosulphonates

In a recent review on the degradation of organosulphonates we stressed the importance of membrane transport in the metabolism of these highly polar compounds [5]. We certainly expect transport to be relevant for the smaller molecules mentioned in this review (e.g. SPC, Fig. 2). It is, however, unclear whether LAS, which sorbs so well to biomass [e.g. 13, 23], can actually penetrate membranes, or whether here also a transport mechanism is necessary.

8. Conclusions

LAS has been in use for nearly 40 years, but only now are the analytical problems involved in working with these complex chemical mixtures being overcome. The complexity led to 'model LAS' being used, effectively without proof that it was a valid model; we suspect that it is not a valid model. Correspondingly, the data indicate that degradation of commercial LAS involves consortia, representative members of which are now under study in our laboratory, and should yield information on the nature of the organisms, enzymes and transport processes involved.

Commercial LAS is subject not only to degradation as a carbon source but also as a sulphur source, and it is likely that we know the family of enzymes involved in the desulphonation. A corresponding desulphonation under strictly anoxic conditions is also known, but we do not even know the nature of the product(s). There is, as yet, no proof of dissimilation of LAS under anoxic conditions.

Acknowledgements

We are grateful to S. Radajewski and J. C. Murrell, University of Warwick, for cooperation with the PCR analyses, to P. Eichhorn and T. Knepper, ESWE, Wiesbaden, for cooperation with the LC-MS analyses, and to M. A. Kertesz, ETH-Zürich and University of Manchester, for making available unpublished data. WD was supported by an exchange studentship from the China-Gesellschaft EV. Our research was funded by the University of Konstanz, the DFG, ECOSOL, DOW, BASF and the Fonds der Chemischen Industrie.

References

1. Andree, H. 1997. Conclusions from the colloquium of the Hauptausschuß Detergenzien 'The state of the debate on the ecological risk assessment of surfactants' on 24th April 1997 in Frankfurt. Tenside, Surfactants, Detergents 34:224.
2. Cain, R. B. 1981. Microbial degradation of surfactants and "builder" components, p. 323-370. *In* T. Leisinger, A. M. Cook, R. Hütter, and J. Nüesch (ed.), Microbial degradation of xenobiotics and recalcitrant compounds. Academic Press, London, United Kingdom.
3. Chien, C.-C., E. R. Leadbetter, and W. Godchaux III. 1995. Sulfonate-sulfur can be assimilated for fermentative growth. FEMS Microbiology Letters 129:189-194.
4. Cook, A. M. 1998. Sulfonated surfactants and related compounds: facets of their desulfonation by aerobic and anaerobic bacteria. Tenside, Surfactants, Detergents 35:52-56.
5. Cook, A. M., H. Laue, and F. Junker. 1998. Microbial desulfonation. FEMS Microbiological Reviews 22:399-419.
6. Denger, K., and A. M. Cook. 1997. Assimilation of sulfur from alkyl- and arylsulfonates by *Clostridium* spp. Archives of Microbiology 167:177-181.
7. Denger, K., and A. M. Cook. 1999. Linear alkylbenzenesulfonate (LAS) bioavailable to anaerobic bacteria as a source of sulfur. Journal of Applied Microbiology 86:165-168.
8. Denger, K., M. A. Kertesz, E. H. Vock, R. Schön, A. Mägli, and A. M. Cook. 1996. Anaerobic desulfonation of 4-tolylsulfonate and 2-(4-sulfophenyl)butyrate by a *Clostridium* sp. Applied and Environmental Microbiology 62:1526-1530.
9. di Corcia, A., R. Samperi, and A. Marcomini. 1994. Monitoring aromatic surfactants and their biodegradation intermediates in raw and treated sewages by solid-phase extraction and liquid chromatography. Environmental Science and Technology 28:850-858.
10. Dong, W., S. Radajewski, P. Eichhorn, K. Denger, T. Knepper, J. C. Murrell, and A. M. Cook. 2000. Linear alkylbenzenesulfonate (LAS) surfactant: bacterial communities initiate catabolism by various ω-oxygenations followed by ß-oxidation, ring cleavage and desulfonation. Submitted.
11. Feigel, B. J., and H.-J. Knackmuss. 1993. Syntrophic interactions during degradation of 4-aminobenzenesulfonic acid by a two species bacterial culture. Archives of Microbiology 159:124-130.
12. Field, J. A., D. J. Miller, T. M. Field, S. B. Hawthorne, and W. Giger. 1992. Quantitative determination of sulfonated aliphatic and aromatic surfactants in sewage sludge by ion-pair/supercritical fluid extraction and derivatization gas chromatography/mass spectrometry. Analytical Chemistry 64:3161-3167.
13. Giger, W., A. C. Alder, P. H. Brunner, A. Marcomini, and H. Siegrist. 1989. Behaviour of LAS in sewage and sludge treatment and in sludge-treated soil. Tenside, Surfactants, Detergents 26:95-100.
14. González-Mazo, E., M. Honing, D. Barceló, and A. Gómez-Parra. 1997. Monitoring long-chain intermediate products from the degradation of linear alkylbenzene sulfonates in the marine environment by solid-phase extraction followed by liquid chromatography/ionspray mass spectrometry. Environmental Science and Technology 31:504-510.
15. Holmes, A. J., A. Costello, M. E. Lidstrom, and J. C. Murrell. 1995. Evidence that particulate methane monooxygenase and ammonia monooxygenase may be evolutionarily related. FEMS Microbiology Letters 132:203-208.
16. Hrsak, D. 1995. Aerobic transformation of linear alkylbenzenesulphonates by mixed methane-utilizing bacteria. Archives of Environmental Contamination and Toxicology 28:265-272.
17. Hrsak, D., and A. Begonja. 1998. Growth characteristics and metabolic activities of the methanotrophic-heterotrophic groundwater community. Journal of Applied Microbiology 85:448-456.
18. Hrsak, D., and D. Grbic-Galic. 1995. Biodegradation of linear alkylbenzenesulfonates (LAS) by mixed methanotrophic-heterotrophic cultures. Journal of Applied Bacteriology 78:487-494.
19. Jiménez, L., A. Breen, N. Thomas, T. W. Federle, and G. S. Sayler. 1991. Mineralization of linear alkylbenzene sulfonate by a four-member aerobic bacterial consortium. Applied and Environmental Microbiology 57:1566-1569.
20. Kahnert, A., P. Vermeij, C. Wietek, P. James, T. Leisinger, and M. A. Kertesz. 2000. The *ssu* locus plays a key role in organosulfur metabolism in *Pseudomonas putida* S-313. Journal of Bacteriology 182:2969-2878.
21. Kanz, C., M. Nölke, T. Fleischmann, H.-P. E. Kohler, and W. Giger. 1998. Separation of chiral biodegradation intermediates of linear alkylbenzenesulfonates by capillary electrophoresis. Analytical Chemistry 70:913-917.
22. Kertesz, M. A. 2000. Riding the sulfur cycle - metabolism of sulfonates and sulfate esters in Gram-negative bacteria. FEMS Microbiological Reviews.

23. Kertesz, M. A., P. Kölbener, H. Stockinger, S. Beil, and A. M. Cook. 1994. Desulfonation of linear alkylbenzenesulfonate surfactants and related compounds by bacteria. Applied and Environmental Microbiology 60:2296-2303.
24. Kölbener, P., U. Baumann, T. Leisinger, and A. M. Cook. 1995. Linear alkylbenzenesulfonate (LAS) surfactants in a simple test to detect refractory organic carbon (ROC): attribution of recalcitrants to impurities in LAS. Environmental Toxicology and Chemistry 14:571-577.
25. Kölbener, P., U. Baumann, T. Leisinger, and A. M. Cook. 1995. Non-degraded metabolites arising from the biodegradation of commercial linear alkylbenzenesulfonate (LAS) surfactants in a laboratory trickling filter. Environmental Toxicology and Chemistry 14:561-569.
26. Kölbener, P., A. Ritter, F. Corradini, U. Baumann, and A. M. Cook. 1996. Refractory organic carbon and sulfur in the biotransformed by-products in commercial linear alkylbenzenesulfonate (LAS): identifications of arylsulfonates. Tenside, Surfactants, Detergents 33:149-156.
27. Kosswig, K. 1994. Surfactants, p. 747-817. In W. Gerhartz and B. Elvers (ed.), Ullmann's encyclopedia of industrial chemistry, 5 ed, vol. A25. VCH, Weinheim.
28. Lie, T. L., J. R. Leadbetter, and E. R. Leadbetter. 1998. Metabolism of sulfonic acids and other organosulfur compounds by sulfate-reducing bacteria. Geomicrobiology Journal 15:135-149.
29. Moir, J. W., L. C. Crossman, S. Spiro, and D. J. Richardson. 1996. The purification of ammonia monooxygenase from *Paracoccus denitrificans*. FEBS Lett 387:71-4.
30. Painter, H. A., and F. E. Mosey. 1992. The anaerobic biodegradability of linear alkyl benzene sulfonate (LAS). 3rd CESIO Internat. Surfact. Cong. pp. 34-43, London, England, June, 01-05, 1992.
31. Sawyer, C. N., and D. W. Ryckman. 1957. Anionic synthetic detergents and water supply problems. Journal of the American Water Works Association 49:480-490.
32. Schleheck, D., W. Dong, K. Denger, E. Heinzle, and A. M. Cook. 2000. An α-proteobacterium converts linear alkylbenzensulfonate (LAS) surfactants into sulfophenylcarboxylates, and linear alkyldiphenyletherdisulfonate surfactants into sulfodiphenylethercarboxylates. Applied and Environmental Microbiology 66:1911-1916.
33. Schöberl, P. 1993. Biologischer Tensid-Abbau, p. 407-464. In K. Kosswig and H. Stache (ed.), Die Tenside. Carl Hanser Verlag, München.
34. Schöberl, P. 1997. Linear alkylbenzenesulphonate (LAS) monitoring in Germany. Tenside, Surfactants, Detergents 34:233-237.
35. Schulz, S., W. Dong, U. Groth, and A. M. Cook. 2000. Enantiomeric degradation of 2-(4-sulfophenyl)butyrate via 4-sulfocatechol in *Delftia acidovorans* SPB1. Applied and Environmental Microbiology 66:1905-1910.
36. Schulze, K. 1996. Der westeuropäische Tensidmarkt 1994/1995. Tenside, Surfactants, Detergents 33:94-95.
37. Sigoillot, J.-C., and M.-H. Nguyen. 1992. Complete oxidation of linear alkylbenzene sulfonate by bacterial communities selected from coastal seawater. Applied and Environmental Microbiology 58:1308-1312.
38. Swisher, R. D. 1987. Surfactant biodegradation, second ed. Marcel Dekker, New York, NY.
39. Swisher, R. D. 1970. Surfactant biodegradation, fifst ed. Marcel Dekker, New York.
40. Tabor, C. F., and L. B. Barber II. 1996. Fate of linear alkylbenzene sulfonate in the Mississippi river. Environmental Science and Technology 30:161-171.
41. White, G. F., and N. J. Russell. 1994. Biodegradation of anionic surfactants and related molecules, p. 143-177. In C. Ratledge (ed.), Biochemistry of microbial degradation. Kluwer, Dordrecht.

MECHANISMS INVOLVING THE AEROBIC BIODEGRADATION OF PCB IN THE ENVIRONMENT

DENNIS D. FOCHT[1], MICHAEL V. MCCULLAR[2], DENISE B. SEARLES[3], SUNG-CHEOL KOH[4]

[1] *University of California, Riverside, CA, USA, fax:909-787-4294*
email: focht@citrus.ucr.edu
[2] *SuperGen, Dublin, CA, USA,*
[3] *Michigan State University, MI, USA,*
[4] *Korea Maritime University, Pusan, KOREA*

1. Introduction

Polychlorinated biphenyls (PCBs) are among the most persistent environmental contaminants known. Because they are comprised of 60 to 80 congeners, they present a more formidable challenge to microorganisms than a single compound. Remediation strategies are still in the research and development phase although promise has been demonstrated with a few limited field trials. The current paradigm is that aerobic biodegradation is generally limited to PCB mixtures having an average mass percentage of 42% chlorine (e.g. Aroclor 1242). Although metabolism of higher chlorinated congeners (e.g. hexachlorobiphenyls) has been demonstrated in cultures grown with biphenyl, the process has, so far, not been implemented in soils and sediments contaminated by Aroclors 1254 (54% Cl) or 1260 (60% Cl). These more highly chlorinated congeners, nevertheless, undergo slow reductive dehalogenation in flooded soils and sediments to less chlorinated congeners, which would be susceptible to aerobic biodegradation. Coupled anaerobic and aerobic metabolism in sediments is probably how biodegradation of highly chlorinated congeners occurs in sediments. Recent studies have shown that addition of brominated analogs (Wu *et al.*, 1999) or ferrous sulphate (Zwiernik *et al.*, 1998) enhances dehalogenation of PCBs. However, the organisms involved in anaerobic dehalogenation have not been isolated, and this has complicated the task of understanding and manipulating the process to our advantage.

Despite the current paradigm on the limits of aerobic biodegradation of PCBs, our knowledge has, for the most part, been focused on a single pathway, namely the catabolism of biphenyl (Figure 1). Discoveries of reactions not common to the biphenyl pathway raise the question as to whether or not alternative aerobic pathways exist in the environment that can be induced to effect greater metabolism towards more highly chlorinated congeners (Ahmad *et al.*, 1991; Massé *et al.*, 1989; Omori *et al.*, 1988). This review will focus on aerobic biodegradation of PCBs and the factors that contribute to their persistence in the environment.

S.N. Agathos and W. Reineke (eds.),
Biotechnology for the Environment: Strategy and Fundamentals, 185–203.

2. Bioavailability

Microorganisms take up substrates more rapidly from fluids than from solid or sorbed states. Biphenyl, the common growth substrate for PCB-degrading bacteria, has an aqueous solubility of 8.0 mg/L, which is sufficiently low to effect slow growth of these bacteria in liquid media. Growth on biphenyl is much more rapid in the vapour phase when cultures are grown on solid media on sealed agar plates. Havel and Reineke (1995) observed that growth on biphenyl and 2-chlorobiphenyl could be greatly enhanced in liquid medium by supplying the substrate on a folded filter to provide a large surface that dispensed the vapour in the culture flask

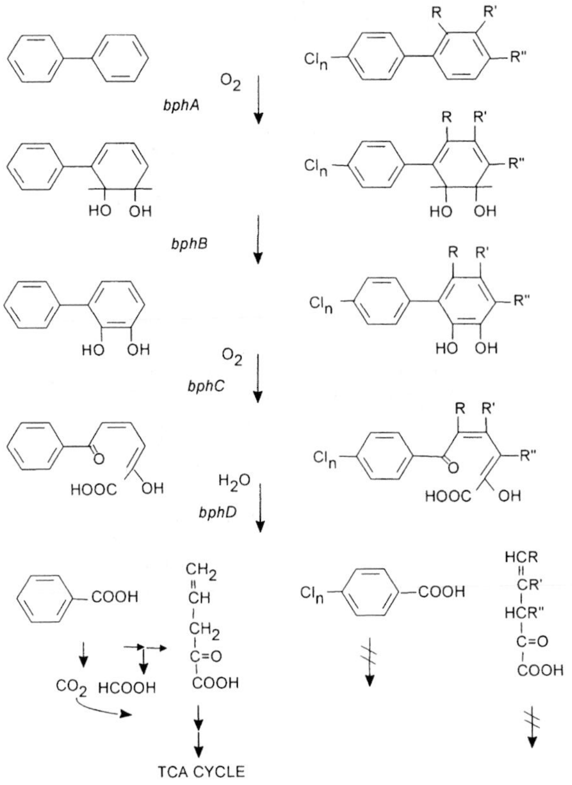

Figure 1. Schematic of the aerobic biphenyl/PCB degradation pathway. Bph refers to the genes that encode the respective enzymes. and R,R', R" refer to either Cl or H substitution.

This method did not work with 3,5-dichlorobiphenyl, which had a higher vapour pressure. Nevertheless, growth on 3,5-dichlorobiphenyl accelerated after accumulation of the yellow meta ring cleavage product (See Figure 1). By measuring the surface tension, they showed that the yellow compound possessed detergent-like activities, and increased the amount of 3,5-dichlorobiphenyl dissolved in the medium. The yellow ring fission product is a common observation among researchers involved in PCB degradation. It is occasionally observed with biphenyl, but always noted with 4-chlorobiphenyl as the growth substrate. Thus, the observations by Havel et al indicate an important role of the ring fission product as a factor enhancing PCB uptake.

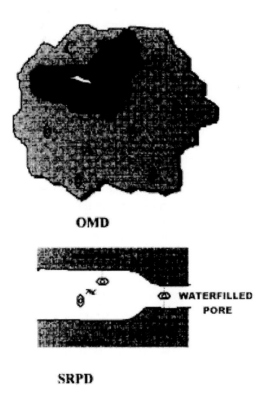

Figure 2. Entrapment of contaminants in soil by organic matter coating (OMD) and by clay lattices (SRPD). From (Pignatello and Xing, 1996). Reproduced by courtesy copyright ACS publication, not profit organzation.

Solubility of mono- and di-chlorobiphenyls are much higher than what is observed for commercial Aroclors 1242, 1248, and 1260, which are 0.2, 0.04, and 0.025 mg/L respectively (Waid, 1986). Addition of specific surfactants to promote growth of PCB-metabolising bacteria has effected enhanced degradation of Aroclors 1242 and 1248 in field trials (Lajoie *et al.*, 1994; Lajoie *et al.*, 1993). Populations of the inoculant strains increased to 10^9 cells/ml, which resulted in a half-fold reduction of the indicator congener 2,3,2',5'-tetrachlorobiphenyl over 50 to 60 days. Although surfactants show promise for enhancing bioavailability of PCBs, not all of them support growth of bacteria that co-metabolise PCBs. Moreover, they may also increase the availability of other industrial oils associated with PCB contaminated soils. This could be beneficial or detrimental depending on whether or not the oils select for or against PCB degraders. A more recent approach has been to select for PCB degrading bacteria that produce surfactants (Shi *et al.*, 1998) or clone genes from surfactant degraders into PCB degraders. The former approach is more desirable because of governmental regulations regarding the use of genetically engineered microorganisms in the environment.

Although it is reasonable that addition of specific surfactants would enhance biodegradation of PCBs, it is less clear how effectively it works with aged contaminants, in contrast to soils spiked with known amounts of PCBs. "Bound residues" has been a term used for many years to describe that fraction of a xenobiotic chemical that is not readily extracted from soil with solvents. Steinberg *et al.* (1987) were the first to demonstrate this phenomenon with EDB (ethylene dibromide) that had been a soil contaminant many years prior to analysis. They periodically extracted the same soil samples by Soxhlet procedures over 200 days, yet continued to recover diminishing quantities of the contaminant. The conclusion of their studies was that equilibrium between "free" and sorbed or bound contaminant existed in soil, and that extractions shifted the equilibrium continuously in the direction of the free state. The two primary mechanisms accounting for slow release of chemicals from soil are sorption to and coating by organic matter and entrapment in the clay mineral pores (Figure 2).

3. Spatial variability

Laboratory investigations characteristically involve the addition (spiking) of known quantities of PCBs to cultures, soil, sediments, or sludge. PCB-spiked microcosms are useful for determining differences in rates of PCB transformation as effected by substrates, microbial inoculants, or environmental parameters. Demonstration of remediation on a small field scale, however, is far more difficult, due to the inherent spatial variability of the contaminant. Few soil parameters indeed can be found which follow a normal arithmetic distribution. An example of this is shown with concentrations of dieldrin, a chlorocyclodiene insecticide having similar chemical properties as PCBs, over a small field grid (Figure 3). Most soil parameters tend to be log-normally distributed. The consequence of a log normal distribution means that three samples replicates are vary unreliable in making valid statistical comparisons between different treatments. Thus, marked differences that may be seen in uniformly mixed laboratory experiments, where two or three replicates are sufficient, will invariably require more samples for noticing differences between control and remedied field plots. Aside from the statistical difficulties in assessing remediation, spatial variability of concentrations may influence the rates of degradation due to kinetic effects. Uniform degradation rates in the field cannot be assumed to occur throughout the soil matrix. "Hot spots" (areas of high concentration) presumably would not exert toxic effects due to the low water solubility of PCBs. Nevertheless, the generalised effect of oils, such as PCBs, on membrane disruption of bacteria cannot be discounted.

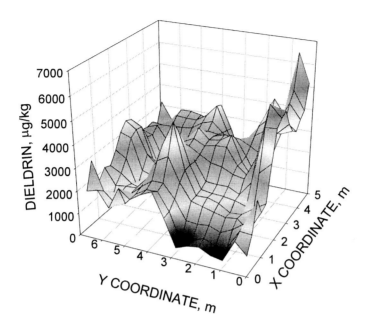

Figure 3. Distribution of dieldrin concentrations over a grid surface in the field.

4. Halogen substitution

It has been well documented that organochlorine compounds are more resistant than their non-chlorinated analogs to microbial degradation since Carson (1962) noted that organochlorine insecticides (e.g. DDT) persisted in the environment. Tucker *et al.* (1975) observed a continual decline in transformation of PCBs from Aroclor 1221 through 1254 with sludge cultures. Similarly, the rate and extent of mineralization between Aroclors 1242 and 1254 is quite marked (Figure 4).

It is unclear exactly how Cl substituents impede biodegradation of PCBs. In terms of aerobic metabolism, it was generally believed that steric effects, specifically, the lack of an adjacent *ortho, meta* site containing no chlorine substituents precluded dioxygenase attack of the aromatic ring. However, Furukawa *et al.* (1979) noted 20 years ago that unidentified chlorinated metabolites were produced from congeners with substituted *ortho* or *meta* sites, and Bedard *et al.* (1987), similarly noted that dioxygenase attack at *ortho, p* sites was more predominate with *Alcaligenes eutrophus* H850 than with Furukawa's isolate.

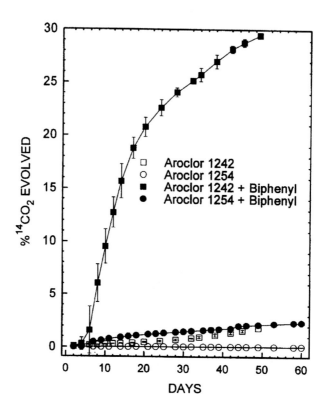

*Figure 4. Comparative mineralization of Aroclor 1242 and 1254 in an Altamont soil.
Aroclor 1242 data redrawn from Focht and Brunner (1985).*

The enigmatic 2,5,2'5' congener, which is not normally found in commercial Aroclors, represents an exception to the dogma of unsubstituted adjacent *ortho, meta* sites. Haddock *et al.* (1995) have established two modes of attack by *Alcaligenes eutrophus* H850. Both modes refute the dogma of an obligatory 2,3-dioxygenase attack on both unsubstituted sites. The 2,3-dioxygenase attack results in spontaneous elimination of the *ortho* chlorine to give the corresponding catechol. This mechanism appears similar to the dioxygenolytic attack of chlorobenzenes. In the case of 1,2,4-trichlorobenzene, dioxygenase attack at the 5,6 position (equivalent *ortho, meta* site to a 2,3 substitution) was preferred (Sander *et al.*, 1991). Substitution of either a nitro group or another Cl atom at the 5 site, resulted in dioxygenolytic loss of the nitro group or chlorine atom with subsequent formation of the corresponding catechols. Thus, formation of a dihydrodiol (first reaction in Figure 1) is bypassed.

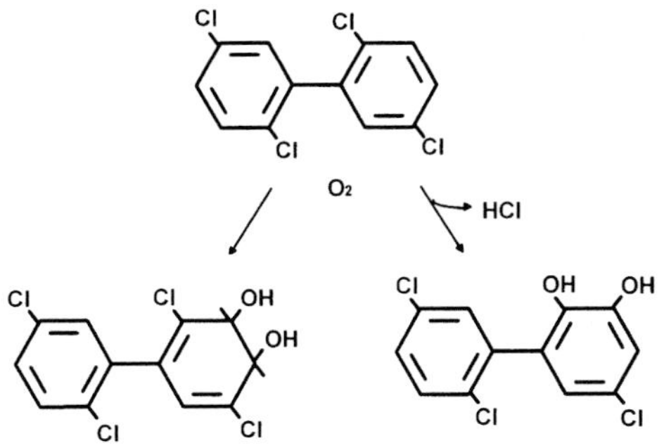

Figure 5. Dioxygenation of 2,5,2',5' tetrachlorobiphenyl. Redrawn from (Haddock et al., 1995)

Dioxygenation of 2,5,2',5'-tetrachlorobiphenyl at the 3,4 site was the first conclusive demonstration of an alternative oxidation of PCBs (Haddock *et al.*, 1995). However, further metabolism of the diol was not demonstrated. Moreover, the likely catecholic intermediate, bearing a 3,4,- instead of a 2,3,- dihydroxy substitution would not likely be metabolised by biphenyl oxidising bacteria, because 3,4-dihydroxybiphenyl (4-phenylcatechol) is not metabolised by biphenyl degraders (Adams *et al.*, 1992; Taira *et al.*, 1988). Thus, the significance of this alternative pathway is unknown and warrants further investigation in finding alternative pathways to PCB degradation.

5. Pathway construction

Although spurious claims have appeared for cultures purported to use PCBs as growth substrates, no such isolates have yet been demonstrated to grow on any of the commercial Aroclor mixtures, that are free of biphenyl. Such claims do not stand up to rigorous proof demonstrating increased cell yields with the concomitant consumption of substrate and release of chloride. Many isolates able to grow on monochlorobiphenyl and a few dichlorobiphenyls have been reported. 4-Chlorobiphenyl is the most common congener that supports growth. In all cases, growth is less extensive with 4-chlorobiphenyl than with biphenyl, and most strains produce 4-chlorobenzoate as a final product. Utilisation of the other two monochloro isomers as sole carbon sources is less common from environmental isolates (Bedard, 1990; Focht, 1993). It has been observed that 2- and 3-chlorobiphenyl are more slowly metabolised than 4-chlorobiphenyl, which has been attributed to steric interference with the 2,3-biphenyl dioxygenase -- particularly with the more recalcitrant 2-chlorobiphenyl. However, because the non-chlorinated ring is oxidised, the steric concept is difficult to support.
The reason for the uncommon usage of 2 or 3-chlorobiphenyls as growth substrates is related to the production of suicidal intermediates, specifically 3-chlorocatechol. A profound concept, advanced many years ago by Reineke and Knackmuss (1979) established that all methyaromatics degrading bacteria cause rupture of the benzene ring by an extradiol dioxygenase, in contrast to chloroaromatic-utilising bacteria, which

cause an intradiol cleavage. Thus, extradiol dioxygenation of 3-chlorocatechol in *B. cepacia* P166 (Arensdorf and Focht, 1994) causes production of an acyl halide, which irreversibly inactivates the enzyme (Figure 6). A similar, but more complex, phenomenon was observed with growth of recombinant strains of *B. cepacia* in the presence of several different congeners (Stratford *et al.*, 1996). Suicidal intermediates were formed from congeners not utilised for growth. These studies suggest that another underlying factor in the slow degradation of environmental PCBs may be the production of suicidal products.

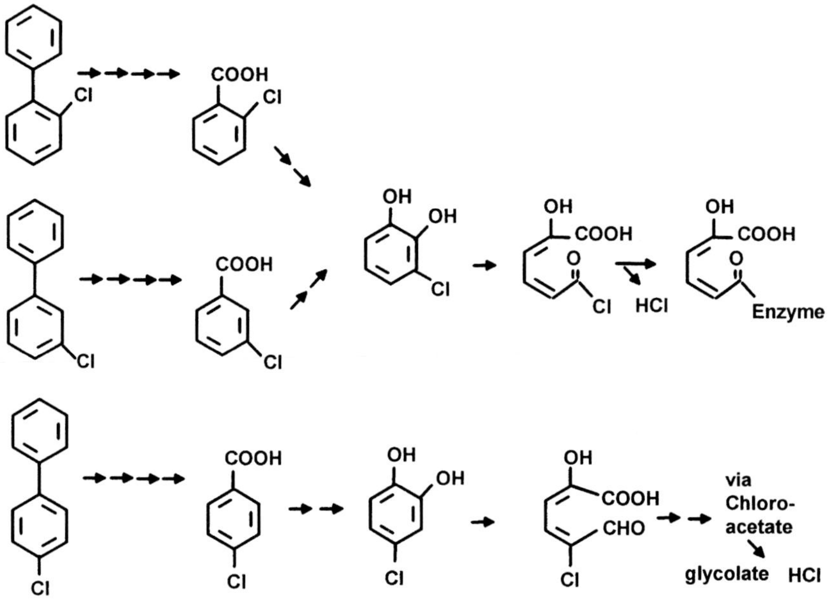

Figure 6. Degradation pathway of monochlorobiphenyls utilised for growth by Burkholderia (Pseudomonas) cepacia P166. From (Arensdorf and Focht, 1994. Reproduced by courtesy copyright ASM, non-profit organiszation

Hence, Reineke and Knackmuss (Reineke and Knackmuss, 1984) advanced the concept of patchwork construction of catabolic pathways, in which a benzene-utilising isolate was converted, after 9 months of adaptation in a chemostat, to a chlorobenzene-degrader. The isolate lacked any *meta* (extradiol) fission activity, but had acquired *ortho* (intradiol) fission, which it used in metabolism of chlorobenzene. Kröckel and Focht (1987) also obtained a chlorobenzene-utiliser by mating two parental strains, a benzene and a 3-chlorobenzoate degrader, through a continuous multi-chemostat system. This recombinant strain differed slightly from that of Reineke and Knackmuss in that it still retained meta fission activity when grown on benzene or toluene. Although it contained a modified TOL (pWWO) plasmid similar to its recipient parental strain, it would not grow on xylenes or toluene, both of which are stronger inducers for *meta* fission activity than benzene. This activity was not expressed when the culture was grown on chlorobenzene, which was metabolised through the *ortho* fission pathway. Additional experiments revealed that the recombinant organism was more similar to the benzene degrader and received the chlorocatechol (*clc*) genes from the chlorobenzoate degrader (Carney *et al.*, 1989).

The isolation and characterisation of other chlorobenzene degrading bacteria confirmed the Reineke-Knackmuss concept: all strains utilised the *ortho* fission pathway (de Bont *et al.*, 1986; Haigler and Spain, 1989; Haigler *et al.*, 1988; Nishino *et al.*, 1992; Pettigrew *et al.*, 1991; Schraa *et al.*, 1986; Spain, 1990; Spain and Nishino, 1987; van der Meer *et al.*, 1991). Exceptions to this rule have only recently been noted. *Pseudomonas putida* GJ31 contains an unusual catechol 2,3-dioxygenase that converts 3-chlorocatechol and 3-methylcatechol, which enables the organism to use both chloroaromatics and methylaromatics simultaneously for growth (Mars *et al.*, 1997; Mars *et al.*, 1999). The enzymes of the modified *ortho* cleavage pathway were never present, while the enzymes of the *meta* cleavage pathway were detected in all cultures. The authors concluded that *P. putida* GJ31 has a *meta* cleavage enzyme which is resistant to inactivation by the acylchloride. These recent findings may have relevance to similar observations noted with metabolism of chlorobiphenyls. The first oxidation of the aromatic ring leads to extradiol cleavage (Figure 1), and no alternative pathway (e.g. intradiol cleavage) is known. In the case of 3-chlorobiphenyl, chlorobenzoate is oxidised to 3-chlorocatechol: this leads to suicidal inactivation of the catechol dioxygenase, as it acts non-lethally on 2,3-dihydroxy-3'-chlorobiphenyl (Adams, et al., 1992). Nevertheless, the organism still is able to grow on 3-chlorobiphenyl.

Pseudomonas acidovorans M3GY (McCullar *et al.*,1994) is a recombinant strain containing genes from three parental strains: *Rhodococcus globerulus* (Asturias *et al.*, 1994), *Pseudomonas* sp. HF1 (Adams, et al., 1992), and *Pseudomonas acidovorans* CC1 (Kohler-Staub and Kohler, 1989; McCullar, 1996), which contribute the biphenyl, chlorocatechol, and chloroacetate genes, respectively. The organism grows on 3,4'-dichlorobiphenyl (3,4'-DCBP) productively by oxidising either aromatic ring, which causes production of 3- and 4-chorobenzoates. Oxidation of both chlorobenzoates occurs through both *ortho* and *meta* fission pathways, yet is suicidal only with 3-chlorobenzoate (Figure 7). Despite suicidal meta cleavage of 3-chlorocatechol, growth of strain M3GY on 3,4'-DCBP occurs. In contrast, *Pseudomonas* sp. CB15, which contains the *bph* and *clc* genes from *Rhodococcus globerulus* and *Pseudomonas* sp. HF1, cannot grow on 3,4'-DCBP. The difference between the two strains is that strain M3GY, unlike strain CB15, metabolises chloroacetate, a degradation product of 4-chlorobenzoate.

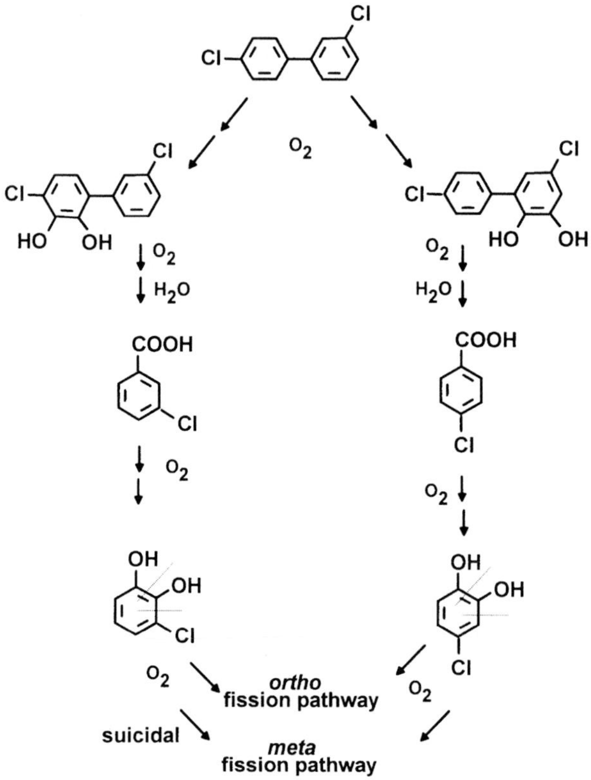

Figure 7. Metabolism of 3,4'-dichlorobiphenyl by recombinant strain Pseudomonas acidovorans *M3GY. From McCullar (1996).*

Vertical pathway construction is focused on assembling genes that encode catabolic enzymes from complementary pathways, as discussed so far. In many cases, the structural genes may be present, but the regulatory genes may be absent or defective. *P. acidovorans* M3GY has all the catabolic enzymes for complete mineralization of 4-chlorobenzoate, yet it is unable to grow on this substrate as a sole carbon source because the substrate alone does not induce for its uptake or oxidation. Only benzoate, 3-chlorobenzoate, and 4-fluorobenzoate can be utilised as sole carbon sources. Thus, when strain M3GY is grown on 3,4'-DCBP, the inducer for 4-chlorobenzoate oxidation is 3-chlorobenzoate, which is produced by oxidation of the 4-Cl substituted ring (Figure 7). Transposon mutagenesis of M3GY produced a mutant strain, designated as M3GY-I, which was able to grow on 4-fluoro-, 4-chloro- and 4-bromobenzoate, and showed rapid oxygen consumption rates in comparison to the wild type strain (Figure 8). Broadening of an existing function, specifically the benzoate dioxygenase, is referred to as horizontal pathway construction.

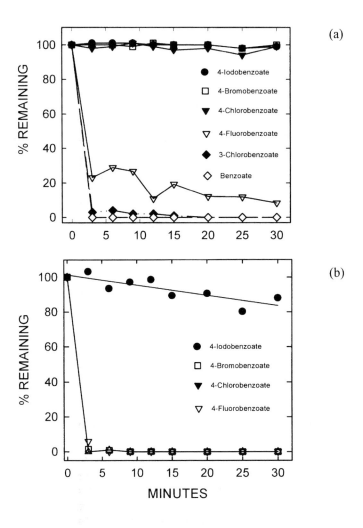

Figure 8. Oxidation of halobenzoates by wild type strain Pseudomonas acidovorans *M3GY (a) and EMS mutagenised strain M3GY-I (b). Substrates that were oxidised were also utilised as sole carbon sources for growth. From McCullar (1996)).*

6. Consortia

Three catabolic types of bacteria have been identified so far in having the complete catabolic pathway for PCBs: 1) biphenyl degraders, 2) chlorobenzoate degraders, 3) chloroacetate degraders. It is uncommon to find all three of these traits in a single organism, which is why the focus has been on strain construction. For example, forty two biphenyl degrading strains, representing nine genera, were isolated from soil, and only one was able to utilise chloroacetate as a growth substrate (Hernandez *et al.*, 1995). In contrast to the ubiquitous nature of biphenyl-degrading bacteria in soil, chlorobenzoate degraders are not readily isolated. Thus, the addition of chlorobenzoate degraders to soil lacking these organisms facilitated transformation and mineralization of Aroclor 1242 (Hickey *et al.*, 1993). Additional studies showed that genetic exchange in soil was brought about through transfer of the *bph* genes from an indigenous soil biphenyl degrader to the inoculant strain (Focht *et al.*, 1996).

Spatial and temporal limitation restrict optimisation of microbial consortia for enhanced degradation of PCBs in the environment. In considering the three types of bacteria that have been so far identified, complete mineralization would require diffusion of chlorobenzoate metabolites from the biphenyl degrader to the chlorobenzoate degrader, and finally chloroacetate to the chloroacetate degraders, under idealised conditions. In reality, growth in soil is very transient and the stationary phase is not long lasting (Figure 9); (Focht and Brunner 1985). Moreover, PCB degraders might literally kill themselves by formation of suicidal products while metabolising chlorobiphenyls. Arensdorf and Focht (1994) noted similar turbidity of *B. cepacia* cells during growth on biphenyl and the three monochlorobiphenyls. However, they found viable cell counts to be 2 orders of magnitude lower with 2- and 3-chlorobiphenyl, which was attributed to acyl halide formation. Thus, it is unlikely that a simple diffusion model of metabolites flowing through soil to other organisms is applicable. More likely, cells grow and die, akin to a steady state system, while effecting dehalogenation through acyl halides (3-chlorocatechol) or the TCA cycle (chloroacetate). Other microorganisms probably exploit the products of their autolysis.

One reason why the addition of chlorobenzoate-degrading bacteria to soil enhances PCB degradation may be in minimising suicidal inactivation of biphenyl degraders through genetic exchange. Enhancement of natural genetic exchange -- in contrast to the addition of recombinants -- has three distinct advantages, not necessarily related. First, the process does not fall under governmental regulations involving recombinant organisms in nature if the chlorobenzoate degrader was isolated from another environment. Second, the constructed strains may not be very good PCB degraders because they are unstable. Third, the indigenous bacteria that co-metabolise PCBs may be more fit than introduced strains, particularly with aged contaminants that are not readily available. An example of the latter two cases involved a study comparing degradation rates of aged PCBs (Aroclor 1248) and spiked PCBs (Aroclor 1242) between inoculants and indigenous bacteria. The recombinants, isolated from a previous study (Focht, et al., 1996), rapidly lost the ability, after inoculation, to utilise chlorobenzoates as none were recovered after 28 days (Searles, 1995). Moreover, the recombinants were no more effective in metabolising PCBs in vitro than their presumed parental strain AW. Although strain AW, isolated from another soil, was dominant over the indigenous flora in utilising biphenyl (Figure 9), it was less effective in metabolising aged PCBs (Aroclor 1248) and spiked PCBs (Aroclor 1242) (Table 1). The obvious point here is that rapid growth at the expense of biphenyl is not necessarily correlated with better oxidation of PCB congeners. Nevertheless, this is precisely how we isolate PCB degraders.

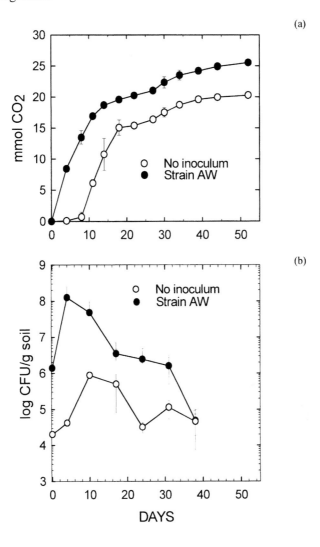

Figure 9. Comparison of biphenyl mineralization in a PCB contaminated industrial soil (a) and growth of biphenyl utilising bacteria (b): 1.0 g biphenyl were added per kg soil (dry mass). (Focht and Brunner 1985). Reproduced by courtesy of ASM, non-profit organiszation.

Table 1. *Comparison of PCB transformation (Aroclors remaining) and mineralization ($^{14}CO_2$) by indigenous biphenyl degraders and inoculant strain AW in an industrial soil (50 days). From Searles (1995)).*

Treatment	Aroclor 1248 (1)	^{14}C-Aroclor 1242 (2)	$^{14}CO_2$ produced (3)
no inoculum + BP	77.2 ± 1.7b[3]	60.3 ± 1.7c	2.93 ±0.65b
strain AW + BP	89.4 ± 2.2a	70.8 ± 2.1b	0.66 ± 0.17a
no inoculum, no BP	100 ± 1.1a	83.7 ± 5.1a	0.98 ± 0.65a

1. *% recovery and analysis by gas chromatography of Aroclor originally present in soil.*
2. *% of spiked ^{14}C-Aroclor 1242 extracted from soil.*
3. *Average values (± standard deviation) followed by a common letter (within a column) were not significantly different (95% level) by Student's t test.*

All cases in which PCB degradation has been significantly enhanced in soil requires the addition of biphenyl, or a related substrate, to raise these organisms from their indigenous level of around 10^4 /g soil to 10^8/g or higher (Focht and Brunner, 1985; Focht, *et al.*, 1996). The addition of biphenyl to dredged sediments contaminated with Aroclor 1242 has been successful on a field scale level as well (Harkness *et al.*, 1993). Since the first studies on PCB degradation (Ahmed and Focht, 1973), biphenyl has characteristically been the substrate used for isolation of PCB degraders. Although research on PCBs has been served well by the use of its non chlorinated analog biphenyl, we should expand our search for alternative substrates and pathways because biphenyl shows little promise for aerobic degradation of highly chlorinated PCBs. Moreover, the US EPA lists biphenyl as a priority pollutant.

The addition of biphenyl to soil demonstrates the validity of the competitive exclusion principle: those bacteria with the faster growth rate on biphenyl will be dominant. In the previous example (Figure 9, Table 1), we showed that fast growth may not necessarily be correlated with broad spread activity for oxidation of PCB congeners. Current isolation methods select for rapid growing prototrophic strains able to utilise biphenyl as a sole carbon source. It is not inconceivable, that slower growing, more fastidious microorganisms exist in the environment that may be more efficacious in oxidising PCBs. This begs the question of how well we have characterised the microflora in soil for this specific function.

7. Alternative strategies: is mineralization necessary?

Addition of biphenyl to soil has resulted in enhanced degradation of PCBs, yet this strategy is limited by the organisms it selects and by governmental agencies that have listed biphenyl as a priority pollutant. Alternative natural substrates, such as terpenes

(Hernandez et al., 1997) and flavinoids (Donnelly et al., 1994), have been shown to enhance PCB degradation to an equal or greater extent than biphenyl. The mechanisms by which natural plant products enhance degradation of PCBs are unclear. The addition of plant residues, rich in terpenes, resulted in complete disappearance of Aroclor 1242 after 6 months with a concomitant several log fold increase in bacteria on dilution plates containing biphenyl as the sole carbon source, even though biphenyl was not added to the soil (Hernandez et al., 1997). However terpenes, for the most part, are oxidised by monooxygenases (Trudgill, 1994). Flavinoids (e.g. ubiquinone) themselves are constituents of many mono- and dioxygenase reactions. It is unclear what role terpenes or flavinoids play in stimulating PCB degradation. It may be likely that both mono- and dioxygenase reactions are favoured as a result of a greater diversity of selection than what occurs with biphenyl.

This has been assumed since the early work on PCB degradation that the rate-limiting step is the initial oxidation of the aromatic ring. Despite more studies that have demonstrated that product inhibition is also an impediment, this early concept is still valid. Clearly, the oxidation of a very insoluble compound to a hydroxylated, more soluble, compound makes it more bioavailable for other organisms to attack.

Some doubt might be raised about the persistence and environmental safety of multichlorinated phenols. The question of metabolite stability is an old one that continues to surface since Bartha et al. (1968) demonstrated that the dimerised product of chloroaniline hydrolysis was far more persistent in soil. A more recent review (Dec and Bollag, 1994) puts this question in its proper form by showing that chlorophenols can become incorporated into soil humus where they become part of the normal constituent of soil organic matter. The fate of chlorocatechols in soil has been well documented to become an innocuous part of the soil organic matter. As they are either broken down by microorganisms or polymerised into humus, chlorophenols are eventually mineralised in a similar fashion to nonchlorinated catechols and phenols (Stott et al., 1982).

The schematic in Figure 10 shows the possible fate of 2,2'4,4',5,5'-hexachlorobiphenyl, a PCB congener that is not readily metabolised by Rhodococcus globerulus or Alcaligenes eutrophus H850, two of the best characterised PCB degraders. Dioxygenolytic attack would effectively route the compound through the biphenyl pathway. Monooxygenolytic dehalogenation would produce a hydroxylated PCB. The fate of these compounds is unknown, as they are not oxidised by biphenyl degrading bacteria (Higson and Focht, 1989; Kohler et al., 1988). Thus, the product is likely to be excreted by the organism effecting its catalysis. Two fates are possible, namely further metabolism by some, as yet, unidentified microorganism, or polymerisation through free radical addition and soil enzymatic coupling to humus. Excretion of diols and catechols from the biphenyl pathway could also become incorporated through the same mechanism. Brown coloured oxidation products resembling humic precursors have been noted in cultures growing on chlorobiphenyls. A dimer, 3-chloro-5(2-hydroxy-3-chlorophenyl)-1,2-benzoquinone was found from the oxidation of 3-chlorobiphenyl, and proposed to be the precursor to the brown humic-like substance (Adams et al., 1992).

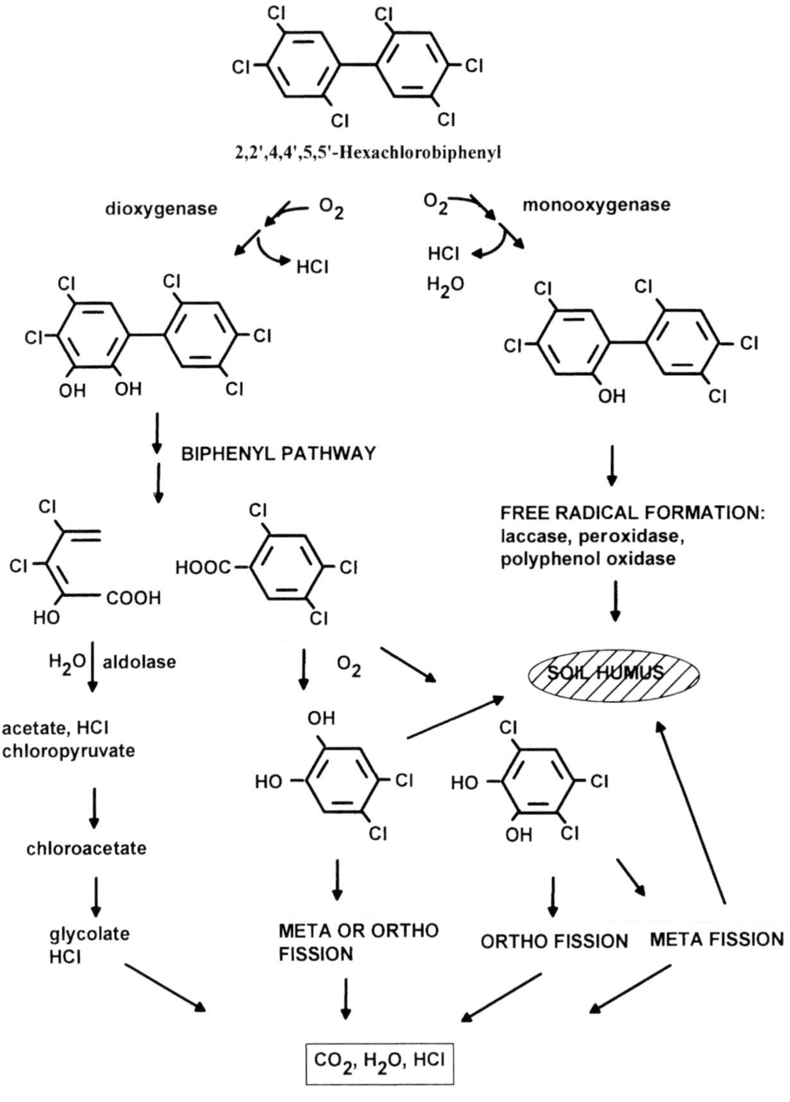

Figure 10. Possible pathways and fate of aerobic degradation of PCBs. For mechanisms on chlorophenol dimerization, see Dec and Bollag, J.-M. (1994 and Bolla & Liu S-Y (1985).

Research on plant callus tissue has shown that PCB congeners are metabolised by monooxygenase attack to produce hydroxylated PCBs (Mackova *et al.*, 1996). Thus, the mechanism and the potential for phytoremediation of PCBs as a long term strategy should not be overlooked. As noted by Hegde and Fletcher (1996), the extensive penetration of roots and accompanying rhizosphere is the easiest and most natural method to ensure that plants or their associated microflora get to the contaminant.

Nevertheless, reaction kinetics of PCB uptake and metabolism by plants is not likely to be any faster than by bacteria, and is likewise limited by low solubility. Finally, it should be recalled that many halogenated natural products (e.g. antibiotics) are produced by the action of soil bacteria and marine algae.

With the recent introduction of molecular methods to study microbial communities without having to culture the individual members, it should be possible to study the substrates and microorganisms in the soil -- other than biphenyl and the fast-growing bacteria that utilise it -- that act on PCBs and their metabolites. We know that biphenyl utilising bacteria are ubiquitous in soil, but that biphenyl is not. This raises the obvious questions of what substrates these organisms subsist on, and for what purpose the enzymes, encoded by the so-called *bph* genes, are really designed? Our knowledge is scant on this topic, and even less so, in knowing for certain what other catabolic pathways may offer more promise for oxidation of higher chlorinated PCBs. We are not likely to discover alternatives as long as we continue to rely so heavily upon the enrichment culture with biphenyl.

Acknowledgements

We are grateful to the National Science Foundation for partial funding of this work and to Greg Jenkins for providing us with the data for Figure 3.

References

Adams, R. H., Huang, C. M., Higson, F. K., Brenner, V. and Focht, D. D. (1992). Construction of a 3-chlorobiphenyl-utilizing recombinant from an intergeneric mating. Appl. Environ. Microbiol. 58, 647-654.

Ahmad, D., Sylvestre, M. and Masse, R. (1991). Bioconversion of 2-hydroxy-6-oxo-6-(4'-chlorophenyl)hexa-2,4-dienoic acid, the meta-cleavage product of 4-chlorobiophenyl. J. Gen. Microbiol. 137, 1375-1385.

Ahmed, M. and Focht, D. D. (1973). Degradation of polychlorinated biphenyls by two species of Achromobacter. Can. J. Microbiol. 19, 47-52.

Arensdorf, J. J. and Focht, D. D. (1994). Formation of chlorocatechol meta cleavage products by a pseudomonad during metabolism of monochlorobiphenyls. Appl. Environ. Microbiol. 60, 2884-2889.

Asturias, J. A., Moore, E., Yakimov, M., Klatte, S. and Timmis, K. N. (1994). Reclassification of the polychlorinated biphenyl-degraders Acinetobacter sp. strain P6 and Corynebacterium sp. strain MB1 as Rhodococcus globerulus. Syst. Appl. Microbiol. 17, 226-231.

Bartha, R., Linke, H. A. B. and Pramer, D. (1968). Pesticide transformations: production of chloroazobenzenes from chloroanilines. Science 161, 582-583.

Bedard, D. L. (1990). Bacterial transformation of polychlorinated biphenyls. In Biotechnology and biodegradation, pp. 369-388. Edited by D. Kamely, A. Chakrabarty and G. Omenn. Woodlands, TX and Houston: Portfolio Publishing Co. and Gulf Publishing Co.

Bedard, D. L., Haberl, M. L., May, R. J. and Brennan, M. J. (1987). Evidence for novel mechanisms of polychlorinated biphenyl metabolism in Alcaligenes eutrophus H850. Appl. Environ. Microbiol. 53, 1103-1112.

Bollag, JM & Liu S-Y (1985) Copolymerization of halogenated phenols and syringic acid. Pesticide Biochemistry and Physiology 23:261-272

Carney, B. F., Krockel, L., Leary, J. V. and Focht, D. D. (1989). Identification of Pseudomonas alcaligenes chromosomal DNA in the plasmid of the chlorobenzene-degrading recombinant Pseudomonas putida strain CB1-9. Appl. Environ. Microbiol. 55, 1037-1039.

Carson, R. (1962). Silent Spring. Boston: Houghton Miflin Co.

deBont, J. A. M., Vorage, M. J. A. W., Hartmans, S. and van den Tweel, W. J. J. (1986). Microbial degradation of 1,3-dichlorobenzene. Appl. Environ. Microbiol. 52, 677-680.

Dec, J. and Bollag, J.-M. (1994). Dehalogenation of chlorinated phenols during binding to humus. In Bioremediation through rhizosphere technology, pp. 102-111. Edited by T. A. Anderson and J. R. Coats. Washington: American Chemical Society.

Donnelly, P. K., Hegde, R. S. and Fletcher, J. S. (1994). Growth of PCB-degrading bacteria on compounds from photosynthetic plants. Chemosphere 28, 981-988.

Focht, D. D. (1993). Microbial degradation of chlorinated biphenyls. In Soil Biochemistry, pp. 341-407. Edited by J.-M. Bollag and G. Stozky. New York: Marcel Dekker

Focht, D. D. and Brunner, W. (1985). Kinetics of biphenyl and polychlorinated biphenyl metabolism in soil. Appl. Environ. Microbiol. 50, 1058-1063.

Focht, D. D., Searles, D. B. and Koh, S.-C. (1996). Genetic exchange in soil between introduced chlorobenzoate degraders and indigenous biphenyl degraders. Appl. Environ. Microbiol. 62, 3910-3913.

Furukawa, K., Tonomura, K. and Kamibayashi, A. (1979). Effect of chlorine substitution on the bacterial metabolism of various polychlorinated biphenyls. Appl. Environ. Microbiol. 38, 301-310.

Haddock, J. D., Horton, J. R. and Gibson, D. T. (1995). Dihydroxylation and dechlorination of chlorinated biphenyls by purified biphenyl 2,3-dioxygenases from Pseudomonas sp. strain LB400. J. Bacteriol. 177, 20-26.

Haigler, B. and Spain, J. C. (1989). Degradation of para-chlorotoluene by a mutant of Pseudomonas sp. strain JS6. Appl. Environ. Microbiol. 55, 372-379.

Haigler, B. E., Nishino, S. F. and Spain, J. C. (1988). Degradation of 1,2-dichlorobenzene by a Pseudomonas sp. Appl. Environ. Microbiol. 54, 294-301.

Harkness, M. R., McDermott, J. B., Abramowicz, D. A., Salvo, J. J., Flanagan, W. P., Stephens, M. L., Mondello, F. J., May, R. J., Lobos, J. H., Carroll, K. M., M.J., B., Bracco, A. A., Fish, K. M., WArner, G. L., Wilson, P. R., Dietrich, D. K., Lin, D. T., Morgan, C. B. and Gately, W. L. (1993). In situ stimulation of aerobic PCB biodegradation in Hudson River sediments. Science 259, 503-507.

Havel, J. and Reineke, W. (1995). The influence of physicochemical effects on the microbial degradation of chlorinated biphenyls. Appl. Microbiol. Biotechnol. 43, 914-919.

Hegde, R. S. and Fletcher, J. S. (1996). Influence of plant growth stage and season on the release of root phenolics by mulberry as related to development of phytoremediation technology. Chemosphere 32, 2471-2479.

Hernandez, B. S., Arensdorf, J. J. and Focht, D. D. (1995). Catabolic characteristics of biphenyl-utilising isolates which cometabolize PCBs. Biodegradation 6, 75-82.

Hernandez, B. S., Koh, S.-C., Chial, M. and Focht, D. D. (1997). Terpene-utilizing isolates and their relevance to enhanced biotransformation of polychlorinated biphenyls in soil. Biodegradation 8, 153-158.

Hickey, W. J., Searles, D. B. and Focht, D. D. (1993). Enhanced mineralization of polychlorinated biphenyls in soil inoculated with chlorobenzoate-degrading bacteria. Appl. Environ. Microbiol. 59, 1194-1200.

Higson, F. K. and Focht, D. D. (1989). Bacterial metabolism of hydroxylated biphenyls. Appl. Environ. Microbiol. 55, 946-952.

Kohler, H.-P. E., Kohler-Staub, D. and Focht, D. D. (1988). Degradation of 2-hydroxybiphenyl and 2,2'-dihydroxybiphenyl by Pseudomonas sp. strain HBP1. Appl. Environ. Microbiol. 54, 2683-2688.

Kohler-Staub, D. and Kohler, H.-P. E. (1989). Microbial degradation of B-chlorinated four-carbon aliphatic acids. J. Bacteriol. 171, 1428-1434.

Kröckel, L. and Focht, D. D. (1987). Construction of chlorobenzene-utilising recombinants by progenitive manifestation of a rare event. Appl. Environ. Microbiol. 53, 2470-2475.

Lajoie, C. A., Layton, A. C. and Sayler, G. S. (1994). Cometabolic oxidation of polychlorinated biphenyls in soil with a surfactant-based field application vector. Appl. Environ. Microbiol. 60, 2826-2833.

Lajoie, C. A., Zylstra, G. J., Deflaun, M. F. and Strom, P. F. (1993). Development of field application vectors for bioremediation of soils contaminated with polychlorinated biphenyls. Appl. Environ. Microbiol. 59, 1735-1741.

Mackova, M., Macek, T., Ocenaskova, J., Burkhard, J., Demnerova, K. and Pazlarova, J. (1996). Selection of the potential plant degraders of PCB. Chemicke Listy 90, 712-713.

Mars, A. E., Kasberg, T., Kaschabek, S. R., van Agteren, M. H., Janssen, D. B. and Reineke, W. (1997). Microbial degradation of chloroaromatics: Use of the meta-cleavage pathway for mineralization of chlorobenzene. J. Bacteriol. 179, 4530-4537.

Mars, A. E., Kingma, J., Kaschabek, S. R., Reineke, W. and Janssen, D. B. (1999). Conversion of 3-chlorocatechol by various catechol 2,3-dioxygenases and sequence analysis of the chlorocatechol dioxygenase region of Pseudomonas putida GJ31. J. Bacteriol. 181, 1309-1318.

Massé, R., Messier, F., Ayotte, C., Lévesque, M.-F. and Sylvestre, M. (1989). A comprehensive gas chromatographic/mass spectrometric analysis of 4-chlorobiphenyl bacterial degradation products. Biomed. Environ. Mass Spectrom. 18, 27-47.

McCullar (1996). Metabolism of chlorinated biphenyls and chlorobenzoates by Pseudomonas acidovorans M3GY. PhD Thesis. Riverside: University of California.

Nishino, S. F., Spain, J. C., Belcher, L. A. and Litchfield, C. D. (1992). Chlorobenzene degradation by bacteria isolated from contaminated groundwater. Appl. Environ. Microbiol. 58, 1719-1726.

Omori, T., Ishigooka, H. and Minoda, Y. (1988). A new metabolic pathway for meta ring fission compounds of biphenyl. Agric. Biol. Chem. 52, 503-509.

Pettigrew, C. A., Haigler, B. E. and Spain, J. C. (1991). Simultaneous biodegradation of chlorobenzene and toluene by a Pseudomonas strain. Appl. Environ. Microbiol. 57, 157-162.

Pignatello, J. J. and Xing, B. (1996). Mechanisms of slow sorption of organic chemicals to natural particles. Environ. Sci. Technol. 30, 1-11.

Reineke, W. and Knackmuss, H.-J. (1979). Construction of haloaromatics utilising bacteria. Nature 277, 385-386.

Reineke, W. and Knackmuss, H.-J. (1984). Microbial metabolism of haloaromatics: Isolation and properties of a chlorobenzene-degrading bacterium. Appl. Environ. Microbiol. 47, 395-402.

Sander, P., Wittich, R. M., Fortnagel, P., Wilkes, H. and W., F. (1991). Degradation of 1,2,4-trichlorobenzene and 1,2,4,5-tetrachlorobenzene by Pseudomonas strains. Appl. Environ. Microbiol. 57, 1430-1440.

Schraa, G., Boone, M. L., Jetten, M. S. M., vanNeerven, A. R. W., Colberg, P. J. and Zehnder, A. J. B. (1986). Degradation of 1,4-dichlorobenzene by Alcaligenes sp. strain A175. Appl. Environ. Microbiol. 52, 1374-1381.

Searles, D.B. (1995). Biodegradation of PCBs by biphenyl- and chlorobenzoate-utilising recombinants isolated from soil inoculated with a chlorobenzoate degrader. PhD Thesis. Riverside: University of California.

Shi, Z., LaTorre, K. A., Ghosh, M. M., Layton, A. C., Luna, S. H., Bowles, L. and Sayler, G. S. (1998). Biodegradation of UV-irradiated polychlorinated biphenyls in surfactant micelles. Water Sci. Technol. 38, 25-32.

Spain, J. C. (1990). Metabolic pathways for biodegradation of chlorobenzenes. In Pseudomonas biotransformations, pathogenesis, and evolving biotechnology, pp. 197-206. Edited by S. Silver, A. M. Chakrabarty, B. Iglewski and S. Kaplan. Washington, DC: American Society for Microbiology.

Spain, J. C. and Nishino, S. F. (1987). Degradation of 1,4-dichlorobenzene by a Pseudomonas sp. Appl. Environ. Microbiol. 53, 1010-1019.

Steinberg, S. M., Pignatello, J. J. and Sawhney, B. L. (1987). Persistence of 1,2-dibromoethane in soils: entrapment in intraparticle micropores. Environ. Sci. Technol. 21, 1201-1208.

Stott, D. E., Martin, J. P., Focht, D. D. and Haider, K. (1982). Biodegradation, stabilization in humus, and incorporation into soil biomass of 2,4-D and chlorocatechol carbons. Soil Sci. Soc. Am. J. 47, 66-70.

Stratford, J., Wright, M. A., Reineke, W., Mokross, H., Havel, J., Knowles, C. J. and Robinson, G. K. (1996). Influence of chlorobenzoates on the utilisation of chlorobiphenyls and chlorobenzoate mixtures by chlorobiphenyl/chlorobenzoate-mineralising hybrid bacterial strains. Arch. Microbiol. 165, 213-218.

Taira, K., Hayase, N., Arimura, N., Yamashita, S., Miyazaki, T. and Furukawa, K. (1988). Cloning and nucleotide sequence of the 2,3-dihydroxybiphenyl dioxygenase gene from the PCB-degrading strain of Pseudomonas paucimobilis Q1. Biochemistry 27, 3990-3996.

Trudgill, P. W. (1994). Microbial metabolism and transformation of selected monoterpenes. In Biochemistry of microbial degradation, pp. 33-62. Edited by C. Ratledge. Dordrecht: Kluwer Academic Publishers.

Tucker, E. S., Saeger, V. W. and Hicks, O. (1975). Activated sludge primary biodegradation of polychlorinated biphenyls. Bull. Environ. Contam. Toxicol. 14, 705-713.

van der Meer, J. R., van Neerven, A. R. W., de Vries, E. J., de Vos, W. M. and Zehnder, A. J. B. (1991). Cloning and characterisation of plasmid-encoded genes for the degradation of 1,2-dichloro-, 1,4-dichloro-, and 1,2,4-trichlorobenzene of Pseudomonas sp. strain P51. J. Bacteriol. 173, 6-15.

Waid, J. S. (1986). PCBs and the environment. Boca Raton, FL: CRC Press.

Wu, Q. Z., Bedard, D. L. and Wiegel, J. (1999). 2,6-dibromobiphenyl primes extensive dechlorination of Aroclor 1260 in contaminated sediment at 8-30°C by stimulating growth of PCB-dehalogenating microorganisms. Environ. Sci. Technol. 33, 595-602.

Zwiernik, M. J., Quensen III, J. F. and Boyd, S. A. (1998). FeSO$_4$ amendments stimulate extensive anaerobic PCB dechlorination. Environ. Sci. Technol. 32, 3360-3365

ENZYMOLOGY OF THE BREAKDOWN OF SYNTHETIC CHELATING AGENTS

THOMAS EGLI AND MARGARETE WITSCHEL

Swiss Federal Institute for Environmental Science and Technology
EAWAG), CH-8600 Dübendorf, Switzerland

1. Introduction

Organic metal-complexing agents are able to form water-soluble complexes with metal ions. These compounds are included in many different industrial and domestic processes and products to

- prevent the formation of metal precipitates,

- hinder metal ion catalysis of unwanted reactions,

- remove metal ions from systems,

- make metal ions more available by keeping them in solution.

Their areas of application are very diverse (Bell, 1977; Egli, 1988; Witschel and Egli, 1999). It includes their use in household and industrial cleaning agents to avoid precipitation of metal salts at high pHs during the washing process, in photographic industry as an iron-carrier, in bleaching processes in the pulp and paper industry, in electroplating of metal surfaces, for the decontamination of nuclear reactors and equipment, the remediation of metal-contaminated soils or sediments, in fertilisers, but also as additives in food, pharmaceutics or cosmetics to prevent spoilage of the products (Wolf and Gilbert, 1992; Pothoff-Karl *et al.*, 1996).

In all these applications the most widely applied synthetic organic complexing agents are aminopolycarboxylic acids (APCAs). The chemical structure of the most important APCAs is shown in Fig. 1. Other metal-complexing synthetic compounds synthetically produced are the phosphonic acid analogues of APCAs (Fig. 1). However, they are used in much lower quantities. Because most applications of APCAs are water based these compounds generally end up in wastewater or in the aqueous environment. Therefore, their behaviour in technical and environmental compartments and in particular their (bio) degradability has received much attention. A number of concerns

S.N. Agathos and W. Reineke (eds.),
Biotechnology for the Environment: Strategy and Fundamentals, 205–217.

have been raised with respect to the possible effects of APCAs in the environment (Anderson *et al.*, 1985), including adverse affects on the operation of wastewater treatment plants, toxic effects on aquatic organisms, their contribution to eutrophication due to nitrogen release, and the potential of APCAs to mobilise heavy metals. Presently, after more than 30 years of practical experience it is mainly the ability of APCAs to complex and mobilise heavy metals which may become critical (Alder *et al.*, 1990; Nowack *et al.*, 1997; Twachtmann *et al.*, 1998), whereas at the concentrations of APCAs prevailing in aquatic environments the other issues have been shown to be of little importance (Anderson *et al.*, 1985; Wolf and Gilbert, 1992).

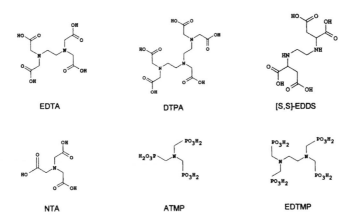

Figure 1. *Chemical structure of the commercially most important synthetic aminopolycarboxylic acids. EDTA, ethylenediaminetetraacetic acid, DTPA, diethylenediaminepentaacetic acid, [S,S]-EDDS, [S,S]-ethylenediaminedisuccinic acid, NTA, nitrilotriacetic acid, ATMP, aminotrimethylphosphonic acid, EDTMP, Ethylenediaminetetramethylphosphonic acid*

Both chemical and biological processes bring about elimination of different APCAs from the environment. Surprisingly, despite their structural and chemical similarity, it has turned out that the mechanism primarily responsible for the elimination of individual APCAs can be quite different. For example, whereas biodegradation is in the first place responsible for the elimination of NTA (Anderson *et al.*, 1985) photodegradation of the iron complex appears to be the key mechanism for the elimination of EDTA in surface waters (Kari and Giger, 1995). In this contribution we will concentrate on the biochemistry of the breakdown of APCAs but for a overview on the abiotic elimination processes for APCAs in different environmental compartments the reader is referred to our recent review (Witschel and Egli, 1999).

Based on consumption, EDTA and NTA are certainly the most important APCAs and they have therefore received most attention. In comparison, little is known about the environmental fate and biological breakdown of other metal-chelating agents such as DTPA or HEDTA (summarised in Witschel and Egli, 1999). Only recently, some attention has also been given to the biological degradation of [S,S]-EDDS, a structural analogue of EDTA, because it was found to be well biodegradable during aerobic wastewater treatment (Schowanek et al., 1997; Witschel and Egli, 1998). Also phosphonic acids are beginning to receive more attention (Schowanek and Verstraete, 1990a,b; Nowack, 1998).

2. The microorganisms able to degrade complexing agents

Elucidation of catabolic pathways for the degradation of chelating agents requires that pure cultures of microbial strains able to break down these compounds are available. Presently, a number of different NTA- and three EDTA-degrading microbes have been isolated in pure culture and characterised (Witschel and Egli, 1999). Furthermore, a number of strains able to attack organophosphonates (Piepke and Amrhein, 1988) and metal-complexing phosphonates, including ATMP and EDTMP, have been isolated and described (Schowanek and Verstraete, 1990a,b). However, all these strains were able to utilise the phosphonates only as a source of phosphorus and the fate of the carbon and nitrogen remained unclear. Only recently, the isolation of strains able to use aminoalkylphosphonates as a source of carbon and nitrogen has been reported (McGrath et al., 1997) but they still remain to be thoroughly studied. To date, interest in the degradation of organic phosphonates has concentrated on the herbicide glyphosate and the authors are not aware of any biochemical studies concerning the cleavage reaction of the C-P bond of metal-complexing phosphonic acids.

Obligatory aerobic NTA-degrading strains can be easily isolated (Egli et al., 1988; Wehrli and Egli, 1988) and there is good evidence that they are present in most aerobic environments (Wilberg et al., 1993). Most of the known well-characterised isolates are obligatory aerobic Gram-negative rods. They are members the α-subgroup of Proteobacteria and two new genera, Chelatobacter and Chelatococcus, were established to house them (Auling et al., 1993). A denitrifying isolate able to grow with NTA was found to be a member of the γ-subgroup of Proteobacteria; its exact taxonomic position still has to be elucidated (Wanner et al., 1990).

In contrast, it is still tedious to enrich EDTA-degrading strains. This indicates that their environmental distribution is probably limited or that we know still too little about their nutritional requirements for successful isolation. Presently, isolation of three different strains has been reported in the literature (an Agrobacterium sp. by Lauff et al., 1990; strain BNC1 by Nörtemann, 1992; and our isolate assigned the DSM number 9103, Witschel et al., 1995). All of them are members of the Agrobacterium-Rhizobium branch in the α-subgroup of Proteobacteria (Witschel and Egli, 1999). It is very likely that they are members of a new genus in this branch.

In the course of our investigations we found that the NTA-degrading Chelatobacter strains and the EDTA-degrading strain isolated in our laboratory are also able to utilise [S,S]-EDDS (Witschel and Egli, 1998). However, it is not yet known how widespread this ability is. Nevertheless, the relative fast degradation of [S,S]-EDDS in activated sludge and the environment suggests that EDDS-degrading microorganism are present in relatively large numbers.

3. Biochemistry of the breakdown of chelating agents

3.1. THE KEY ENZYMES INITIATING THE BREAKDOWN

3.1.1. Transport across the cell membrane

The first step in the metabolism of chelating agents is their transport across the cytoplasmic membrane into the cytoplasm. In the early seventies preliminary data were presented which indicated that transport of NTA into the cell is an energy-dependent process (Wong et al., 1973) and similar observations were made in our laboratory (Wilberg, 1989). From the data given one cannot deduce whether the free form or a metal-complexed NTA species was transported and we are not aware of any further attempts to study the transport of NTA in NTA-degrading microorganisms.

Recently, the transport of EDTA was investigated in the EDTA-degrading strain DSM 9103 in some detail (Witschel et al., 1999). It was clearly demonstrated that EDTA transport is energy-dependent and probably driven by the proton gradient. EDTA-grown cells readily took up free ^{14}C-EDTA, as well as Ca^{2+}-, Ba^{2+}-, and Mg^{2+}-complexed EDTA at similar rates. Uptake of Mn^{2+}-EDTA proceeded at a slower rate and complexes with stability constants higher than 10^{16} (Zn^{2+}, Co^{2+}, Cu^{2+}, Ni^{2+}, or Fe^{3+}) were not transported. The transport system exhibited a high affinity for EDTA (affinity constant $K_t = 0.39$ μM) and was rather specific for this chelating agent because, from several structurally related compounds, only diethylenetriamine-pentaacetate (DTPA) inhibited its transport. Also it was shown that the rate of uptake of Ca^{2+} was considerably enhanced in the presence of EDTA, indicating a transport of Ca^{2+} together with EDTA. A further indication for the transport of Ca^{2+}-complexed EDTA in this strain comes from electron microscopic studies. In EDTA-grown cells the presence of electron-dense bodies was observed, whereas they were absent in fumarate-grown cells. Electron disperse X-ray revealed enhanced concentrations of Ca, Mg and P in these inclusion bodies, indicating that they consist of Ca-Mg-phosphate precipitate. The dynamics of Ca^{2+}-EDTA uptake and Ca^{2+}-release after EDTA exhaustion indicates that the cells transport at least a part of the EDTA metabolised in the metal-complexed form. Hence, to cope with the incoming metal-flux a part of it metal is precipitated inside the cell whereas the remainder is most likely excreted again. We are not aware of a similar presence of electron-dense bodies in NTA-grown cells of Chelatobacter or Chelatococcus strains.

In summary, transport studies at the whole cell level indicate that probably free NTA or EDTA (or one of the relatively unstable complexes with Mg^{2+} or Ca^{2+}) is primarily transported. This observation could one of the reasons for the apparent differences in the biodegradability of NTA and EDTA in the environment. In the environment, EDTA is mainly present in the form of the highly stable Fe^{3+} and Zn^{2+} complexes that cannot be transported into the cell, whereas a large part of NTA is present in the labile and transportable Ca^{2+}-complexed form (Nowack et al., 1997).

3.1.2. NTA and EDTA monooxygenase

In the different obligatory aerobic NTA-utilising strains of *Chelatobacter* and *Chelatococcus*, as well as the Gram-negative isolate DSM 9103 able to grow with EDTA, monooxygenases are responsible for the initial enzymatic reaction in the catabolism of the chelating agent. The two monooxygenases, NTA MO and EDTA MO, have been purified and characterised (Uetz *et al.*, 1992; Witschel *et al.*, 1997). Here they will be discussed together because - although distinctly different enzymes - they share a number of interesting properties.

NTA monooxygenase. Involvement of a monooxygenase in catalysing the two initial breakdown reaction leading from NTA to iminodiacetate (IDA) then to glycine was proposed from information obtained with cell-free extracts (Cripps and Noble, 1973; Firestone and Tiedje, 1978). A functional NTA MO from *Chelatobacter heintzii* ATCC 29600 was finally purified and characterised by Uetz and co-workers (Uetz *et al.*, 1992) and later more information was obtained by sequencing the corresponding genes (Knobel *et al.*, 1996; Xu *et al.*, 1997). NTA MO catalyses the $NADH/O_2$-dependent oxidation of NTA to IDA and glyoxylate. Furthermore, the reaction was strictly dependent on the presence of Mg^{2+}, Co^{2+}, or Mn^{2+} ions, indicating that not free NTA but the metal-complexed compound acts as the true substrate. NTA complexed with Ca^{2+}, Fe^{2+}, Fe^{3+}, Zn^{2+}, Cu^{2+}, or Ni^{2+} were not accepted (Uetz *et al.*, 1992; but compare also Xun *et al.*, 1996). The substrate spectrum of this enzyme appears to be restricted to NTA and it does not catalyse the subsequent cleavage of IDA to glycine and glyoxylate. Based on the characterisation of the functional enzyme NTA MO was originally proposed to consist of two weakly associated components, cA and cB. The presence of both of them was thought to be mandatory for the reaction. More recently, it was demonstrated that cA and cB are two distinct enzymes, with cA being a true monooxygenase catalysing the oxidative cleavage of NTA and consuming $FMNH_2$ and molecular oxygen, and cB being an oxidoreductase providing $FMNH_2$ for the MO by transferring reducing equivalents from NADH to FMN (Fig. 2). The oxidoreductase can be replaced by other NADH:FMN oxidoreductases, such as the oxidoreductase from *Photobacterium fischeri*, which normally delivers $FMNH_2$ to luciferase (Xu *et al.*, 1997). Also from *Chelatococcus asaccharovorans* a protein equivalent to cA has been isolated and was shown to share many similarities with cA from *C. heintzii*, whereas no protein equivalent to cB has yet been found in *C. asaccharovorans* (Uetz, 1992). However, cB from *C. heintzii* was shown to form a functional enzyme complex with cA from *C. asaccharovorans*. This confirms the crucial role of cA, i.e., the NTA MO, in the breakdown of NTA, whereas cB can probably be replaced by any other enzyme providing reduced FMN.

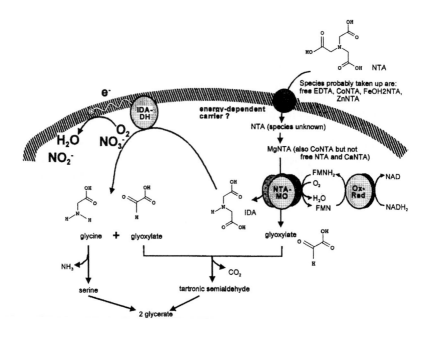

Figure 2. *Catabolism of NTA in Chelatobacter heintzii ATCC 29600 and related strains.*

Recently, the genes for the two enzymes cA and cB were cloned and sequenced (Knobel *et al.*, 1996; Knobel, 1997; Xu *et al.*, 1997). They are oriented divergently with an intergenic region of 307 bp, a rather unusual organisation for two enzymes acting so closely together. The amino acid sequence deduced from the gene sequences of cA and cB indicated some homology with other monooxygenases and NADH:FMN oxidoreductases involved in the oxidation of some structurally quite unrelated compounds, namely a sulphur-containing compounds and three antibiotics (see also Kertesz *et al.*, 1999).

The regulation of NTA MO at the physiological level was studied in some detail for *C. heintzii* ATCC 29600 during growth in carbon-limited continuous culture (Egli *et al.*, 1988; Uetz, 1992; Bally *et al.*, 1994; Bally and Egli, 1996). The enzyme system was expressed only during growth with NTA and IDA, whereas it is absent during growth with a range of other substrates, including glycine, succinate, citrate, glucose, or nutrient broth (Uetz, 1992). However, no significant repression of NTA MO by other substrates seems to occur because during batch growth with mixtures of NTA plus another carbon source the cells are able to utilise both substrates simultaneously. Under such conditions several NTA-utilising strains were shown to achieve even higher specific growth rates than during growth with NTA alone (Egli *et al.*, 1988). Also in carbon-limited chemostat culture the cells were able to utilise NTA simultaneously with glucose (Bally *et al.*, 1994; Bally and Egli, 1996) and both, the steady-state levels of expression and the dynamics of the synthesis of NTA MO was found to depend strongly on the ratio of glucose to NTA fed.

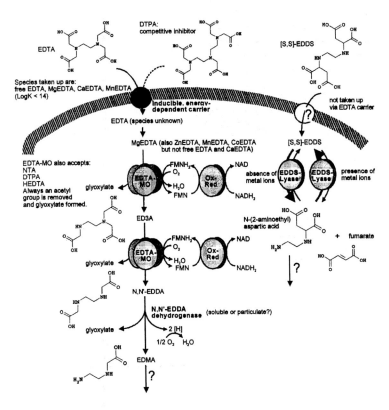

Figure 3. *Degradation pathway for EDTA and [S,S]-EDDS in the Gram-negative strain DSM 9103 (from Witschel, 1999). EDTA MO, EDTA monooxygenase; Ox-Red, NADH:FMN oxydo-reductase.*

EDTA monooxygenase. EDTA MO was purified and characterised from strain DSM 9103 (Witschel *et al.*, 1997). Also here a two-enzyme system (cA' and cB', in analogy to NTA MO) was found to catalyse the oxidative cleavage of EDTA in a reaction dependent on the presence of FMN and consuming NADH and molecular oxygen. In this reaction two acetyl groups were removed from EDTA and the products formed were two molecules of glyoxylate and N,N'-EDDA (no formation of N,N-EDDA observed). It was demonstrated that ED3A was intermediately formed in this reaction (Fig. 3). Free EDTA was not accepted as a substrate; only when complexed with Mg^{2+}, Zn^{2+}, Mn^{2+}, Co^{2+}, or Cu^{2+} glyoxylate was formed. Ca^{2+}-, Fe^{3+}-, or Ni^{2+}-complexed EDTA was not transformed. The enzyme system exhibited broad substrate specificity.

Not only EDTA and ED3A acted as substrates, but also NTA, DTPA, HEDTA, diethylenetriamine-pentaacetate, and 1,3-diaminopropanoltetraacetic acid were transformed. Other NADH:FMN oxidoreductases such as cB from NTA MO or the oxidoreductase from P. fischeri could also substitute component B' from this enzyme complex. Interestingly, the N-terminal amino acid sequences of cA' from EDTA MO and cA from NTA MO showed no significant homology (Witschel et al., 1997). Moreover, immunoblotting with polyclonal antibodies raised against cA and cB from C. heintzii did not reveal a similarity at the protein level between corresponding components of the two enzyme systems.

3.1.3. NTA Dehydrogenase

Strain TE11 is able to grow with NTA as the only source of carbon and energy under denitrifying conditions. Obviously, the initial NTA breakdown has to proceed via an oxygen-independent step. From cells of this bacterium grown under either denitrifying (Jenal-Wanner et al., 1993) or aerobic conditions (Kemmler, 1992) a dehydrogenase was isolated that catalysed the oxidation of NTA resulting in the formation of glyoxylate and IDA. This enzyme is able to deduce electrons from NTA and, with the help of the artificial redox-compound PMS (phenazine methosulfate), to transfer them to either a nitrate reductase or via the electron transfer chain to molecular oxygen. The enzyme is completely independent of the presence of metal ions, strongly indicating that NTA DH accepts that only free NTA and no complexes with bivalent metal ions (Kemmler, 1992). The substrate spectrum of NTA DH seems to be restricted to NTA because none out of some 20 structurally related compounds were transformed. The enzyme contains a covalently bound FAD moiety and probably four iron atoms. Difference spectra argue against the presence of a heam chromatophore and, therefore, one has to assume that the enzyme contains two 2Fe-2S clusters. Thus, the electrons derived from NTA are presumably first transferred to FAD, then to the iron-sulphur cluster, and from there to a still unknown in-vivo electron carrier (or in vitro to PMS). EM immunochemical labelling indicated a cytosolic location of NTA DH, whereas nitrate reductase was found to be associated with or integrated into the cytoplasmatic membrane (Knobel, 1997). This is an additional strong indication for the involvement of a redox-compound in the transfer of electrons from NTA DH to nitrate reductase or to other redox components of the membrane. The N-terminal amino acid sequence of NTA DH had a high similarity to other flavin-containing enzymes (Kemmler, 1992), all exhibiting in their N-terminus a FAD-binding segment ($\beta\alpha\beta$-fold). It is therefore not surprising that attempts made to clone the NTA DH gene failed because the information contained in the terminus was too low and the probes synthesised hybridised with many different DNA fragments (Knobel, 1997).

3.1.4. IDA Dehydrogenase

The products of the enzymatic breakdown of NTA by both NTA MO and NTA DH results are glyoxylate and IDA. The former is a common intermediate of the central metabolism. IDA, however, is so far still considered a xenobiotic compound. Neither NTA MO nor NTA DH was found to accept IDA as a substrate, indicating the presence of an additional enzyme in the catabolism of NTA. A membrane-bound dehydrogenase (IDA DH) was subsequently identified to catalyse the breakdown of IDA. It was possible to enrich this enzyme in artificial membrane vesicles and characterise it (Uetz and Egli, 1993). NTA DH catalyses the oxidative cleavage of IDA to glycine and glyoxylate and it is distinctly different from other membrane-associated dehydrogenases such as succinate dehydrogenase. Also this enzyme exhibited high substrate specificity for IDA and no other substrates for IDA DH are presently known. Integration of NTA DH into soybean phospholipid vesicles resulted in significant oxygen consumption only when the vesicles contained ubiquinone Q-10. This indicates that *in vivo* the enzyme is feeding the electrons from IDA into the electron transfer chain at the level of ubiquinone Q-10. IDA DH activity was found to be present in both *Chelatobacter* and *Chelatococcus* strains, as well as in the denitrifying strain TE11.

3.1.5. EDDS Lyase

The EDTA-degrading bacterial strain DSM 9103 isolated in our laboratory is also able to grow with [S,S]-ethylenediaminedisuccinate (EDDS), a structural isomer of EDTA. Despite its structural similarity, this complexing agent was not a substrate for EDTA MO. In cell-free extract [S,S]-EDDS was consumed without addition of co-factors and subsequently an EDDS lyase was purified which catalysed the non-hydrolytic splitting of [S,S]-EDDS leading to the formation of fumarate and N-(2-aminoethyl) aspartic acid (Witschel and Egli, 1997). Whereas [R,R]-EDDS was not a substrate for this lyase [R,S]-EDDS was accepted. The enzyme catalysed the transformation of free [S,S]-EDDS and of metal-[S,S]-EDDS complexes with stability constants lower than 10^{10}, namely of Mg^{2+}, Ca^{2+}, Ba^{2+}, and at a low rate also Mn^{2+}-[S,S]-EDDS, whereas Fe^{3+}, Ni^{2+}, Cu^{2+}, Co^{2+}, and Zn^{2+}-[S,S]- EDDS complexes were not transformed. The reaction was reversible and an equilibrium constant of approximately $43*10^{-3}$ M was determined. A stereospecific degradation pathway has obviously evolved for this biotic complexing agent. This is probably due to the fact that the [S,S]-form is the only stereoisomer which is produced biologically by the actinomycete *Amycolatopsis orientalis* (and perhaps also other organisms) in response to Zn-deficiency (Nishikiori et al., 1984; Cebulla, 1995).

3.2. THE LOWER PART OF THE PATHWAYS

So far only incomplete information is available on the fate of the metabolites produced in the initial reactions of the breakdown of the different chelating agents.

In the case of NTA the subsequent metabolism of the identified products, glycine and glyoxylate, should present no major problem. Cripps and Noble (1973) proposed a pathway for the complete degradation of NTA based on a comparison of enzyme levels determined in NTA- and glucose-grown cells of their NTA-utilising strain T23 (Fig. 2). However, confirmation of this pathway, e.g., from mutant studies, is still lacking.

In contrast, the pathways for the complete degradation of EDTA and EDDS still remain to be elucidated (Fig. 3). There is good evidence for the degradation of N,N'-EDDS to EDMA (ethylenediaminemonoacetate) but no further (Witschel, 1999). Also the fate of N-(2-aminoethyl)aspartic acid produced by EDDS lyase from [S,S]-EDDS is still unknown.

3.3. INFLUENCE OF METAL SPECIATION

In both growth media and the natural environment APCAs occur in a metal-complexed form and only to a very limited extent freely (see Witschel and Egli, 1999). Therefore, metals must be expected to have a major influence on the biodegradation of these compounds. Some of these aspects have already been discussed above for the transport of EDTA into cells of strain DSM 9103. Unfortunately, little conclusive data exists in this field because in such studies a number of difficulties are met that are difficult to avoid. These problems include, e.g., metal toxicity, the presence of metal-complexing ligands at the bacterial cell surface, or the dynamics of metal exchange during assays. Hence, there is much work still to be done in this area and, therefore, this aspect will be touched here only shortly.

Experimental data reported for NTA consumption rates of whole cells in combination with calculations suggested that either free NTA or NTA complexed with Co^{2+}, $FeOH^{2+}$ or Zn^{2+} is transported into the cell (Bolton et al., 1996). However, under environmental conditions neither of these species is abundant. Furthermore, once transported, NTA MO accepts none of them as a substrate (Uetz et al., 1992). Hence, one must assume that the species accepted by NTA MO, most likely MgNTA, is generated intracellularly. All this indicates that for clear-cut information transport of complexing agents has to be studied in membrane vesicles. However, attempts made in our laboratory to produce vesicles from the cytoplasmic membrane of C. heintzii have met little success so far, probably because of the rigid cell wall of this organism.

Also in the case of EDTA utilisation the information concerning the degradability of different metal complexes is far from complete. Whereas the strain isolated by Lauff and co-workers (Lauff et al., 1990) was reported to utilise EDTA only when complexed with Fe^{3+}, strains BNC1 and DSM 9103 prefer free EDTA and metal complexes with stability constants lower than 10^{14}. The information available today indicates that for strains BNC1 and DSM 9103 Zn-, Cu-, Ni-, Co-, and Fe(III)-complexed EDTA is not directly accessible but can only be degraded after dissociation. However, dissociation rates for most of these complexes are rather low (Nowack and Sigg, 1997). As in the case of NTA, an intracellular exchange of the metal associated with EDTA must be expected before the degradation is possible. This is suggested from a comparison of the species transported across the cell membrane (free, Ca-, Mg-complexed EDTA, but not Fe(III)-, Co-, or Ni-species) and those accepted as substrates by EDTA MO (Mg-, Zn- or Mn- complexed EDTA, but not free, or Ca-complexes).

4. Outlook

In the past interest in the biodegradation of chelating agents was focussed to NTA and EDTA. However, the search for easily biodegradable chelating agents with good metal-complexing characteristics is increasingly attracting interest (Potthoff-Karl et al., 1996; Schowanek et al., 1997). Although aminopolycarboxylic acids seem to us similar in their chemical structure and properties, investigation of the biological breakdown of these compounds has proven to be far from boring. Isolation of the key enzymes for the breakdown of NTA, EDTA and [S,S]-EDDS have demonstrated that not one type of enzyme but (mechanistically completely) different enzymes are involved in the catabolism of the apparently so similar APCAs. Furthermore, the authors are convinced that the challenging investigation of transport and metabolism of different metal-species will have a number of surprises waiting for us.

5. References

Alder, A.C., Siegrist, H, Gujer, W., and Giger, W. (1990) Behaviour of NTA and EDTA in biological wastewater treatment, Water Research 24, 733-742.

Anderson, R.L., Bishop, W.E., and Campbell, R.L. (1985) A review of the environmental and mammalian toxicology of nitrilotriacetic acid, CRC Critical Reviews in Toxicology 15, 1-102.

Auling, G., Busse, H.J., Egli, T., El-Banna, T., and Stackebrandt, E. (1993) Description of the Gram-negative, obligatory aerobic, nitrilotriacetate (NTA)-utilizing bacteria as Chelatobacter heintzii, gen. nov., sp. nov., and Chelatococcus asaccharovorans, gen. nov., sp. nov.,. Systematic and Applied Microbiology 16, 104-112.

Bally, M., Wilberg, E., Kühni, M., and Egli, T. (1994) Growth and regulation of enzyme synthesis in the nitrilotriacetic acid (NTA) degrading bacterium Chelatobacter heintzii sp. ATCC 29600, Microbiology 140, 1927-1936.

Bally, M., and Egli, T. (1996) Dynamics of substrate consumption and enzyme synthesis in Chelatobacter heintzii during growth in carbon-limited continuous culture with different mixtures of glucose and nitrilotriacetate, Applied and Environmental Microbiology 62, 133-140.

Bell, C.F. (1977) Principles and applications of metal chelation, Claredon Press, Oxford.

Bolton, H.J., Girvin, D.C., Plymale, A.E., Harvey, S.D., and Workman, D.J. (1996) Degradation of metal-nitrilotriacetate complexes by Chelatobacter heintzii, Environmental Science and Technology 30, 931-938.

Cebulla, I. (1995) Gewinnung komplexbildender Substanzen mittels Amycolatopsis orientalis, PhD thesis, Eberhard-Karls-University Tübingen, Germany.

Cripps, R.E., and Noble, A.S. (1973) The metabolism of nitrilotriacetate by a Pseudomonad, Biochemical Journal 136, 1059-1086.

Egli, T. (1988) (An)aerobic breakdown of chelating agents used in household detergents, Microbiological Sciences 5, 36-41.

Egli, T., Weilenmann, H.-U., El-Banna, T., and Auling, G. (1988) Gram negative, aerobic nitrilotriacetate-utilizing bacteria from wastewater and soil, Systematic and Applied Microbiology 10, 297-305.

Firestone, M.K., and Tiedje, J.M. (1978) Pathway of degradation of nitrilotriacetate by a Pseudomonas species, Applied and Environmental Microbiology 35, 955-961.

McGrath, J.-W., Ternan, N.G., and Quinn, J.P. (1997) Utilization of organophosphonates by environmental micro-organisms, Letters in Applied Microbiology 24, 69-73.

Jenal-Wanner, U. (1991) Anaerobic degradatio of nitrilotriacetate in a denitrifying bacterium: Purification and characterization of the nitrilotriacetate dehydrogenae / nitrate reductase enzyme complex, PhD thesis No 9531, Swiss Federal Institute of Technology, Zürich, Switzerland.

Jenal-Wanner, U., and Egli, T. (1993) Anaerobic degradation of nitrilotriacetate (NTA) in a denitrifying bacterium: Purification and characterization of a NTA dehydrogenase / nitrate reductase enzyme complex, Applied and Environmental Microbiology 59, 3350-3359.

Kari, F.G., and Giger, W. (1995) Modeling the photochemical degradation of ethylenediaminetetracetate in the river Glatt, Environmental Science and Technology 29, 2814-2827.

Kemmler, J. (1992) Biochemistry of nitrilotriacetate degradation in the facultatively deitrifying bacterium TE 11, PhD thesis No 9983, Swiss Federal Institute of Technology, Zürich, Switzerland.

Kertesz, M.A., Schmidt-Larbig, K., and Wüest, T. (1999) A novel reduced flavin mononucleotide-dependent methanesulfonate sulfonatase encoded by the sulphur-regulated msu operon of Pseudomonas aeruginosa, Journal of Bacteriology 181, 1464-1473.

Knobel, H.-R. (1997) Genetic study of bacterial nitrilotriacetate degrading enzymes, PhD thesis No 12146, Swiss Federal Institute of Technology, Zürich, Switzerland.

Knobel, H.-R., Egli, T., and van der Meer, J.R. (1996) Cloning and characterization of the genes encoding nitrilotriacetate monooxygenase of Chelatobacter heintzii ATCC 2960, Journal of Bacteriology 178, 6123-6132.

Lauff, J.J., Steele, D.B., Coogan, L.A., and Breitfeller, J.M. (1990) Degradation of the ferric chelate of EDTA by a pure culture of an Agrobacterium sp. Applied and Environmental Microbiology 56, 3346-3353.

Nishikiori, T., Okuyama, A., Naganawa, T., Takita, T., Hamada, M, Takeuchi, T., Aoyagi, t., and Umezawa, H. (1984) Production by actinomycetes of (S,S)-N,N'-ethylenediamine-disuccinic acid, an inhibitor of phospholipase C, Antibiotics 37, 426-427.

Nörtemann, B. (1992). Total degradation of EDTA by mixed cultures of a bacterial isolate, Applied and Environmental Microbiology 58, 671-676.

Nowack, B. (1998) The behaviour of phosphonates in wastewater treatment plants of Switzerland, Water Research 32, 1271-1279.

Nowack, B., and Sigg, L. (1997) Dissolution of Fe(III)(hydr)oxydes by metal-EDTA complexes, Geochimica et Cosmochimica Acta 61, 951-963.

Nowack, B., Xue, H., and Sigg, L. (1997) Influence of natural and anthropogenic ligands on metal transport during infiltration of river water to groundwater, Environmental Science and Technology 31, 866-872.

Payne, J.W., Bolton, H.J., Campbell, J.A., and Xun, L. (1998) Purification and characterization of EDTA monooxygenase from the EDTA-degrading bacterium BNC1, Journal of Bacteriology 180, 3823-3827.

Piepke, R., and Amrhein, N. (1988) Degradation of phosphonate herbicide gyphosate by *Arthrobacter atrocyaneus* ATCC 13752, Applied and Environmental Microbiology 54, 1293-1296.

Potthoff-Karl, B., Greindl, T., and Oftring, A. (1996) Synthese abbaubarer Komplexbildner und ihre Anwendung in Waschmittel- und Reinigungsformulierungen, SÖFW-Journal 122, 392-397.

Schowanek, D., and Verstraete, W. (1990a) Phosphonate utilization by bacterial cultures and enrichments from environmental samples, Applied and Environmental Microbiology 56, 895-903.

Schowanek, D., and Verstraete, W. (1990b) Phosphonate utilization by bacteria in the presence of alternative phosphorus sources, Biodegradation 1, 43-53.

Schowanek, D., Feijtel, T.C.J., Perkins, C.M., Hartman, F.A., Federle, T.W., and Larson, R.J. (1997) Biodegradation of [S,S], [R,R] and mixed stereoisomers of ethylene diamine disuccinic acid (EDDS), a transition metal chelator, Chemosphere 34, 2375-2391.

Twachtmann, U., Petrick, S., Merz, W., and Metzger, J.W. (1998) Zum Einfluss umweltrelevanter Konzentrationen des Komplexbildners EDTA auf die Remobilisierung von Schwermetallen im Belebungsverfahren, Vom Wasser 91, 101-120.

Uetz, T. (1992) .Biochemistry of nitrilotriacetate degradation in obligatory aerobic, Gram-negative bacteria, PhD thesis No 9722, Swiss Federal Institute of Technology, Zürich, Switzerland.

Uetz, T., and Egli, T. (1993) Characterization of an inducible, membrane-bound iminodiacetate dehydrogenase from *Chelatobacter heintzii* ATCC 29600, Biodegradation 3, 423-434.

Uetz, T., Schneider, R., Snozzi, M., and Egli, T. (1992) Purification and characterization of a two component monooxygenase that hydroxylates nitrilotriacetate (NTA) from "*Chelatobacter*" strain ATCC 29600, Journal of Bacteriology 174, 1179-1188.

Wanner, U., Kemmler, J., Egli, T., Weilenmann, H.-U., El-Banna, T., and Auling, G. (1990) Isolation and growth of a bacterium able to degrade nitrilotriacetic acid under denitrifying conditions, Biodegradation 1, 31-41.

Wehrli, E., and Egli, T. (1988) Morphology of nitrilotriacetate-utilizing bacteria, Systematic and Applied Microbiology 10, 306-312.

Wilberg, E. (1989) Zur Physiologie und Ökologie Nitrilotriacetat (NTA) abbauender Bakterien. PhD thesis No 9015, Swiss Federal Institute of Technology, Zürich, Switzerland.

Wilberg, E., El-Banna, T., Auling, G., and Egli, T. (1993) Serological studies on nitrilotriacetic acid (NTA)-utilizing bacteria: Distribution of *Chelatobacter heintzii* and *Chelatococcus asaccharovorans* in sewage treatment plants and aquatic ecosystems, Systematic and Applied Microbiology 16, 147-152.

Witschel, M. (1999) Biochemical and physiological characterisation of a bacterial isolate able to grow with EDTA and other aminopolycarboxylic acids, PhD thesis No 12967, Swiss Federal Institute of Technology, Zürich, Switzerland.

Witschel, M., Weilenmann, H.U., and Egli, T. (1995) Degradation of EDTA by a bacterial isolate, 54th Annual Meeting of the Swiss Society for Microbiology, Lugano, Switzerland.

Witschel, M., Nagel, S., and Egli, T. (1997) Identification and characterization of the two-enzyme system catalyzing oxidation of EDTA in the EDTA-degrading bacterial strain DSM 9103, Journal of Bacteriology 179, 6937-6943.

Witschel, M., and Egli T. (1998) Purification and characterization of a lyase from the EDA-degrading bacterial strain DSM 9103 that catalyzes the splitting of [S,S]-ethylenediaminedisuccinate, a structural isomer of EDTA, Biodegradation 8, 419-428.

Witschel, M., and Egli, T. (1999) Fate of aminopolycarboxylic acids in the environment, FEMS Microbiology Reviews (submitted).

Witschel, M., Egli, T., Zehnder, A.J.B., Wehrli, E., Spycher, M. (1999) Transport of EDTA into cells of EDTA-degrading strain DSM 9103, Microbiology 145, 973-983.

Wolf, K., and Gilbert, P.A. (1992) EDTA-Ethylenediaminetetraacetic acid, in O. Hutzinger (ed.), The Handbook of Environmental Chemistry Springer Verlag, Berlin-Heidelberg, pp.241-259.

Wong, P.T.S., Liu, D., and McGirr, D.J. (1973) Mechanism of NTA degradation by a bacterial mutant, Water Research 7, 1367-1374.

Xu, Y., Mortimer, M.W., Fisher, T.S., Kahn, M.L., Brockman, F.J., and Xun, L. (1997). Cloning, sequencing, and analysis of a gene cluster from *Chelatobacter heintzii* ATCC 29600 encoding nitrilotriacetate monooxygenase and NADH:flavin mononucleotide oxidoreductase, Journal of Bacteriology 179, 1112-1116.

Xun, L., , Reeder, R.B., Plymale, A.E., Girvin, D.C., and Bolton, H. (1996) Degradation of metal nitrilotriacetate complexes by nitrilotriacetate monooxygenase, Environmental Science and Technology 30, 1752-1755

PART 4
ECOTOXICOLOGY

TRANSGENIC NEMATODES AS BIOSENSORS OF ENVIRONMENTAL STRESS

DAVID I. DE POMERAI, HELEN E. DAVID, ROWENA S. POWER, MOHAMMED H.A.Z MUTWAKIL AND CLARE DANIELLS

Molecular Toxicology Division, School of Biological Sciences, the University of Nottingham, University Park, Nottingham NG7 2RD, United Kingdom.

Summary

Three transgenic strains of the soil nematode, *Caenorhabditis elegans*, have been evaluated as biosensors of environmental stress. These strains carry *lacZ* and/or GFP reporter genes under the control of stress-inducible heat-shock promoters. The activation of heat-shock genes is a universal cellular response to protein damage, whether caused by heat, toxic chemicals or radiation. Using the transgenic reporter strains, reporter expression is inducible in aquatic or soil media by a wide range of environmental stressors including heavy metals, pesticides and microwave radiation. Induced reporter products can be detected very simply and cheaply (much more so than authentic heat-shock proteins), but *C. elegans* itself poses intrinsic limitations of sensitivity as a test organism.

1. Introduction

1.1. THE HEAT-SHOCK RESPONSE

All organisms respond to a sudden rise in ambient temperature (heat shock) by activating a small set of emergency-response genes encoding heat-shock proteins (HSPs). Collectively, these serve a variety of protective functions within the cell, by refolding and reactivating heat-denatured proteins, breaking up protein aggregates, and helping to dispose of irreparably damaged proteins [1]. Proteins encoded by constitutively expressed members of the HSP gene families perform similar functions in non-stressed cells. However, many of the heat-shock genes are strictly stress-inducible or at least stress-modulated, the extent of their expression being proportionate to the stress applied. A heat-shock transcription factor (HSF) is present in the form of cytoplasmic monomers in unstressed cells, but under stress these trimerise and

S.N. Agathos and W. Reineke (eds.),
Biotechnology for the Environment: Strategy and Fundamentals, 221–236.
© 2002 *Kluwer Academic Publishers. Printed in the Netherlands.*

translocate into the nucleus, where they bind co-operatively to arrays of heat-shock elements (HSEs; consensus NGAAN) prefacing the stress-inducible HSP genes, and thereby activate their transcription [2]. HSF is normally held in the cytoplasm through interactions with constitutive HSPs (such as HSP90; [3]), but these complexes dissociate when HSPs are needed to interact with denatured cellular proteins.

1.2. BIOMONITORING USES OF HEAT-SHOCK PROTEINS

The major signal for HSP gene induction is protein damage (proteotoxicity), thus any chemical or physical stressor, which damages proteins, will also induce a heat-shock response. Although such damage may be incidental rather than a primary route of toxicity, HSPs are nevertheless inducible by several heavy metals (especially Cd), pesticides, oxidative stressors and radiation, as well as by heat itself [4]. The generality of this response suggests that it could be applied for biomonitoring purposes, providing a non-specific indicator of stress in the test organism [4, 5]. The main drawback with this approach is the expense and technical difficulty of HSP detection, whether by means of metabolic labelling or immunoblotting. The former involves radioactive amino-acids (applicable only to small organisms; [6]) and autoradiography, while the latter requires blotting with specific anti-HSP antibodies [7] and second-antibody detection systems; both also necessitate protein separation by gel electrophoresis in one or two dimensions [8]. Neither of these methods is conducive to routine biomonitoring, although protein dot-blotting or gene-specific oligonucleotide probes may provide less laborious alternatives.

1.3. *CAENORHABDITIS ELEGANS* IN TOXICOLOGY

The free-living soil nematode *Caenorhabditis elegans* is uniquely well characterised among metazoan animals; its complete somatic cell lineage was elucidated over 15 years ago [9], while its genome sequence was completed in 1998 [10]. Its short lifecycle (3 days at 25°C), ease of laboratory culture (feeding off bacterial lawns on agar plates), and predominant mode of reproduction by hermaphrodite self-fertilisation (conducive to the maintenance of mutant stocks) have all contributed to its popularity as an experimental system. Transgenic strains are also relatively easy to produce, by microinjecting DNA constructs into the gonad along with a selectable marker [11, 12]. Despite this, nematodes such as *C. elegans* have been little used in toxicological studies until recently. There is a widespread perception that nematodes are relatively insensitive as compared to other test species, and this is particularly true of short-lived, opportunistic species such as *C. elegans* [13]. Its very robustness, however useful in other contexts, is something of a drawback in toxicology. Nevertheless, a number of *C. elegans*-based assays have been developed for ecotoxicological purposes, including LC50 tests in aquatic [14, 15] and soil [16] media, and growth/fecundity measurements [17].

1.4. TRANSGENIC *C. ELEGANS* AS AN ALTERNATIVE FOR HSP BIOMONITORING

Although the induction of particular HSPs in *C. elegans* can be investigated by means of immunoblotting, e.g. for HSP70 [8] or HSP16 (this paper), such an approach is time-consuming, expensive and inconvenient. A much simpler alternative is to place a conventional reporter gene (e.g. *lacZ* encoding β-galactosidase, or Green Fluorescent Protein) under the control of a stress-inducible heat-shock promoter, and then to integrate this construct into the genome of *C. elegans*. These strains show strictly stress-inducible expression of the reporter product, in parallel with HSP induction. The first such transgenic strain, CB4027, carried multiple copies of a *Drosophila hsp70* promoter fused to a *lacZ* reporter [11]. This strain was first used for biomonitoring purposes in our laboratory [18], but requires an inconveniently high exposure temperature of 31-32°C for optimal sensitivity to added stressors, because the *C. elegans hsp70* gene is heat-activated only at 34°C. This strain will not be considered further here. A second strain, PC72, comprising a *lacZ* reporter gene fused in-frame into the 2nd exon of the *C. elegans hsp16-1* gene, was independently developed by Peter Candido and co-workers in Vancouver [19], and was subsequently adopted in our laboratory [20]. Ecotoxicological tests on a series of metal-polluted water samples from southwest England using both of these strains (PC72 and CB4027) found them to be of comparable sensitivity. The advantage of strain PC72 is that it can be used at lower exposure temperatures (24-25°C, at the upper end of the normal growth range), because the *hsp16* genes are heat-inducible at 28°C. Note that both strains display optimal sensitivity to added stressors at temperatures 2-3°C below those required for heat activation of the promoter used, suggesting co-operability between the effects of heat and other stressors. The Vancouver and Nottingham laboratories have also collaborated to produce a third transgenic strain, PC161, which is essentially similar to PC72 except that it incorporates a GFP reporter gene in addition to *lacZ* (both under the control of the *hsp16-1* promoter). In all such strains, the reporter response is inducible by stressors ranging from anti-surface antibodies [22] and microwave radiation [23], to pesticides [24, 25] and heavy metals both in aqueous [18, 19, 21] and soil [26, 27] media.

2. Materials and methods

2.1. MATERIALS.

The PC72 transgenic strain of *C. elegans* and anti-HSP16 antibody were generous gifts from Professor E.P.M. Candido (University of British Columbia, Vancouver, Canada). $CdCl_2.2\frac{1}{2}H_2O$ was ACS grade from Sigma Ltd (Poole, Dorset). The Transverse Electromagnetic (TEM) cell used in this work was the same as that described previously [23], and all microwave exposures were carried out overnight (18 h) at 750 MHz and 0.5 W (27 dBm), giving an E-field of approximately 45 V m^{-1} in the centre of the TEM cell with an estimated SAR of approximately 0.002 W kg^{-1}. Soil samples contaminated with heavy metals (to the current UK limits for dumping sewage sludge) were kindly supplied by Drs A. Tye and S. Young of the Environmental Science Division in this School.

2.2. WORM CULTURE

PC72 and PC161 worms were cultured at 15°C on NGM agar plates with a lawn of *E. coli* P90C food bacteria [18]. Worms were washed off actively growing cultures and size-fractionated by filtration through a 5 μm mesh (Wilson Sieves, c/o Mr D. Wilson, School of Biological Sciences, University of Nottingham), as described previously [28]. L1 plus a few L2 larvae were recovered from the filtrate by centrifugation, then grown for a further 3 days on fresh NGM agar plates at 15°C. This gives a synchronous population of late L4 larvae plus young adults, which were used at once for exposure to stressors, since young adults are the most sensitive stage in terms of stress-responsiveness [18].

2.3. WORM EXPOSURE TO STRESSORS

L4/adult worms were exposed overnight to microwave radiation (see 3.1 above) at temperatures of between 21 and 27°C; controls were kept shielded (wrapped in Al foil) outside the TEM cell, but within the same incubator and for the same period of time. This ensures that control and exposed populations within each run experience identical temperature regimes, and differ only in respect of microwave exposure. Such exposures were normally conducted in 12- or 24-well multiwell plates, in which identical aliquots of worm suspension were added to 0.2 or 0.1 ml of K medium (53 mM NaCl, 32 mM KCl [14]) plus food bacteria (approximately 10^7 CFU) in each test well. In this way, multiple replicates could be conducted within each test run. In one set of experiments, cadmium was added to 0.5 μg ml^{-1} in some of the test wells. Exposure to metal-contaminated freshwater samples (those analysed previously in [21]) was conducted at 25°C over 7 h, in 6- or 12-well multiwell plates containing 1.5 ml of test sample or ultra pure water (negative control) plus food bacteria (10^8 CFU) per well. Exposures were also conducted with a range of different soil types (pH 4-8), all of which had been contaminated 8 months previously with a mixture of heavy metals (as nitrates) to the limits currently specified by UK legislation controlling the dumping of sewage sludge (further details in [29]). For these experiments, food bacteria (10^8 CFU g^{-1}) and test nematodes (about 10^4 g^{-1}) were mixed with 5 g soil samples at 50% moisture-holding capacity (MHC) and incubated at 25°C for 24 h [26, 29]. Following exposure to all stressors, worms were immediately frozen, then thawed and washed several times in ultra pure water for reporter assays. In the case of soil exposures, worms were recovered by flotation on colloidal silica (Ludox HS40 [16]), and washed with ultra pure water prior to assay [26]. The cadmium (and other heavy metal) contents of freshwater and soil water samples were determined by atomic absorption spectrometry (AAS; 18).

2.4. REPORTER ASSAY PROCEDURES

Expression of GFP in strain PC161 was measured in fluorescence units on a Perkin-Elmer HTS7000 Bioassay reader set for FITC. This is sub-optimal for GFP detection in *C. elegans*, since these wavelength parameters also detect significant background fluorescence originating from gut granules that are strongly autofluorescent. Nevertheless, by normalising the results against those from controls maintained at 15°C (100%), significant stress-induction of GFP could be demonstrated. Assays for the other reporter product (β-galactosidase) were essentially as described previously for strain PC72 [20, 24], and involved permeabilising the worms with acetone, allowing them to dry, then incubating them for 30 min at 37°C with the non-fluorescent substrate, 4-methylumbelliferyl-βD-galactopyranoside (MUG). This is cleaved by β-galactosidase activity to yield the highly fluorescent product, 4-methylumbelliferone (MU), which can be quantified (against MU standards) on the HTS7000 Bioassay reader, using non-fluorescent black microplates for optimal sensitivity. Once again, results obtained at different temperatures or in response to different stressors have been normalised relative to controls maintained at 15°C or 25°C. Welch's alternate t test (which does not assume equal variances) was used to assess the significance of differences in reporter expression between the exposed and control populations (both in multiple replicates).

2.5. HSP16 EXPRESSION INDUCED BY EXPOSURE TO CADMIUM IN SOIL WATER

A standard sandy loam (Lufa 2.2, details in [26]) was admixed with food bacteria (10^8 CFU g^{-1}) plus CdCl$_2$.2½H$_2$O at 250 µg g^{-1}. After preincubation for 24 h at 25°C, soil water was extracted by centrifugation at 3000 x g for 1 h [26]. A similar procedure was applied to Lufa 2.2 soil without Cd additions (control). PC72 worms were exposed for 24 h at 25°C to 0.1 ml samples of both soil-water extracts in 24-well microplates. Worms from several such exposures were pooled for each condition, then frozen, and processed for 2D gel electrophoresis as described previously [8]. First dimension IEF was performed using pH3-10 Immobiline DryStrips (LKB-Pharmacia Ltd) and second dimersion gels were run using precast ExcelGel 8-18% polyacrylamide gels (same source). After blotting onto ECL nitrocellulose membranes (Amersham International Ltd), the blots were stained with ProtoGold (British BioCell International), then blocked overnight with 5% (w/v) Bovine Serum Albumin in TBS (154 mM NaCl, 10 mM Tris-HCl pH 7.5). Blots were then probed for 1 h with a 1:2500 dilution of a rabbit polyclonal antibody directed against *C. elegans* HSP16, washed 3 times in TBS, and incubated for 1 h with a 1:5000 dilution of peroxidase-linked goat anti-rabbit IgG (Sigma). After 5 further washes in TBS, the location of bound antibodies was determined using the ECL chemiluminescent detection system (Amersham International Ltd) according to the manufacturer's instructions.

David I. De Pomerai, Helen E. David, Rowena S. Power, Mohammed H.A.Z Mutwakil and Clare Daniells

3. Results

3.1. REPORTER RESPONSES TO MICROWAVE EXPOSURE

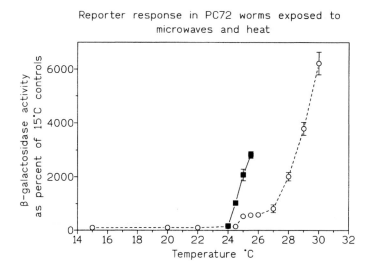

Figure 1. *Reporter response in PC72 worms exposed to microwaves and heat. L4/adult PC72 worms were exposed for 18 h at various temperatures to microwave radiation (750 MHz, 0.5 W), or were kept shielded in the same incubator. All worms were assayed for reporter enzyme (β-galactosidase) activity as described in Methods. Each data point shows the mean and standard error (SEM) derived from 12 (24.0-25.5 °C) or 6 (controls outside this range) replicates. Enzyme activities from different runs have been normalised relative to 15 °C controls (as 100%). Open circles, dashed line: reporter response to heat alone. Filled squares, solid line: reporter response to heat plus microwave radiation.*

We have previously demonstrated reporter induction in PC72 worms following prolonged exposure to microwave radiation (750 MHz, 0.5 W) at 25°C [26, 29]. As shown in Figures 1 and 2, this effect is highly temperature-dependant. Relative to 15°C controls (100%), controls exposed to heat alone show little sign of increased reporter activity up to 24.5°C (Figure 1). Between 25 and 27°C, there is modest induction (about 5-600% relative to 15°C controls), followed by a steep rise across the 28-30°C range (Figure 1) as the *hsp16* heat-shock response is activated. For PC161 worms, control expression levels for both reporters remain low across the range 21-25°C, but increase sharply at 27°C, especially in the case of the *lacZ* reporter (Figure 2). There is minimal response to microwave exposure at 24.0°C in the case of PC72 (Figure 1), or at 21°C in the case of PC161 (Figure 2); other evidence (not shown) confirms that reporter induction is slightly more temperature sensitive in PC161 than in PC72, even though the constructs carried by both strains employ the same *hsp16-1* promoter. Across the range 24.5 to 25.5°C, the reporter carried by PC72 becomes increasingly responsive to microwave exposure, to the extent that the microwave induction curve parallels that of heat-only controls at temperatures some 2-3°C higher (Figure 1). Similarly, in PC161 worms, microwave exposure induces clear responses for both reporters at 23°C and much larger responses at 25°C (Figure 2). The response of GFP appears much less than that of *lacZ*, largely because of high backgrounds from gut autofluorescence when measuring GFP expression using FITC settings. There is much less net effect of microwave treatment at 27°C (Figure 2), as heat induction becomes the dominant factor controlling reporter expression. These findings strongly suggest synergism between the effects of heat and of microwave exposure.

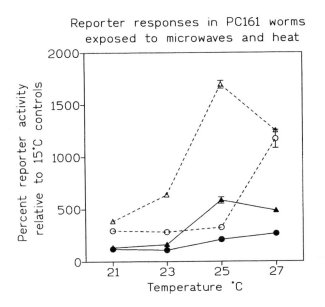

Figure 2. Reporter responses in PC161 worms exposed to microwaves and heat. PC161 worms were exposed to microwave radiation, or kept shielded, at four different temperatures. Reporter induction was measured as GFP fluorescence and β-galactosidase activity (as in Figure 1). Open symbols, dashed lines: β-galactosidase activity. Filled symbols, solid lines: GFP fluorescence. Circles show controls (heat only), whereas triangles show the corresponding effects of microwaves. All points show mean and SEM derived from 6 replica assays, normalised relative to 15°C controls.

To determine whether there is any similar synergism between microwave exposure and other stressors capable of inducing the reporter response, we exposed PC72 worms overnight to 0.5 μg ml^{-1} Cd, or to microwave (radio-frequency, RF) radiation as above, or to both. The concentration of Cd chosen is one which is normally just sufficient to induce a clear stress response [20, 21], and the experiment was conducted at three temperatures up to and including those required to see microwave-induced responses (Figure 3). At 23.5°C, there is no significant response either to Cd or microwave exposure, whereas at 24.0°C there is significant reporter induction by Cd and a smaller effect of microwaves (Figure 3). Although the response to Cd plus microwaves is somewhat greater than that to Cd alone, this difference is no more than additive and is not statistically significant (p = 0.09; n = 12).

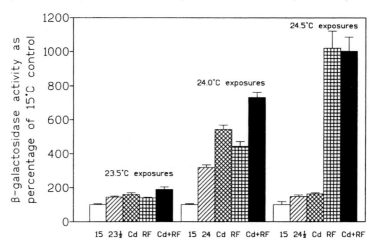

Figure 3. Temperature response of PC72 worms exposed to cadmium and microwaves. PC72 worms were exposed overnight to cadmium (0.5 μg ml⁻¹), or microwave radiation (as for Figures 1 and 2), or both together, or neither (temperature control). This experiment was conducted at 23.5, 24.0 and 24.5 °C. The induction of β-galactosidase activity was determined under each test condition, together with a set of 15 °C controls. All results have been normalised relative to the 15 °C controls as 100%, so as to ensure comparability between different temperatures. Left-hand group: 23.5 °C. Centre group: 24.0 °C. Right-hand group: 24.5 °C. Within each group (from left to right):- open bar, 15 °C control (100%); hatched bar, temperature control; cross-hatched bar, 0.5 μg ml⁻¹ Cd treatment; chequered bar, radio-frequency (RF) microwave treatment; filled bar, simultaneous 0.5 μg ml⁻¹ Cd plus microwave treatment. All bars show mean and SEM derived from 12 replicates.

At 24.5°C, there is very little effect of Cd, but a strong effect of microwave exposure (as in Figure 1); the effect of both stressors combined does not differ from that of microwaves alone. Although the apparent lack of effect of Cd at 23.5 and 24.5°C is somewhat anomalous, these results do not suggest any marked synergism between Cd (a classic oxidative stressor) and microwave radiation. It may be noted in passing that Cd induction of the reporter response is detectable right across the temperature range from 20 to 26°C [19, 30]; thus there is no clear synergism between Cd and heat paralleling that observed here between microwave radiation and heat (Figures 1 and 2).

3.2. REPORTER RESPONSES TO CADMIUM IN FRESHWATER AND SOIL

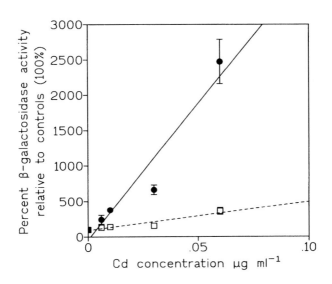

Figure 4. PC72 reporter responses to metal-contaminated water samples. C72 worms were exposed for 7 h at 25 °C to water samples from 4 sites in the Carnon river basin in southwest England. These samples are ranked according to their cadmium content (see text). Two data sets are shown, using the same samples, which had been kept frozen over the intervening year. All β-galactosidase activities have been normalised relative to 25 °C controls (ultra pure water = 100%) run in parallel with each set of samples (n = 6), and are shown as means with SEM. For comparison between data sets, the mean reporter activity induced by 2 µg ml[-1] Cd (positive control) is also shown. Filled circles, solid line: 1997 data set. Open squares, dashed line: 1998 data set.

Water samples taken from the Carnon river basin in south-west England are all contaminated to varying extents with a range of heavy metals (Cd, As, Cu, Mn, Zn, Al, Fe etc [21]), even though at least one of these sites supports a thriving macroinvertebrate fauna. A detailed study of transgenic nematode responses to these samples has been published previously [21], and for the purposes of this report, the responses of PC72 worms to four of these samples are simply plotted against the Cd concentration determined by AAS. In order of increasing Cd concentration, these samples correspond to sites 1, 3, 4 and leakage from 5, as described by Mutwakil *et al.* [21], although the samples used here were taken from these sites on a later occasion and have been reanalysed for Cd content. These samples have been stored frozen at -20°C for over 2 years, and reporter responses (normalised relative to ultra pure water controls as 100%) were determined as part of a class practical experiment in two successive years (1997 and 1998). As can be seen from Figure 4, there is an apparently large difference in response between the 1997 and 1998 data sets, mainly due to much lower control values in the earlier year. Responses to 2 μg ml^{-1} Cd alone (positive control) are over 3-fold higher in the earlier year, although this is less than the 7-fold difference in slope between the two regression lines in Figure 4. It is also likely that dissolved metals may have been lost through precipitation and adsorption during prolonged storage. Nevertheless, reporter responses to all test samples in both years were significantly different from the corresponding controls (one matched set for each sample set tested), and both data sets give the same rank order of responses. Indeed, in both data sets, the responses from site 4 fall somewhat below the regression line while those from site 5 fall slightly above this line. Despite the fact that these assays were conducted by undergraduate students with no previous experience of using this technique, the responses observed are unambiguous and the error levels are acceptable even when the overall response appears smaller (1998). It is notable that worm responses to metal mixtures (present in all these samples) are far more sensitive than to any of the component metals singly [21]. Thus the regression lines in Figure 4 denote responses to Cd plus other metals rather than to Cd alone. Even so, there is no clear linear relationship between reporter response and the concentration of any other single metal present in these mixtures. It may be that such metals merely exacerbate the effect of Cd. Using Cd alone, reporter responses are sometimes detectable at concentrations of 0.1-0.5 μg ml^{-1}, but higher concentrations (2-16 μg ml^{-1}) are used routinely in aqueous tests as positive controls [18].

A similar trend can be observed among a range of different soil types, all contaminated with the same mixture of metals including Cd. The observed reporter responses among 22 such soils are shown in Figure 5a. Broadly speaking, they fall into two categories:- group B with low levels of Cd in the soil water compartment, and group A with higher levels (circled in Figure 5a). For the group A soils, there is again a close relationship (this time exponential rather than linear) between the soil-water Cd concentration and the observed reporter response (Figure 5b). Figure 5b also includes a 23rd soil with much higher levels of soil-water Cd, which gave a response beyond the range covered in Figure 5a. Once again, when Cd is present in significant amounts, its effect appears to predominate over that of the other metals present in these mixtures. Experiments using paired metal inputs confirm that Cu exacerbates the toxic effect of Cd, both in soil and in aqueous solution, whereas Zn has an opposite ameliorating influence on Cd toxicity [26].

Part a shows the mean reporter response to 22 metal-contaminated soils (see text), ranked according to the soil-water cadmium concentration. Most of these soils contain very low concentrations of soil-water Cd (group B), but for those with higher concentrations (group A) there is a clear dose-response relationship between [Cd] and reporter response. This point is amplified in part b, which shows the mean reporter responses (mean $\pm$ SEM) for group A soils (plus one additional soil whose soil-water concentration of Cd is beyond the range covered in part a), plotted on a logarithmic scale. For all points, the reporter response is derived from 4 replicate 24 h soil tests at 25°C (see Methods).

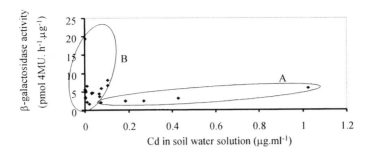

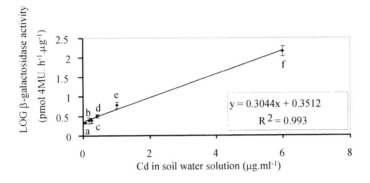

Figure 5. PC72 reporter responses to metal-contaminated soils.

Finally, we have checked for co-induction of HSP16 itself in parallel with the reporter response. This experiment used soil water extracted from Lufa 2.2 standard soil spiked with 250 μg g^{-1} Cd, as compared to control soil water from Lufa 2.2 soil without Cd. Measurements of the actual Cd concentration present in the former case suggest that > 95% of the added Cd becomes adsorbed to clay and organic components of the soil during the 24 h preincubation, such that the extracted soil water contains less than 10 μg ml^{-1} of Cd. This is nevertheless sufficient to induce a strong and highly significant reporter response in PC72 worms [26]. As shown in Figure 6, exposure to Cd in soil water induces numerous changes in the proteome (total protein spot pattern), including the appearance of several spots at around 70 kD (right-hand edge of Figure 6b, not present in Figure 6a control) which probably represent HSP70 isoforms [8]. Immunoblotting with an anti-HSP16 antibody confirms that HSP16 is also induced by Cd in soil water (dark streak and smaller spot in lower left-hand corner of Figure 6d, not present in Figure 6c control). HSP16 spots are not apparent in the corresponding stained proteome (Figure 6b), suggesting that these are not abundant proteins. There are other smaller spots apparently reacting with the anti-HSP16 antibody in both Figure 6c and 6d; however, these are all much larger than 16 kD and also correspond to strong protein spots on the corresponding proteome (Figures 6a and 6b, respectively), suggesting that they are merely artefacts.

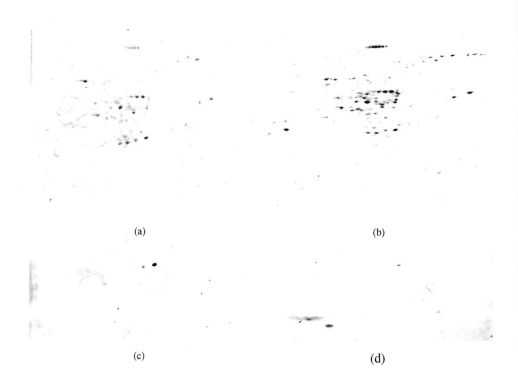

(a)   (b)

(c)   (d)

Figure 6. Co-induction of HSP16 by cadmium in soil water. PC72 worms were exposed for 24 h to soil water extracted from Lufa 2.2 soil supplemented with 250 μg g⁻¹ Cd (see Methods), or else to control soil-water from untreated Lufa 2.2 soil. Proteins were extracted, separated by 2D gel electrophoresis, blotted onto ECL membranes, stained with ProtoGold (parts a and b), and probed with an anti-HSP16 antibody followed by secondary-antibody ECL detection (parts c and d). Molecular weight markers are shown on the right-hand side of the figure. Part a, stained proteome after control exposure; part c, anti-HSP16 blot after control exposure. Part b, stained proteome after Cd exposure; part d, anti-HSP16 blot after Cd exposure. The dark streak and smaller spot in the lower left corner of part d correspond in size and pI to authentic HSP16. Note also a series of spots at the upper right of part b, which correspond in size and pI to HSP70s. Neither of these HSP groups is detectable in controls (parts c and a, respectively).

4. Discussion

4.1. REPORTER INDUCIBILITY BY MICROWAVE RADIATION

There is increasing evidence that electromagnetic fields (EMFs) are capable of inducing HSP genes, particularly in the case of extremely low-frequency (50-60 Hz [31]) and radio-frequency (RF; 200-2000 MHz [23]) ranges. Although microwave ovens exert their heating effect by operating at much higher power levels and at 2.4 GHz (a resonant frequency for water), generalised heating cannot account for the findings reported here and elsewhere [23, 26, 31]. Microwave exposure limits are set such that any temperature increase in biological tissue is much less than 1°C. We calculate an E-field of approximately 45 V m^{-1} in the centre of our TEM cell. Based on the electrical properties of comparable biological materials [32], we estimate an SAR of about 0.002 W kg^{-1}, which is much lower than the estimated SARs for both analogue and digital mobile phones [33]. In addition, we have measured the temperature of agar wells after prolonged exposure both to microwave and control conditions, and found no detectable difference between them (24.70 versus 24.75°C, respectively [26]). We have also used a Luxtron fibre-optic probe to measure the medium temperature during microwave exposure within the TEM cell, and again found no indication of any inflection in the temperature record. Thus we can eliminate the possibility that generalised microwave heating might be responsible for the effects observed here (Figures 1 and 2). Highly localised heating confined to biological tissue (i.e. the worms) is much harder to eliminate, although it would seem difficult to maintain a temperature differential of 2-3°C between such a small (1 mm) worm and its surroundings. Direct thermal imaging of worms during exposure would be needed to address this point. Even if this differential heating explanation proves to be correct, this would still raise significant health concerns about excessive human use of mobile phones.

If localised heating cannot account for the microwave effects reported here, there is clearly a need for some alternative, non-thermal mechanism through which microwave radiation could induce a heat-shock response. If microwave effects are mediated through protein damage (the classic proteotoxicity signal for a heat-shock response), it is conceivable that such radiation might somehow disrupt the hydrophobic or other weak interactions that maintain proteins in their active 3D conformations. Even very limited protein denaturation and/or aggregation might be sufficient to induce a heat-shock response. Alternatively, microwaves might interact with other signalling pathways (e.g. kinases and phosphatases) that also feed into the regulation of the heat-shock response. C. elegans is the ideal organism in which to explore these pathways by means of mutants, since its genome has now been completely sequenced [10] and a programme of systematic gene knockouts is now under way. Although the stressing effects of microwaves have been documented only in a nematode [23, 26; this paper], given the universality of the heat-shock response, it seems likely that similar effects will be detectable in vertebrates and indeed in human cell cultures. The close dependance on temperature documented here suggests that this hitherto-neglected variable deserves much closer attention from experimenters. The possibility that microwave exposure might cause oxidative stress (which can also induce a heat-shock response) seems less plausible in the light of our inability to demonstrate any synergism between the effects of microwaves and of Cd (a classic oxidative stressor; Figure 3).

4.2. REPORTER INDUCIBILITY BY HEAVY METALS IN WATER AND IN SOIL

Figure 3 also demonstrates that the reporter construct used in these transgenic worms is not especially sensitive to low concentrations of Cd (below 1 μg ml^{-1}) when used as a single stressor. In some experiments, we have detected responses (not usually statistically significant) at concentrations down to 0.1 μg ml^{-1}, but 0.5 μg ml^{-1} (as used in Figure 3) is normally at the lower limit of sensitivity. Similar considerations apply to a range of less toxic metals, such as Mn, Fe, Zn or Cu; each of these, when applied singly, fails to induce a detectable reporter response at concentrations below 10 μg ml^{-1} (5 μg ml^{-1} in the case of Cu [20]). This apparent lack of sensitivity may arise in part from barriers to metal entry into the worms. Given the toughness and relative impermeability of the nematode cuticle, the most likely major route of entry is through the gut. There are also innate defence mechanisms against metal contaminants, such as constitutively expressed metallothioneins in the larval pharynx [34]. In passing, we may note that these worms' apparent sensitivity to microwave-induced stress may result from the whole body being irradiated simultaneously throughout exposure, with no requirement for ingestion or uptake of the stressor into cells. Be that as it may, it is clear from this and previous studies [21, 26, 27, 29] that transgenic nematodes are far more sensitive to metal mixtures than to single metals. This is true even for mixtures containing a single highly toxic metal (such as Cd) in combination with others (such as Zn and Mn), which elicit reporter responses only at extremely high concentrations. Furthermore, as documented here (Figures 4 and 5), there is an apparent dose-response relationship between the concentration of the most toxic metal present (Cd) and the overall reporter response. Yet both for metal-polluted water samples (Figure 4) and for most of the soil water samples (Figure 5), the measured Cd concentrations are far less than 0.5 μg ml^{-1}. Since no such dose-response relationship is apparent for any of the other metals present in these mixtures, it seems plausible to suggest that Cd availability may be a primary determinant of the reporter response. If this is the case, then why is such sensitivity to Cd only displayed in metal mixtures, and not when Cd is used as a single stressor? One possible explanation is that the other metals present may overwhelm defence mechanisms (such as metallothioneins) that would otherwise sequester Cd and prevent it from inducing a reporter response. Other important interactions may occur at the level of uptake, with metal ions competing with each other and with Ca for entry into cells (principally via Ca channels [35]). Studies with paired metal inputs have confirmed that Cu can indeed exacerbate Cd toxicity, although Zn has the converse effect of reducing Cd toxicity [26, 35]. Clearly such interactions are of major importance when attempting to assess the overall environmental impact of complex mixtures containing multiple toxicants. One important advantage of the transgenic nematode approach described here is that the heat-shock response is essentially summative, such that the extent of reporter expression reflects the overall level of protein damage within the organism. In general, the rank order of reporter responses to different toxicants (7 h exposure) is the same as that obtained from longer-term 24 or 48 h LC50 tests [30] on *C. elegans*. Figure 4 also demonstrates that the reporter assay methodology is sufficiently simple and robust to give acceptable results in inexperienced hands, and that despite large differences in the extent of reporter activity between runs (1997 versus 1998), the overall rank order remains unchanged for the same samples reassayed on different occasions. We conclude that transgenic nematodes such as PC72 or PC161, if developed in a kit form suitable for semi-automatic microplate assays, could provide rapid and reliable ecotoxicological assessments of both water and soil pollution, and well as some forms of radiation. PC161 has the additional advantage that GFP responses can be monitored in living worms, allowing the kinetics of the stress response to be followed over time, rather than focussing on one or a few arbitrary snapshots.

Acknowledgements

The authors would like to thank BBSRC, the Wellcome Trust and DERA for financial support in various aspects of this work. MHAZM was in receipt of a postgraduate studentship from the government of Saudi Arabia, while HED is a Wellcome Ecotoxicology Prize student.

References

1. Parsell, D.A. and Lindquist, S. (1993). The function of heat-shock proteins in stress tolerance: degradation and reactivation of damaged proteins. Ann. Rev. Genet. 27, 437-496.
2. Lis, J. and Wu, C. (1993). Protein traffic on the heat-shock promoter: parking, stalling, and trucking along. Cell 74, 1-4.
3. Zou, J., Guo, Y., Guettouche, T., Smith, D.F. and Voellmy, R. (1998). Repression of heat shock transcription factor HSF1 activation by HSP90 (HSP90 complex) that forms a stress-sensitive complex with HSF1. Cell 94, 471-480.
4. Sanders, B.M. (1993). Stress proteins in aquatic organisms: an environmental perspective. Crit. Rev. Toxicol. 23, 49-75.
5. de Pomerai, D.I. (1996). Heat-shock proteins as biomarkers of pollution. Human Exper. Toxicol. 15, 279-285.
6. Hakimzadeh, R. and Bradley, B.P. (1990). The heat shock response in the copepod Eurytemora affinis (Poppe). J. Thermal Biol. 15, 67-77.
7. Sanders, B.M., Martin, L.S., Nakagawa, P.A., Hunter, D.A., Miller, S. and Ullrich, S.J. (1994). Specific cross-reactivity of antibodies raised against two major stress proteins, stress 70 and chaperonin 60, in diverse species. Environ. Toxicol. Chem. 13, 1241-1249.
8. Guven, K. and de Pomerai, D.I. (1995). Differential expression of HSP70 proteins in response to heat and cadmium in Caenorhabditis elegans. J. Thermal Biol. 20, 355-365.
9. Sulston, J.E., Schierenberg, E., White, J.G. and Thomson, J.N. (1983). The embryonic cell lineage of the nematode Caenorhabditis elegans,. Devl. Biol. 100, 64-119.
10. The C. elegans Sequencing Consortium (1998). Genome sequence of the nematode C. elegans: a platform for investigating biology. Science 282, 2012-2018.
11. Fire, A. (1986). Integrative transformation in Caenorhabditis elegans. EMBO J. 5, 2673-2680.
12. Mello, C.C., Kramer, J.M., Stinchcomb, D. & Ambros, V. (1991). Efficient gene transfer in C. elegans: extrachromosomal maintenance and integration of transforming sequences. EMBO J. 10, 3959-3970.
13. Bongers, T. (1990). The maturity index: an ecological measure of environmental disturbance based on nematode species composition. Oecologia 83, 14-19.
14. Williams, P. and Dusenbery, D. (1990). Acute toxicity testing using the nematode C. elegans. Environ. Toxicol. Chem. 9, 1285-1290.
15. Tatara, C., Newman, N., McCloskey, J. and Williams, P. (1998). Use of ion characteristics to predict relative toxicity of mono-, di- and trivalent metal ions: Caenorhabditis elegans LC50. Aquatic Toxicol. 42, 255-269.
16. Donkin, S. and Dusenbery, D. (1993). A soil toxicity test using the nematode Caenorhabditis elegans and an effective method of recovery. Arch. Environ. Contam. Toxicol. 25, 145-151.
17. Traunspurger, W., Haitzer, M., Hoss, S., Beier, S., Ahlf, W. and Steinberg, C. (1997). Ecotoxicological assessment of aquatic sediments with C. elegans (Nematoda): a method for testing liquid medium and whole sediment samples. Environ. Toxicol. Chem. 16, 245-250.
18. Guven, K., Duce, J.A. and de Pomerai. D.I. (1994). Evaluation of a stress-inducible transgenic nematode strain for rapid aquatic toxicity testing. Aquatic Toxicol. 29, 119-137.
19. Stringham, E.G. and Candido, E.P.M. (1994). Transgenic hsp16-lacZ strains of the soil nematode Caenorhabditis elegans as biological monitors of environmental stress. Environ. Toxicol. Chem. 13, 1211-1220.
20. Dennis, J.L., Mutawakil, M.H.A.Z., Lowe, K.C. and de Pomerai, D.I. (1997). Effects of metal ions in combination with a non-ionic surfactant on stress responses in a transgenic nematode. Aquatic Toxicol. 40, 37-50.

21. Mutwakil, M.H.A.Z, Reader, J.P., Holdich, D.M., Smithurst, P.R., Candido, E.P.M., Jones, D., Stringham, E.G. and de Pomerai, D.I. (1997). Use of stress-inducible transgenic nematodes as biomarkers of heavy metal pollution in water samples from an English river system. Arch. Environ. Contam. Toxicol. 32, 146-153.

22. Nowell, M.A., Wardlaw, A., de Pomerai, D.I. and Pritchard, D.I. (1997). The measurement of immunological stress in nematodes. J. Helminthol. 71, 119-123.

23. Daniells, C., Duce, I., Thomas, D., Sewell, P., Tattersall, J. and de Pomerai, D. (1998). Transgenic nematodes as biomonitors of microwave-induced stress. Mutation Res. 399, 55-64.

24. Guven, K., Power, R.S., Avramides, S., Allender, R. and de Pomerai, D.I. (1999). The toxicity of dithiocarbamate fungicides to soil nematodes, assessed using a stress-inducible transgenic strain of Caenorhabditis elegans. J. Biochem. Molec. Toxicol.13, in press.

25. Jones, D., Stringham, E.G., Babich, S.L. and Candido, E.P.M. (1996). Transgenic strains of the nematode C. elegans in biomonitoring and toxicology: effects of captan and related compounds on the stress response. Toxicol. 109, 119-127.

26. Power, R.S. and de Pomerai, D.I. (1999). Effect of single and paired metal inputs in soil on a stress-inducible transgenic nematode. Arch. Environ. Contam. Toxicol., in press.

27. Power, R.S., David, H.E., Mutwakil, M.H.A.Z., Fletcher, K., Daniells, C., Nowell. M.A.. Dennis, J.L, Martinelli, A., Wiseman, R., Wharf, E. and de Pomerai, D.I. (1998). Stress-inducible transgenic nematodes as biomonitors of soil and water pollution. J. Biosci. 23, 513-526.

28. Mutwakil, M.H.A.Z., Steele, T.J.G., Lowe, K.C. & de Pomerai, D.I. (1997). Surfactant stimulation of growth in the nematode Caenorhabditis elegans. Enzyme Microb. Technol. 20, 462-470.

29. Power, R.S. and de Pomerai, D.I. (1999). The application of a transgenic stress-inducible nematode to soil biomonitoring, in P. Rainbow, S. Hopkin and M. Crane (eds.), Forecasting the Environmental Fate and Effects of Chemicals (Proceedings of the 19998 SETAC-UK Conference), in press.

30. Mutwakil, M.H.A.Z. (1998). The nematode Caenorhabditis elegans as a bioindicator of environmental stress. PhD Thesis, the University of Nottingham.

31. Smith, O. (1996). Cells, stress and EMFs. Nature Medicine 2, 23-24.

32. Gabriel, S., Lau, R.W. & Gabriel, C. (1996). The dielectric properties of biological tissues: III parametric models for the dielectric spectrum of tissues. Phys. Med. Biol. 41, 2271-2293.

33. Gandhi, O.P., Lazzi, G. & Furse, C.M. (1996). Electromagnetic absorption in the human head and neck for mobile telephones at 835 and 1900 MHz. IEEE Trans. Microwave Theor. Tech. 44, 1884-1897.

34. Freedman, J.H., Slice, L.W., Dixon, D., Fire, A. and Rubin, C.S. (1993). The novel metallothionein genes of Caenorhabditis elegans: structural organisation and inducible, cell-specific expression. J. Biol. Chem. 268, 2554-2564.

35. Guven, K., Duce, J.A. and de Pomerai, D.I. (1995). Calcium moderation of cadmium stress explored using a stress-inducible transgenic strain of Caenorhabditis elegans. Comp. Biochem. Physiol. 110C, 61-70.

INDEX